JANET
THE FIRST 25 YEARS

Christopher S Cooper

ja.net

First published in 2010 in the United Kingdom by The JNT Association.

© 2010 The JNT Association.

ISBN 978-0-9549207-2-2

This document is copyright The JNT Association trading as JANET(UK). Parts of it, as appropriate, may be freely copied and incorporated unaltered into another document unless produced for commercial gain, subject to the source being appropriately acknowledged and the copyright preserved. The reproduction of logos without permission is expressly forbidden. Permission should be sought from JANET Service Desk.

The right of Christopher S. Cooper to be identified as the author of this work has been asserted by him in accordance with the Copyright, Designs and Patents Acts 1988.

Opinions expressed, where not otherwise attributed, are solely those of the author and do not necessarily represent those of either The JNT Association or JISC.

JANET(UK) manages the operation and development of JANET, the United Kingdom's education and research network, on behalf of JISC (Joint Information Systems Committee) for the UK Higher and Further Education Funding Councils.

For further information please contact:

JANET Service Desk
Lumen House, Library Avenue, Harwell Science and Innovation Campus,
Didcot, Oxfordshire, OX11 0SG
Tel: 0300 300 2212
Fax: 0300 300 2213
E-mail: service@ja.net

053 (10/10)

*To CMC
(aka CM²)*

'As gold which he cannot spend will make no man rich, so knowledge which he cannot apply will make no man wise.'
　　　　　　　　　　Samuel Johnson (1709–84)
　　　　　　　　The Idler No.84, 24 November 1759.

'Those who cannot remember the past are condemned to repeat it.'
　　　　　　　　　　George Santayana (1863–1952)
　　　　　　　　　　　　Life of Reason (vol.1).

CONTENTS

Foreword by Peter Kirstein   i

Preface   v

PART A   BEGINNINGS: THE CREATION OF A NETWORK

Introduction   1

1   Background and Early Networking in the UK   3
    1.1   Regionalisation   4
    1.2   Formative networks   6
    1.3   Beginnings   8

2   Origins of a National Network   13
    2.1   The Wells Reports   13
    2.2   Regional Networks   23
    2.3   Research Council activities   28
    2.4   The Network Unit   31
    2.5   Strategy for a network   35

3   Original JANET Protocols   39
    3.1   Architecture of a network   41
    3.2   Terminal access   46
    3.3   High-level protocols   49

4   JANET   57
    4.1   The Joint Network Team   57
    4.2   Components of a network   62
    4.3   Where to get a backbone   66
    4.4   1984   70

PART B   DEVELOPMENT AND TRANSITION TO INTERNET PROTOCOLS

Introduction   77

5   Early Evolution of JANET   79
    5.1   New services   80

	5.2 Campus access	85
	5.3 Coping with growth	92
6	Protocol Debates and Emergence of the Internet	105
	6.1 Internetworking	106
	6.2 Birth of the Internet	108
	6.3 Adoption of IP by JANET	109
7	International Dimensions	123
	7.1 Transatlantic connections	124
	7.2 European angle	126
	7.3 Other community networks	130

PART C: SUPERJANET ERA

	Introduction	133
8	Multiservice and Ubiquitous Networking	135
	8.1 Evolving horizons	135
	8.2 Technological developments	142
	8.3 Effect of community growth	154
9	Organisational Evolution	157
	9.1 Time for change	158
	9.2 Community stakeholders	160
	9.3 The creation of a new organisation	162
	9.4 The JNT Association	165
10	SuperJANET: Birth of a Concept and ATM trials	171
	10.1 SuperJANET: breaking the log-jam 1989–1994	172
	10.2 SuperJANET II: continued expansion and development 1994–1998	188
	10.3 SuperJANET III: ATM 1997–2001	200
11	Devolved Networks	211
	11.1 Regional networks (MANs)	211
	11.2 Schools networks	216

12	SuperJANET: Realisation with WDM	221
	12.1 SuperJANET4: gigabit routers 2001–2006	221
	12.2 SuperJANET5: meeting a broader remit 2007	232
	12.3 SuperJANET applications and services	244

PART D: MULTISERVICE JANET

Introduction		255
13	Regulation and Security	257
	13.1 Regulation	258
	13.2 Security	264
	13.3 Protection and enforcement	267
14	What Now?	277
	14.1 Recent developments	277
	14.2 Retrospective	285
	14.3 Last words	299

Appendices

A	Computer Board Network Working Parties	305
B	Regional Networks	306
C	Heads of JANET(UK) and its predecessors	308
D	On Layering	310
E	The Coloured Books	312
F	Networkshops	319
G	Digital Transmission Hierarchies	321
H	School Networks	323
I	Computers, Communication and the Law	325

Glossary	329
Sources	371
References	373
Chronology	380
Index	399
Biographical Note	411

List of Figures

2.1	University networks, September 1973 (after Wells, 1973).	14
2.2	Proposed 3-node backbone network (after Wells, 1973).	16
2.3	Lines funded by Computer Board as at 1975 (after Wells, 1975).	20
2.4	Network hierarchy (after Network Unit, 1979).	35
4.1	The JNT PAD.	67
4.2	GEC 4000 PSE.	73
4.3	Initial JANET backbone.	74
4.4	JANET backbone, showing geographical locations of and number of connections to each PSE (after Wells, 1986).	75
5.1	JANET Mk II transition topology.	95
5.2	JANET in 1990.	96
6.1	Shoestring pilot connectivity, August 1991.	112
9.1	Organisational structure of UKERNA in 1996 (after market testing).	169
10.1	SuperJANET pilot network.	185
10.2	14-site ATM network, late summer 1994.	191
10.3	SuperJANET Video Network, showing MCU locations and PVC configuration.	192
10.4	SuperJANET ATM topology from March 1996.	194
10.5	SuperJANET IP VP trunk, 1996.	194
10.6	SuperJANET sites 1993/94.	196
10.7	SuperJANET III backbone.	202
11.1	Regional Networks (MANs).	214
12.1	SuperJANET III backbone with 'band-aid' upgrades, September 2000.	222
12.2	SuperJANET4: boring under the river under the Cam.	223
12.3	SuperJANET4, June 2001.	224
12.4.	GSR 12016 routers in a SuperJANET4 PoP.	225
12.5	Upgraded SuperJANET4 configuration, 2002.	226
12.6	JANET backbone, geographic view, March 2003.	228
12.7	SuperJANET5 flexible transmission platform.	237
12.8	SuperJANET5 transmission architecture.	238
12.9	SuperJANET5 transmission platform topology (schematic).	240
12.10	SuperJANET5 transmission platform topology (geographic).	241
12.11	The JANET Videoconferencing Switching Service.	247

List of Tables

Table 2.1. Key to site numbers in Figure 2.1.	15
Table 10.1. SuperJANET sites 1993/94.	197

Foreword

Peter Kirstein

It is a privilege to write a Foreword to this book on the history of JANET. It is particularly broad-minded that I – someone who was not involved in the mainstream of JANET development – was asked to write it. Although the publication date of this book is close to the twenty-fifth anniversary of the founding of JANET, the story really started 20 years previously, as the early chapters make clear.

British computer networking began with the 1964 proposal by Donald Davies of the National Physical Laboratory (NPL) to establish a national data network based on packet switching. Unfortunately his proposal did not find favour with his political masters in the Department of Trade and Industry, nor with the British Post Office – then the monopoly supplier of telecommunications services; Donald was restricted to developing and deploying a single-node network inside NPL. As a result, the US Advanced Project Research Agency (ARPA) deployed the first national computer network in the US. Donald would have liked to collaborate experimentally with ARPANET, and was given the chance to do so by ARPA. Unfortunately the political realities at the time forced him instead to participate only in a European initiative called the European Informatics Network (EIN). There was no significant interest by anyone in using EIN, and the only gain was to the French who used their experience both to further French interests in later European Commission initiatives and to develop the successful French domestic network TRANSPAC and the French Minitel system. In consequence NPL played no significant role in further British, or European, networking. However the weakness of the institutional participation in EIN made it clear that to be successful much broader participation was necessary. The lesson was clearly recognised, and was the background to many of the organisational decisions that eventually led to the formation of JANET.

It fell to a research group at University College London (UCL) to link to the ARPANET, which for the next 15 years gave the British authorities an inside view of what was happening in the US. In the late 1970s there were competing factions in the US. The high energy physics community under the US Department of Energy (DOE) and the SPAN space community under NASA were embracing VAX computers and DECNET. The Information Services community mainly used IBM mainframes, initially with SNA and later BITNET; the latter internationally becoming EARN. The ARPA community and the Department of Defense (DOD) used the ARPANET protocols, and the Computer Science interests under the National Science Foundation (NSF) used UUCP. In the early eighties, both the

ARPA and NSF communities moved to the Internet protocols, but the others remained with DECNET and BITNET for a considerable time.

The remit of ARPA meant that ARPANET could only be an experimental service and it fell to others, such as the Defense Communications Agency, to run permanent services. The British authorities wanted to avoid the fractionalisation they saw in the US community and strove to have a unified British approach to research and educational computer networks. To achieve this they developed a complete and consistent family of protocols called Coloured Books, based partly on international and partly on national standards, designed to intercept in due course with work emanating from ISO. This book well illustrates how the unified network goal was accomplished, although the activity had unintended consequences.

Successful commercial companies frequently emerge from academic activities and the networking field is no exception. The US-sponsored Internet IP family eventually won out internationally and this book shows how JANET was able to make the transition to IP without interrupting services to the research and education community. The transition took until 1993 to complete and was no mean task, with the result that there were few UK commercial spin-offs arising out of the JANET development activities.

For the first decade of its existence, JANET devoted most of its energy to establishing the completeness of the network between educational and research institutions, as well as managing the transition to IP. In this it was not able to capitalise on the advances being made in the US. The book makes clear how much JANET had to devote to political and organisational issues both inside the organisation and in its relationships with other bodies. It is therefore particularly praiseworthy how JANET, in the last 15 years, has succeeded in broadening its remit to cover so many other areas in the educational field, including further education, schools, etc. Moreover it moved quickly to innovative network applications – including, for example, group communication and videoconferencing, distance learning, etc.

One issue that concerns the author of this Foreword has been a perennial question of the role of the network research and development community in the context of JANET. At times JANET has been encouraged to embrace this role, often to find its freedom being removed later. It has also had to grapple with the provision of special facilities for particular research communities (e.g. astronomy or high energy physics), resulting in tension between different areas of interest. With the expansion of JANET to provide even greater bandwidth capacity, these tensions have now abated.

The SuperJANET contract, which was won by BT in 1993, obliged BT to fund substantial academic research; but this lasted only as long as the contract itself. The British Treasury excluded such obligations from the evaluation criteria when the SuperJANET contract was next put out to tender.

In the last 15 years, JANET has had to show flexibility in coping with the demands of new communities, new applications, scaling and new technology – all in the environment of changing governmental organisation, industrial re-organisation and user expectations. The latter part of the book shows how JANET has dealt with these demands and how it has been able, successfully, to balance the relative requirements for national and international capacity.

Recently JANET has played vital roles in several areas. For example there is its pre-eminence, nationally, in network security; its videoconferencing services; universal support for libraries; and its prototyping of very high speed links,

VoIP and SMS services. In each case it has worked with a small but dedicated community to develop facilities that have become universal services.

After a considerable early gestation period resulting from inherited legacy decisions, this book shows how JANET has evolved into the powerhouse it has now become. In its first 25 years JANET has shown that it can maintain a vital role in the fabric of our research and educational life. It clearly has important functions over the next decades. The UK authorities must be congratulated on their unswerving financial support for JANET over the last quarter century. For the sake of the health of UK education and research it is essential that this support continues even in these times of financial stringency.

– Professor Peter T. Kirstein, University College London

IV

Preface

What is JANET? Certainly it is a communications network. But to many, perhaps depending on the perspective of the individual, it is more: a concept, a community, a variety of services, and a part of UK academic, education and research life. In this book we explore the evolution of JANET, the UK's network which serves the education and research community, from its inception in the 1970s to its state in 2009, its 25th anniversary.

Writing towards the end of the first decade of the 21st century, it is sometimes hard to recall just what 'remote computing' was like before ubiquitous access to information and computational services was brought to our metaphorical doorsteps by the pervasive Internet and the World Wide Web. The author recalls, as a research assistant in a university in the North East of England in the early 1970s, perching by a Teletype connected to a modem, conversing with the local telephone operator while making a 'data call' at 110 bits per second to remote South Oxfordshire (actually, Berkshire at the time: the county boundary was redrawn in 1974) to use the new IBM 360/195 at the Rutherford High Energy Laboratory (RHEL). Such calls were usually made in the evening because there were fewer active calls, so less cross-talk between circuits, hence less noise and fewer bit errors on the transmission path to the computer some 250 miles to the south, an important consideration when entering program or data over a connection with no error detection or correction. Of course, this was an advance over the optical telegraphic semaphore communication of the early 19th century used to transmit intelligence between financial markets (as described by Alexandre Dumas in *The Count of Monte Cristo*[1]) – though on a bad night it didn't always feel like it – and it was still time-consuming and error-prone!

A couple of years later, after the installation of an ARPANET node at UCL, it became possible to access a CDC6600 at Berkeley, California – still using the same Teletype – in furtherance of collaborative physics. However, it had to be that particular terminal (because it was the only one in the department with a modem and its own telephone line) and, because there was no particular agreement at that time about how to represent the non-alphabetic, non-numeric characters amongst various manufacturers' computers, one had to type strange, apparently meaningless characters at times to make things work. To be able to access a computer thousands of miles away was a marvel – but it was hard work, quite slow, and required quite a degree of arcane know-how. The significant aspect was the enabling of the first steps in collaborative endeavour (in this case science) mediated by computers and networks – an aspect which continues today, emerging in its most recent form as Grid computing, the surge in e-everything or 'cyber-infrastructure' and videoconferencing among schools.

---

1   *See also* **Chapter 13.**

In that time of some 35 years, computer processors have increased in (clock) speed by a factor of about 10,000; computer memory decreased in price by a factor of a million, from £1 a byte to £1 a megabyte; disk storage capacity increased from 30 Mbyte to 300 Gbyte or more while shrinking from a weighty, stand-alone box about the size of a domestic washing machine to something comparable with the size of a coin and mountable on the printed circuit board of a laptop. And speaking of laptops, by 2007, the processing capacity of such a machine had far outstripped what could have been crammed into a reasonable sized town house in the 1970s, consuming many kilowatts of power and needing water (or sometimes more exotic fluids) to keep it cool.

In similar time, communications moved from analogue to digital, networks moved from the experimental projects of the computer science community into every part of domestic and business life, backbone communications infrastructure was converted from copper to glass fibre, bit error rates improved by around ten thousand million, transmission rates increased by about the same, access developed to exploit both wired and wireless connections, and roaming access was deployed so that users can access the network from wherever they may happen to be situated, using individual portable devices.

While the focus of the story is JANET, which covers a period of over 30 years, the history of computer networking itself only goes back another 10 years or so. In order to give perspective, I have chosen to include a few remarks on one or two salient aspects from those years. For those interested in the history of the Internet, while there is a considerable amount of information on 'the Net', the book by John Naughton (2000) gives an excellent view of both the sociology and technical aspects of its conception and development – and earns an accolade from Donald Davies (2001) who was one of the leading British contributors to the development of packet switching in the 1960s. The book by Davies and Barber (1973) also contains a wealth of information on the early technical development of networking.

While there are many threads running through the story of JANET, this is primarily a technical history, including an account of those technological developments in communication and networking which formed the background to the evolution of JANET. I have also included some background details of the development of JANET's community, funding, management, operation and stakeholders. In relating the story, I have attempted to concentrate on what actually happened, together with the rationale as to how and why. I have not generally attempted to convey all the debates which took place in respect of almost every aspect of JANET's development, from funding, organisation and management to the detailed technical: but that is not to suggest that there were none – quite the reverse, and they were often vigorous, even heated, a sign perhaps of just how much people cared about the enterprise.

During the period covered by this book, all of the organisations associated with JANET have naturally experienced change – and names have altered, in some cases frequently. I have generally adopted the convention of referring to an organisation by its name at the time of the events being described. In order not to break up the narrative flow, detail of this sort has usually been relegated to a combination of footnotes and the Glossary.

The book is in four parts. Simplistically, Part A deals with the time up to the official start of JANET on 1 April 1984; Part B takes up the story of its subsequent evolution and development, until roughly the beginnings of SuperJANET; Part C relates the story of SuperJANET, from its origin in the need for capacity in 1989 to its realisation of the concept in 2007; and, finally in Part D, the way regulation and security have impinged on JANET is described, together with recent developments and what is inevitably a somewhat personal view of JANET in retrospect. Some parts of the story are not so simple to compartmentalise chronologically. Generally, I have placed particular components of the story in the part where most of them occur. So, for example, FDDI deployment in JANET was a natural evolution of wired, shared-media networks and is dealt with in Part B, although chronologically it overlaps the first years of SuperJANET (Part C); likewise, JANET's transition to TCP/IP is covered in Part B, though again, it overlaps in time with the beginning of SuperJANET. In both these cases, the developments are natural evolutions from those begun earlier and I have chosen to maintain the thread of the story, rather than to try to constrain the telling by strict adherence to chronology. Chapter 14 is a combination of recent developments, retrospective and commentary. Among the activities described, some represent a decade or more of effort but only in recent years have they become a significant part of JANET service – like roaming and IPv6; others have a lighter aspect but are nevertheless illustrative of what can now be achieved – like virtual get-togethers at Christmas for military personnel half a world away from their families.

JANET today is part of the world-wide Internet but it was not always so. Coverage of JANET's international aspect is split, essentially along pre- and post-TEN34 lines. Chapter 7 covers the first of these, by the end of which all education and research networks are part of the Internet. The remainder is told as interludes in the SuperJANET story because by then national and international exploitation of technology were tracking each other. The first complete draft for this history was completed towards the end of March 2010. In April, while revising the draft of this history, I received a copy of the book *A History of International Research Networking* edited by Howard Davies and Beatrice Bresson (2010), the major part of which is a detailed account of the history of interconnection of European national research and education networks. Since the purpose of my thumbnail sketches of the international scene included here were only ever intended as part of the background to the story of JANET, they have been left largely as originally written; I am, however, indebted to both Davies and Bressan's book and my reviewers for improvements in accuracy.

This history is, of course, a personal perspective; but I hope others will find it of interest, perhaps even recognizable! It is not a solo effort. Ben Jeapes did a tremendous amount of research in the original documents relating to JANET, as well as interviewing some of those concerned with the origins and development of JANET, including: Bob Day, David Hartley, Peter Linington, Geoff Manning, Malcolm Read, Roland Rosner, Ian Smith and Mike Wells, the results of all of which he made available to me. I am also grateful for conversations with, as well as comments and information from Keith Blow, Roger Bolam, John Burren, Barrie Charles, Chris Cheney, Andrew Cormack, Jon Crowcroft, Bob Day, David Duce, Brian Gilmore, Mark Handley, Phil Harrison, David Hartley, Bob Hopgood, Jack Houldsworth, Henry Hughes, James Hutton, Peter Kirstein, Peter Linington, Linda McCormick, Andrew Moore, David Parish, Iain Phillips, Kit Powell, Roland Rosner, David Salmon, John Seymour, Rob Symberlist, Geoff Tagg, Robin Tasker, Rolly Trice, Mike Wells, Sue Weston and Shirley Wood. I am particularly grateful to Peter Kirstein, Peter Linington and Shirley Wood for reading the book in draft and providing many helpful and detailed comments, elaborating detail and helping to eliminate blunders; others who also read drafts of parts or all of the book, to whom I am similarly grateful, include Chris Cheney, Andrew Cormack, Jon Crowcroft, Bob Day, David Duce, Brian Gilmore, David Hartley, James Hutton, Linda McCormick, Geoff McMullen, Kit Powell, Roland Rosner,

Rob Symberlist, Rolly Trice, Mike Wells and Sue Weston. My thanks also go to Andrew Cormack for providing the text of Appendix I on the legal developments which have affected JANET, particularly during the last 15 years. Finally, I should also like to thank the production team of Ben Jeapes and Nathan Shelton, together with Shirley Wood, at JANET(UK) for their efforts and contributions in producing the book.

However, one thing stands out: JANET from its inception to the present day has been a community collaborative effort at all levels. I have had the privilege of working with many of those who have contributed to the construction of JANET from 1977 to 2008 and it has been a pleasure – but for errors and omissions in this story, I alone am responsible!

Chris Cooper
Ullapool, 2010

Part A

Beginnings:
The Creation of a Network

Introduction

Computers were big and expensive: they had to be shared. Among UK universities, this happened in the 1960s across campuses, within geographic regions typically spanning several counties, and nationally. At first the media were paper and cards, and communication was by foot, bicycle and van. Soon, the telephone network was exploited to improve the communication and dial-in terminals appeared, along with dedicated outstations nearer to users, so the cards and paper didn't have to travel so far. Then the notion of connecting computers to each other occurred to people: a user could connect to the network and use whichever computer was appropriate; and the computers could also access each other's resources to provide a more powerful, more easily accessible and much more general resource to users. With suitably liberal and generalised interpretation, that might still sum up much of the scene at the time of writing in 2010: but the means, the organisation and the technology have changed out of all recognition.

The US ARPANET demonstrated categorically in the early 1970s that such a network could be created. The idea was immediately attractive to many, funders and users alike, but also gave rise to deep misgivings that it might be used merely as a money-saving device, by forcing people to share just a few big machines, rather than opening up new possibilities. Nevertheless, the pressure to create a network mounted and the next issue was whether to copy the ARPANET; to build another network in hopes of exploiting at least some emerging international standards; or to have one supplied by the monopoly national telecommunications provider. Whichever way, support – technology and organisation – would need to be developed for the UK university community and its computers.

By the mid-1970s it was decided to take things further and a small Network Unit was charged with formulating plans for how to go about this, alongside what was already happening in the community essentially represented by Science Research Council laboratories and the Computer Board regions. The major effort was expended in the development and implementation, on a wide range of systems, of an interim family of national protocol standards – interim because there was expectation of international standards to come. Eventually, in 1979, the Joint Network Team was formed; and then, after a hiatus of several years, caused by uncertainty in the regulatory propriety of building a private community national network, JANET finally came into being in 1984, based on pooling existing network efforts and resources in the university education and research community.

We begin in Chapter 1 by looking at the origins of computer packet networking and the early sharing of computers in the UK. It took a long time to reach consensus, raise the funding and organise how to proceed in pursuit of a national network, though 'on the ground' there was activity all over the country: this is told in Chapter 2. The next chapter is devoted to the protocols created by the UK community during the 1970s, generally known as 'the Coloured Books', which became the technical foundations of JANET. And Chapter 4 tells how JANET finally came into being.

1

BACKGROUND AND EARLY NETWORKING IN THE UK

Computer networking is still a comparatively young technology. The computer may have had its origins in the 19th century with the seminal work of Charles Babbage, and computer science with the 20th century work of Alan Turing, John von Neumann, Maurice Wilkes, Donald Michie and others, but it only entered the UK academic curriculum during the latter part of the 1960s. As a topic within the computer science syllabus, computer networking only arrived on the scene in the following decade. And it was at about that time, in the first half of the 1970s, that the investigation and planning began which eventually led to the construction of JANET. However, to appreciate the origins of JANET as a service network, how it was constructed, and the stage of development of the infant packet-switching technology of which it was constructed, we need to make a brief excursion into the previous decade.

It is sometimes hard to recall just how far the world of digital electronics, communications and computers has come in just fifty years – roughly the second half of the 20th century. Although the modern transistor was developed in the late 1940s, during the 1950s the electronics of the UK domestic market was based mostly on the thermionic valve,[1] primarily in radios and television sets. The telephone system was analogue throughout, with the dialling (signalling) and switching being electro-mechanical – where it was not still manual. The release into the public domain of research conducted during the Second World War, particularly that relating to cryptographic computation, coupled with the first of many steps in miniaturisation, improved reliability and affordability all brought about by the transistor, enabled the development of the first practical programmable electronic computers for use in industry in the UK during the 1950s. While the transistor components used were the forerunners of those in an analogue domestic transistor radio of the late 1950s, the circuitry would come to be called 'digital'.

Another significant legacy of the War was the upsurge in funding for science. By the early part of the 1960s, the consequent demand for scientific computing became apparent and led, in the UK, to a strategy for academic computing provision which was to have a profound effect on the way in which UK academic networking would evolve and be funded.

Following hard on the heels of the wide-scale deployment of computers in the 1960s, research into computer communication laid the foundations for modern packet-switching networks which blossomed into computer networking at the beginning of the 1970s. Much of the development of the subject and its technology

---

1   Also known as the vacuum tube.

was intensely practical, building upon the ability to create new functionality by software programming, rather than having to undertake expensive, time-consuming hardware development, as had generally been the case in much of the telecommunications industry (primarily telephony) until then.

Although computers were digital in operation, the telephone transmission system, over which wide area communication took place, was still analogue – and indeed this would remain so until the 1970s, when the core telephone network migrated to digital operation. With the 1980s came the beginnings of a wider recognition that exploitation of digital techniques by both computers and communications offered opportunities to merge techniques and develop new applications at many levels – termed 'convergence' in the argot of the time – which culminated during the 1990s in prototype deployment of multiservice networking as it is being more generally deployed during the first decade of the 21st century.

1.1 Regionalisation

Following recognition early in 1965 by the UK Government that, in order to maintain its international competitiveness, UK scientific research required a substantial injection of funding for powerful computer provision, a working party under the chairmanship of Brian Flowers[2] was set up under the auspices of the Council for Scientific Policy and the University Grants Committee (UGC)[3] *'to assess the probable computer needs, during the next five years, of users in Universities and civil research establishments receiving support from Government funds'*. The working party made its recommendations in 1965 and its report was published in 1966 (Flowers, 1966). The Government generally accepted the recommendations and, as a result, in 1966 the Computer Board of the Universities and Research Councils[4] was created, with Brian Flowers as its first chairman, to co-ordinate the initial five-year programme of university and Research Council computer upgrades. The Computer Board was funded by and reported to the Department of Education and Science.

For the future of networking in the UK, this action had two notable consequences. The first was that it endorsed the idea that facilities at supra-university level, including national, should be funded directly as part of the national research and education budget, not through individual university budgets. This principle of 'top-sliced' funding, as it became known, was already in effect in respect of Research Council support for research in general, and the provision of computing facilities in particular, and paved the way for how funding of the national network would be approached a decade or so later.

The second aspect derived from the detail of the proposals for how the computing facilities were to be provided. It was observed that informal regional consortia of computer users had developed naturally, particularly in London and around Manchester and Edinburgh. The report proposed that a hierarchical arrangement of computer provision should be formalised, whereby in addition

---

2   Professor Sir Brian H Flowers, FRS (later Baron Flowers), then Langworthy Professor of Physics at the University of Manchester.

3   *See* Glossary.

4   *See* Glossary for the relation of the Computer Board to subsequent bodies responsible for UK computing and network funding.

to a university's own provision there would be a larger facility provided at one of the universities in the region on the basis that others in the region would be given access as of right. The groupings already mentioned were suggested as covering, respectively, the South East, the North West, and Scotland. The South West, Midlands, and North East were identified as further incipient regions in this scheme. Northern Ireland was also identified as a region requiring its own funding.

The majority of computers in the 1960s and early 1970s were 'batch job' systems. Programs were written out and then transferred to media readable by a computer: either paper tape or punched cards. Operators fed these into the machine and later delivered fan-fold printout of the results to users. If the program was run on the regional large mainframe, this could involve road transport for both input and output and a 24-hour turnaround for a 'job' – a single instance of the cycle just described. The next development, in the late 1960s, was a capability for jobs to be submitted and the results printed remotely from the mainframe: 'remote job entry' or RJE in the terminology used by IBM and subsequently adopted more generally. The RJE station was connected to the central mainframe by a permanent transmission line leased from the telephone company – the General Post Office (GPO) in the UK at the time.

As may be imagined, the result of the Computer Board policy, coupled with RJE technology, was the formation of a set of regional stars centred around each of the large regional computers. Of course, this is oversimplified. Computers sited at Edinburgh Regional Computer Centre (ERCC), University of London Computer Centre (ULCC) and University of Manchester Regional Computer Centre (UMRCC) were designated national facilities. To these was added the existing Atlas computer facility at the Atlas Laboratory funded by the Science Research Council (SRC). Subsequently, the computing facilities at two further SRC establishments, Rutherford High Energy Laboratory (RHEL) south of Oxford[5] and Daresbury Laboratory (DL)[6] near Manchester and Liverpool, joined the list of national facilities. The national facilities at ERCC, UMRCC and ULCC were funded directly by the Computer Board; those at Atlas, DL and RHEL were funded by SRC, and indeed the staff at all three latter Laboratories were employees of SRC.

During the latter part of the 1960s and much of the 1970s, provision of powerful computing facilities in the UK was dominated by IBM and CDC. IBM (as its name, International Business Machines, suggests) had concentrated primarily on commercial data processing and its System360 architecture, introduced in 1964, included features specifically designed for that field. CDC focused more on engineering and scientific computation: in that context the 6600 (also unveiled in 1964) outperformed all other machines at the time and has some claim to be the first of what was later dubbed a 'supercomputer'. CDC 6600s were installed at ULCC and UMRCC. The one at ULCC began service in 1969, initially offering a batch service.

---

5 Situated in Berkshire at that time but now in Oxfordshire following substantial rearrangement of the county boundaries in 1974.

6 Close to the village of Runcorn in Cheshire.

1.2 Formative networks

High-level networks, that is, those engineered for dedicated support of one specific application have a long history. Dating from the latter part of the 19th century, the telephone network became the most familiar dedicated network. Even earlier, certainly one of the first networks based on electrical transmission, was the telegraph or telex network for sending messages which were generally of a few hundred characters. During the first half of the 20th century both relied on manual or semi-automated operation, including the routing and onward forwarding at message switching centres and telephone call connection at exchanges. During the 1950s the early international network for supporting airline reservations was based on manual message switching using torn paper tape, essentially a derivative of the telex network. By 1960, both this and banking were becoming interested in networking, the latter, in the first instance, as an aid to centralising account maintenance. Both these operations represented early requirements for data networking and both were looking to modernise operations during the first half of the 1960s, following the adoption of computers to support the data processing requirements of the respective industries (Davies and Barber, 1973). In response, essentially proprietary solutions were developed by the computer industry during the 1960s, roughly in parallel with the early development of computer networking as a technology in its own right.

There are several activities which are generally regarded as being the forerunners of computer packet-switched networking (Roberts, 1978; Roberts, 1999; Davies, 2001). In the early 1960s, Paul Baran of Rand Corporation in the USA had been advising the US military on how to fulfil its needs for a resilient communications facility which could handle a variety of communications, including text and voice. His proposal was for a system based on transmitting relatively short 'message blocks' of data of about 1000 bytes which would form the basis of both multiplexing and switching, the resilience coming from having a mesh topology that provided more than one route between any two nodes in the network.

J.C.R. Licklider had been developing concepts about online human-computer interaction which embraced the concept of linking computers together to form a network. This led to the inauguration of the ARPA programme which would lead to the creation of the ARPANET, with Licklider as its first director.

Independently, Leonard Kleinrock (Lincoln Labs, MIT) had begun applying queuing theory to analyse the performance of a message-switched communications system. As already mentioned, message switching in various manual and semi-automated forms had been the basis of the telegraph / telex system as well as airline reservation for many years. During the first half of the 1960s he made a comprehensive study of the behaviour of this type of system, which had considerable influence on the way computer networking would develop.

In the UK, a group led by Donald Davies at the National Physics Laboratory (NPL) independently conceived the same idea for a network for interconnecting computers which would be based on switching and multiplexing short sequences of a few hundred bytes of data. Traditionally, messages had been transmitted one at a time to completion, with the consequence that short messages may be held up by long messages – a problem long familiar in the telegraph industry. By contrast, breaking messages up into shorter, more or less uniform sized 'packets' allows more equitable sharing of a transmission link (at the expense of increased elapsed time for sending each message). Computer communication is intrinsically bursty and message- or packet-based communication is natural. The capability provided by packets for effective sharing of links (multiplexing) and store-and-forward switching without introducing undue delay make packet switching a cost-effective basis for a shared communications architecture, particularly where line costs are high. Davies is generally credited with having coined the term 'packet switching' in 1965 in NPL design documents, and the term appeared in publication at an ACM symposium in Gatlinburg, Tennessee, in 1967 (Bartlett *et al*, 1967). The NPL team continued development of the packet-switched network at NPL, which entered operation in 1971 (Davies and Barber, 1973).

The ACM meeting at Gatlinburg in October 1967 could be said to mark a turning point. Larry Roberts, by now director of the ARPANET programme, knew of Kleinrock's work – indeed is on record in regarding it as seminal in its influence on the nature of the switching and multiplexing architecture which would be adopted for ARPANET (Roberts, 1978). The NPL group knew of Baran's work. ARPA had decided to build a nation-wide network interlinking computers, having conducted tests in 1965 linking a computer at MIT with another at Systems Development Corporation in California using packets. The meeting effectively revealed just how many people in a variety of places were converging on the same notion (Davies, 2001; Roberts, 1978). Ironically, Donald Davies and his group had had as their goal a UK general purpose national network, but the time for this had not yet come and funding was not forthcoming. Originating from a military context of command and control, ARPA also had in mind a network with national coverage, albeit in the context of research projects for which the Agency was responsible, and one effect of the meeting was to confirm the basic principles, including packet switching, to be used in the construction of what came to be called the ARPANET.[7]

In 1968, in a visionary presentation (Engelbart and English, 1968) at the Fall Joint Computer Conference in San Francisco, Doug Engelbart[8] of SRI[9] showed what might one day be possible for all through use of computer networks by linking back to his 'home' site at SRI (a little further south in the Bay Area) and demonstrating simultaneous audio, video, graphics and hypertext access in plenary session. It took another decade for integration to get under way, around two decades for early 'freeware'[10] network videoconferencing and hypertext tools to begin to appear, and only during the mid-1990s did multiservice network deployment begin in earnest: possibly one of the earliest examples of what John O'Reilly[11] would dub in 2004 the 'slow burn', illustrating just how long it takes to evolve from an experimental research demonstration of an idea all the way to deployment. We shall meet several examples of this in the story of JANET.

ARPA, which had begun the design and construction of its experimental, prototype packet switching network in 1967, had installed the first four switching nodes in 1969, and by 1972 had just over 30 nodes spanning the USA. As early as 1970, the need for separate high-level – including application-level – protocols had been recognised and some of the principles of how to achieve this in a system-

---

7   For further information on the origin of the ARPANET, the reader is referred to the authoritative article by Steve Lukasik (2010), Director of ARPA 1970–75. I am grateful to Peter Kirstein for bringing this article to my attention and to Steve Lukasik for early sight of it.

8   Probably best known as inventor of the mouse (with William English).

9   *See* Glossary.

10 *See* Glossary.

11 Professor Sir John J. O'Reilly, KBE, then CEO of EPSRC, in his keynote address at the UK e-Science All-Hands Meeting 2004, Nottingham, UK.

independent, non-proprietary way articulated. In 1973, Peter Kirstein[12] installed the first node outside the USA at the University of London's Institute for Computer Science (ULICS), the forerunner in part of the Department of Computer Science at UCL. Before the end of the year, the UCL node had been connected to the IBM 360/195 mainframe computer at RHEL and the world of international computer networking was opened up within the UK to an initial community of users – and for a short time the most powerful host on the ARPANET was in the UK.

We shall see in the next chapter that it is in this same year, 1973, against a background of computing provision in the university and Research Council sector dominated by large and expensive mainframes, that Mike Wells[13] submitted the first of his reports to the Computer Board (Wells, 1973) advocating the construction of a national network to serve the whole of academia.

1.3 Beginnings

As a result of the Computer Board's strategy, by the beginning of the 1970s the computer services available to a user might be a combination of local, regional and national. The batch model of computer processing has been mentioned, a scheme in which the end-user has no direct interaction with the computer system. While this might be satisfactory for a major computation once the program to do it was working properly, for almost all other purposes it was a highly inefficient use of people's time and was ultimately only justified by the enormous expense of the computers and the accompanying resources to run them. The advantages of what came to be called interactive computing were soon recognised but a way was needed to enable a number of users to 'share' the system simultaneously. So was born the multi-user interactive system, in which a user had a terminal which could be used to type directly into the machine and see the results directly displayed. Initially a terminal was similar to a typewriter; later the output was displayed electronically on a screen rather than being typed out on paper. A part of the computer operating system arranged that each user had a 'slice' or share of the machine, and the appearance to each user was of a dedicated system, although it was slower if there were a lot of people using it. It is evident that such an arrangement is a compromise, forced entirely by economics. Even so, interactive systems were typically more expensive in a number of ways than simple batch systems. Nevertheless, for some purposes the advantages were such that the economics could be justified, and in the early 1970s both forms of service were in operation, often combined.

Thus, during the first half of the 1970s, there grew up around the regional centres funded by the Computer Board a set of regional remote access arrangements, all based initially on the proprietary technology dictated by the regional computer system. An exactly similar evolution took place in respect of the three SRC facilities at RHEL, Atlas and DL, with one difference that, because these were national facilities, the stars had national, if sparse, coverage.

At this point, the plethora of arrangements becomes evident. Two factors dominated. The design of computers at the time was such that they more or less directly controlled all their devices (peripherals), be they punched card or paper tape readers for input, disk and magnetic tape devices for storage, and

---

12 Professor Peter T. Kirstein, CBE, then Head of the Networks Group at ULICS; *see* Glossary.

13 Professor Mike J. Wells, then Head of Computer Services at the University of Leeds.

lineprinters, card punches and paper tape punches for output. In cases where an interactive terminal service was provided, these terminals were also under the direct control of the mainframe computer. When it became clear that users required access to these expensive machines from remote points without the need to visit the machine, the response was to work out ways in which a telecommunication line provided by the telephone company could be interposed between a single terminal or a group of terminals and the central system. Each manufacturer did this in its own way but all such arrangements retained the essential control which the mainframe had over all its peripheral terminal equipment. The proprietary nature of the peripherals and the assumption of dedicated control by the central machine meant that these remote terminals could not be used for shared access, even between two systems of the same type, let alone among systems from different manufacturers.[14]

The other factor arose as a direct consequence of the hierarchical arrangements for provision of computer services, whereby each university had its own computer for the support of its students and the majority of the work of its staff. For somewhat more demanding work it had access to use of its regional centre. And for the most demanding work its staff could make use of the national centres. Those generally available to all were funded by the Computer Board. In addition, research groups in the sciences supported by SRC (and increasingly also the Natural Environment Research Council, NERC) could make use of Research Council facilities at Atlas Laboratory, Daresbury Laboratory and RHEL. So by the early 1970s, a university might find itself with dedicated remote links to its regional centre, possibly a Computer Board national centre, and probably to at least one of the SRC centres. Some of these would be for batch processing based on cards and lineprinter; others might provide interactive terminals; and some provided both. And if there was no remote link then batch computing would be by courier service providing 24-hour job turnaround, possibly supplemented by a Teletype[15] interactive terminal using dial-up over a slow, error-prone line. (Some examples drawn from the author's experience are described in the panel 'Examples of remote access in the late 1960s/early 1970s'.)

The public sector, by the beginning of the 1970s, was also actively considering the possibility of a public packet-switched service for data communications. In 1971 the beginnings of two activities were announced. In the UK, the GPO announced its intention to mount an Experimental Packet Switched Service (EPSS) and invited participation from interested parties in industry and academia. In Europe, agreement was reached to mount an inter-government packet switching network trial, originally known as the COST[16] 11 Project (COST, 1971), later renamed the European Informatics Network (EIN), directed by Derek Barber, one of Donald Davies' colleagues at NPL.

A number of universities signed up to join EPSS, particularly those associated with regional and national centres, as did three of the SRC laboratories: Atlas, Daresbury and RHEL. EPSS was one of the earliest such trials announced by a

---

14 It should not be inferred from this that such sharing was not technically possible: simply that it could not be purchased.

15 A Teletype was a very slow, cumbersome terminal like a clumsy electric typewriter, but with keys requiring an inch of depression for data entry, which printed on a roll of paper like a credit card receipt only wider, using a font reminiscent of an antediluvian telegram, with no lower case. It operated at 10 characters per second (110 bps). Dial-up over the analogue telephone system of the day gave rise to frequent errors owing to noise on the line.

16 European co-operative support organisation. *See* Glossary.

Examples of remote access in the late 1960s/early 1970s

In London in 1969, colleges had access to the CDC 6600 system at ULCC, which initially only provided a batch service. Remote access to this was available via a 2.4 / 4.8Kbps line using a CDC UT200, CDC's RJE terminal. A college might also have access to an IBM360/65 at UCL but in this case by van or courier service across London.

In 1973 Newcastle had an IBM360/67 (the first System 360 machine with virtual memory). It provided both batch and interactive facilities by a combination of IBM system for batch and an interactive system developed at Michigan University, known as the Michigan Terminal System or MTS. The system (known as the Northumbrian Universities Multiple Access Computer, NUMAC) was shared with Durham which had a 48Kbps link to Newcastle supporting a collection of IBM interactive terminals (similar to the IBM 'golf-ball' typewriter), as well as an IBM 1130[1] RJE station with lineprinter, card reader and operator console. (For amusement and comparison with the single chip accompaniment to a laptop in 2007 providing broadband modem facilities at around 8Mbps, the IBM modem supporting the 48Kbps line was only a little smaller than a two-drawer filing cabinet.) In the next room a GEC 2050, programmed by RHEL staff and having its own lineprinter, card reader and operator console, was connected by 4.8Kbps line to the IBM 360/195 and provided RJE facilities using exactly the same set of access protocols[2] as the 1130 in the next room. But to have shared the IBM 1130 RJE between the RHEL 360/195 and the NUMAC 360/67 was not a manufacturer-supported capability and would have required reprogramming the 1130.[3]

Many universities had ICL 1900 systems, a consequence of the UK's 'buy British' policy at the time. As we shall see in Chapter 2, in each of the regions, one of the systems would be larger than the others and designated to provide regional services. ICL also had a proprietary RJE terminal, the 7020, functionally similar to a CDC UT200.

The proprietary protocols for these RJE terminals were typically referred to as 'CDC UT200 protocol', 'ICL 7020 protocol' and 'IBM RJE protocol'.

---

1   IBM also had a more basic RJE terminal, the 2780, directly comparable to the CDC UT200 and the ICL 7020. However, using the IBM 1130 (which was a very early example of what would later be called a mini-computer, albeit with a much simpler, single-user operating system) allowed more control functions from the system console.

2   Protocol: the fundamental procedures by which a network operates. *See* Glossary.

3   That this was technically feasible had been demonstrated at ULICS, for example, by programming emulators in the PDP9 there for both CDC UT200 and IBM RJE support. As the community more widely took on supporting its own communications, this approach was adopted in a number of the regional networks.

Public Network Operator (PNO),[17] and showed in part the influence of Donald Davies and his team who had had briefing meetings with the GPO and the Department of Trade (which was responsible for both the GPO and NPL) in the latter part of the 1960s. Although it had been hoped that EPSS might begin operation in 1975, in the event it would actually begin operation in 1977; it would turn out to be successful and would later lead to the trial introduction of BT's Packet Switched Service (PSS) in 1980, with full service in 1981.[18]

It is against the background of development over a ten-year timescale described in this chapter, the success of ARPANET, and the seeds of development in the public network sector in the UK, that the Computer Board decided to investigate whether a computer network to support university research and education should be set up, partly in the light of the escalating costs by then associated with the provision of computer service access – both the computers and the associated communications.

---

17   Termed a Postal, Telegraph and Telephone (PTT) operator at the time.
18   IPSS was available earlier than PSS: *see* Glossary and Chapter 7.

2

ORIGINS OF
A NATIONAL NETWORK

At the beginning of the 1970s, as another generation of computers was unveiled by the industry, it became clear that the technology was one which was evolving much faster than any previously encountered in supporting university infrastructure. As ever, to remain competitive, researchers needed leading edge computing facilities; and, increasingly, students also required access to at least modest facilities. In view of the escalating costs of this, there was an initial line of thinking within the Computer Board (see Computer Board, 1970) that it might need to concentrate provision in the regional (and national) centres, and that universities might need to collaborate in developing the resources at these centres. At a subsequent meeting in 1972, the Board recognised that it would need to give increased attention in future to sustainable plans for networked communication access to its facilities (Computer Board, 1973a).

By 1972, the success of ARPANET had gained considerable international attention. It already had just over 30 nodes and was expanding rapidly. And indeed there were those in the university community who were advocating that the UK should build its own ARPANET, thereby addressing not only access but also collaboration. In addition, in the previous year, the Post Office had announced its intention to launch its own experimental packet network, EPSS; and under the COST 11 initiative, six (subsequently to rise to ten) Governments in Europe had announced agreement to launch a packet network trial in the form of EIN (COST, 1971).

Following the publication of the Computer Board report in 1973, it was suggested that a delegation visit the ARPANET to assess developments. Accordingly, a team of four, including Mike Wells, visited the USA for this purpose in the Spring of 1973. On its return the team reported to the Board (Computer Board, 1973b), and its lead recommendation was that a working group should be set up to 'examine the technical and economic feasibility of a British network compatible with ARPANET'.

2.1 THE WELLS REPORTS

The period 1973–75 marks the beginnings of JANET. That there would be a decade of political and technical development, together with an exploration of the ill-defined regulatory situation of the day before JANET became a reality, does not alter the significance of the Wells Reports. Through them was promulgated the

14 | 2. Origins of a National Network

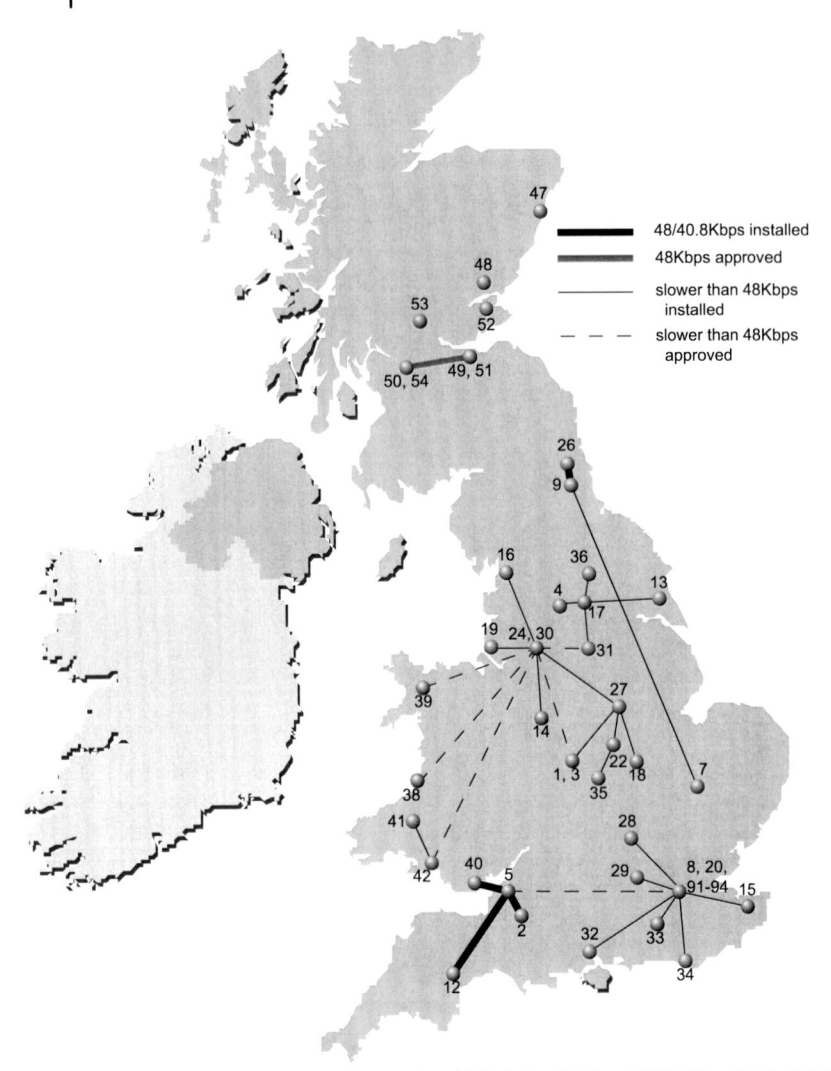

Figure 2.1. University networks, September 1973 (after Wells, 1973). (*See* Table 2.1 for list of sites.)

future potential of networking in higher education and the need for a programme for its support.

The first Wells Report

Following the North American visit, Mike Wells was asked in June 1973 to set up and chair the recommended working party,[1] with the main brief above. It was also asked:

---

1   *See* Appendix A for membership.

Table 2.1. Key to site numbers in Figure 2.1.

No.	Site	No.	Site	No.	Site
1	Aston	20	London Graduate	40	Cardiff U.C.
2	Bath		School of Business	41	St. Davids Lampeter
3	Birmingham		Studies	42	Swansea U.C.
4	Bradford	22	Loughborough	44	University of Wales
5	Bristol	23	Manchester School of		Institute of Science
6	Brunel		Business Studies		and Technology
7	Cambridge	24	Manchester	47	Aberdeen
8	City	26	Newcastle	48	Dundee
9	Durham	27	Nottingham	49	Edinburgh
10	East Anglia	28	Oxford	50	Glasgow
11	Essex	29	Reading	51	Heriot-Watt
12	Exeter	30	Salford	52	St. Andrews
13	Hull	31	Sheffield	53	Stirling
14	Keele	32	Southampton	54	Strathclyde
15	Kent	33	Surrey	91	University of London
16	Lancaster	34	Sussex		Computer Centre
17	Leeds	35	Warwick	92	Imperial College
18	Leicester	36	York		Computer Centre
19	Liverpool	38	Aberystwyth U.C.	93	Queen Mary College
		39	Bangor U.C.	94	University College

- to advise in what ways such a network would be likely to be useful to the universities in helping to meet their future computing requirements; and

- to assess the relevance of existing and proposed (university) computer networks in Britain and abroad.

The Network Working Party reported its findings in October of the same year (Wells, 1973). The report in effect made recommendations in two areas: construction of a national university network and computer provisioning strategy taking account of such a network.

In respect of initiating construction of a network, the report recommended introducing packet switching facilities in parallel with existing arrangements already in existence (shown in Figure 2.1). The proposed network (Figure 2.2) was to be based on three nodes: one at each of Edinburgh, Manchester and London, the three Computer Board national centres. As noted both by the Working Party and in a prior SRC report (SRC, 1973), these three centres were building up substantial (star) networks: it would be sensible to exploit these, as well as being politically expedient.

The design of the switching nodes of the network – termed a BUNNIE (for British UNiversities Network Interface Equipment) – and the protocols to which the network would operate were to be designed by a technical group convened by the Board. The formation of a second group was recommended to determine how the management of the future network and its support services should be organised. The envisaged timescales were for completion of the

Figure 2.2. Proposed 3-node backbone network (after Wells, 1973).

protocol specifications, node design and recommendations on the future form of organisation by April of the following year, 1974, and a projected installation date for the initial network of April 1975.

In terms of recommending how computing facilities provision should be organised in the presence of a network, the Working Party faced a difficult task. There was considerable concern in the community that the Computer Board might take the view it should exploit the network as a tool towards more centralised provision, on the basis that a few large centres might be cheaper to equip and run than the existing situation whereby every institution was entitled to a facility of appropriate size on campus. The Working Party no more shared this view than the community as a whole, and its first recommendation was to endorse the existing principle of a facility at every university, in support of both teaching and research. Its view was that there was much more to be gained by using the network to provide enhanced community-wide access to a wide variety of expertise and facilities – whether data repositories, powerful processors (perhaps of various sorts), computational expertise or specialised computational packages. The network would allow the Board to encourage and increase support of such specialist facilities in a few places, rather than providing diluted support across a wider base. It was suggested that the existence of a network infrastructure would allow the concept of balanced provision of facilities to be achieved over a distributed group, rather than being confined to a single site. So, the network should be viewed as an opportunity to enhance service at reasonable cost, not as a means ultimately to achieve savings – something which in any case it could not do while under construction and before existing communications arrangements had been phased out.

One may perhaps recognise here an instance of what in retrospect is a well-known facet of the introduction of new technology, namely, the tension between viewing the technology merely as a means of saving costs (often through perceived savings in effort, possibly equipment) on the one hand and as an enabler of new capabilities on the other. In practice, the former has seldom realised expectations: more typically there is initially a shift in staff skills required, especially in support of the new technology, and eventually the real benefits turn out to be the latter with the development and adoption of new techniques and services, coupled with a strategic ability to cope with scaling in demand. In the case of JANET, even after it was launched in the following decade, it would be a further decade before the argument of opening up new, 'undreamed horizons' would be fully recognised (though admittedly coupled in the event with a demonstrable need for extra network capacity). The problem is that funding something new with only partially understood or specified objectives is something of a leap in the dark, with risks and costs which are hard to estimate. An apparently simple accounting argument, even if based on only partially articulated underlying assumptions, may appear more acceptable.

The Working Party was fully cognisant of the development activity in packet networking at the time. It included some of those who had visited the ARPANET; it was aware of Peter Kirstein's work at UCL in linking to ARPANET and making it available to university and Ministry of Defence researchers; it had had discussions with the Post Office and knew of its plans for EPSS (which, incidentally, also planned a very similar network of three nodes, situated at London, Manchester and Glasgow); likewise, it was aware of the plans for EIN. Importantly, the Working Party recognised that what was needed was a service network, not another experiment, and that there was a substantial difference between the two.

A major argument against basing the network on either ARPANET or EPSS was that both were experimental. Although EPSS was intended as a precursor to a UK service, the latter was predicated on the outcome of the experiment and there were only estimates of timescales for that and, consequently, for any service. Notwithstanding that ARPANET provided high-level application services,[2] it was not only experimental, it had no remit to evolve into anything else, even in the USA. Basing any proposed service on either of these would prima facie have been inappropriate. In fact, in the case of ARPANET, there was an additional

---

[2] The work at UCL provided terminal access and file transfer, the latter by means of mapping to IBM RJE.

issue. ARPA was a US agency with no formal association with the UK. Basing a UK service network on ARPANET would have left the UK network without control over its own technology – and, indeed, it was already known that there were likely to be changes in the ARPANET protocols. Moreover, expertise in development and operation of its network would be a continuing requirement for the UK community but, with development of ARPANET technology centred in the USA – in spite of the contributions of Peter Kirstein's group at UCL – it was very likely there would be continual leakage of the best UK expertise to the US. This was no idle speculation: the scientific 'brain-drain' to the USA was something which the UK had by then been experiencing for two decades owing to post-war lack of finance to match USA facilities.

Although the technical detail of node design was to be done by a new group, there was a need to address the strategic aspect of what protocols the network should use. Recognising that the network protocols needed to be independent of any manufacturer – since the current position owed not a little to the use of manufacturers' proprietary protocols for interactive and RJE terminal operation – the Working Party recommended that a priority should be the adoption of community-wide standard protocols: indeed, national or international standards if possible. Interconnection with external networks such as ARPANET and EPSS could be accommodated by constructing gateways or 'relays'[3] which could be logically incorporated into the nodes.

With hindsight, it is easy to regard the proposed timescales as aggressively ambitious. However, this was not what in the event contributed to the Report's rejection. There were several issues, all ultimately to do with the novelty of networking. These could be summarised as concerns about the use of packet switching technology; potential loss of local facilities within a university; and uncertainty about the goals of networking.

As we have seen, at this time packet-switching had been very successful in an experimental network, ARPANET, but although there were plans to exploit it in incipient service networks, this had not yet been achieved anywhere. Those in contact with the technology and its development were convinced about its choice but for those concerned with funding and using a service, it remained unproven. Moreover, in the regions, there was already substantial existing investment of equipment and effort based on direct use of circuits (Rosner, 1996): was all this to be wasted? Although the report had recommended the introduction of packet switching facilities alongside the present circuit-switched arrangements, it was clear that the intent was to evolve to using the network once it was operationally proven. Indeed, perhaps in an effort to establish the credibility of the technology, the report included considerable technical detail of the design of the proposed network. For a community and a committee unfamiliar with the whole concept, the attention to technical detail served rather to emphasise the lack of a well-defined strategic objective, instead of encouraging the adoption of a networking approach. Faced with persuading the Board that packet networking was the way to go – something the Working Party had regarded as self-evident and so had not spent time on in the report – it didn't help to have the somewhat jocular acronym BUNNIE as the sobriquet for the main component of the network, estimated at some £125k for the three – a cost of around £1m at 2008 prices.

The real difficulty, recognised at the time by Mike Wells, was that the whole idea of interconnecting computers in open fashion was new territory. Although large commercial enterprises such as the banking and airline businesses had installed

---

3   *See* Glossary.

networks, these were for well-defined purposes, within tightly controlled closed environments, and accessed only by disciplined and trained staff. The networks could be closely tailored to meet detailed targets, supporting specific applications and showing a return on investment. For the research and education community, which is of its essence an open one, the Working Party was proposing a network which would support many different applications, perhaps the majority not yet conceived, accessible to users who were by their very nature unrestrained in originality of use. The network could not have detailed targets and it would be very difficult to show any specific return on the investment. A point which perhaps is only truly capable of appreciation in retrospect was that the real significance of ARPANET was not so much the development of the technology, important as that was, but the beginning of the exploration of what such an ubiquitous infrastructure might eventually mean. Without this, the Computer Board could only fall back on existing experience which, since the Research Council sector had not then quite begun its own network developments, essentially stemmed from the business perspective of well-defined objectives and a clear business case, neither of which were possible.

Comments from the community and Computer Board crystallised into a central concern: why, exactly, should a national network be built? What was the purpose? This was a question which the Working Party had not addressed, perhaps in part because it was sufficiently familiar with the concept as to regard it as a *sine qua non*. Thus the focus of the Report was on how rather than why. Without this underpinning explanation of purpose, the recommendation that every university should continue to be provided with its own computing facilities was substantially weakened. The upshot was that the Board requested a further examination of the issues.

The second Wells Report

The second Network Working Party was set up by the Computer Board in June 1974. The degree of antipathy to the first Wells Report may be judged from the rather blunt summary included in the second Report of the criticisms received: '... the general consensus of the replies received [in response to the Report's publication] was that the approach set out in the Wells Report was wrong'. In particular, a 'major criticism of the Report was that it had not defined the aims of a network development with sufficient clarity'.[4] True, perhaps, if a rather crude summary of what has been discussed above.

The second Working Party had membership[5] which included some of those in the first Working Party but was augmented by representatives from SRC Daresbury Laboratory and Edinburgh University; it reported in October 1975. In the intervening couple of years, things had moved on. People had been using remote access for a few years by then, the number of installed lines had grown (Figure 2.3), and quite a few had been making use of facilities in the USA via the UCL ARPANET link: at the operational and user level, networking was no longer so novel. But the issues of costs, benefits, management and organisation remained.

---

4   Wells (1975), sections 1.1-1.2.

5   *See* Appendix A for membership.

20 | 2. Origins of a National Network

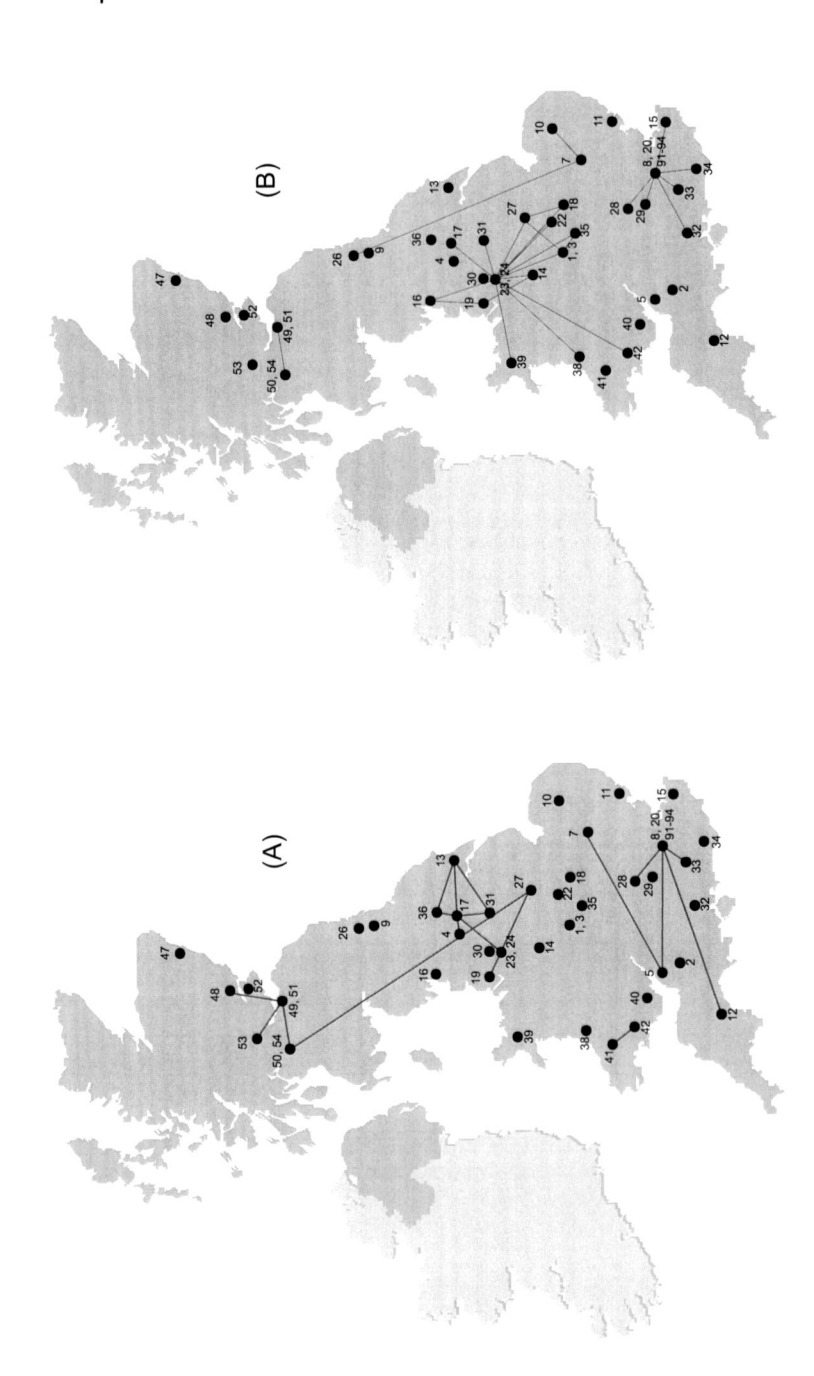

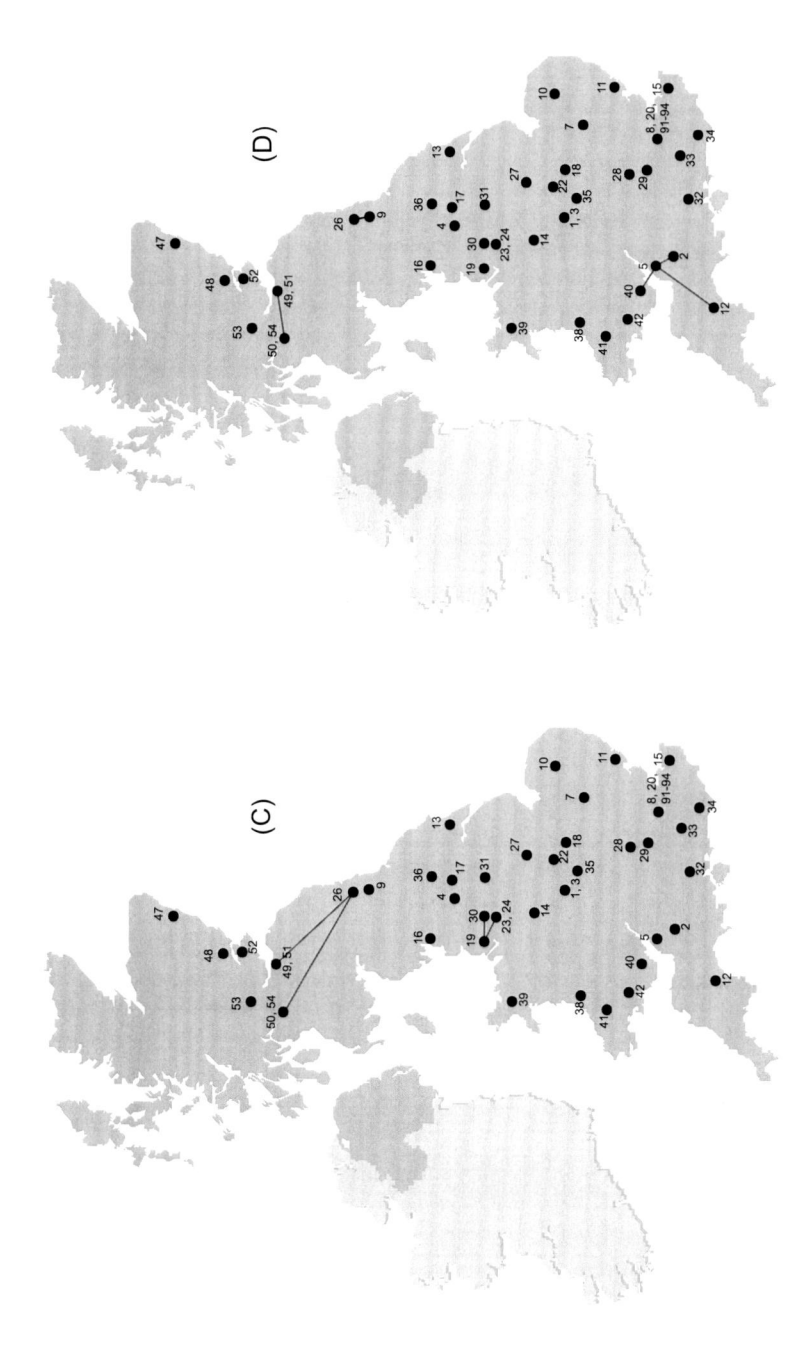

Figure 2.3. Lines funded by Computer Board as at 1975 (after Wells, 1975). (A) 2.4Kbps lines. (B) 4.8Kbps lines. (C) 9.6Kbps lines. (D) 40.8/48Kbps lines. (See Table 2.1 for list of sites.)

The second Wells Report reiterated much of what had been said in the first Report but this time, instead of focusing on the technical, it concentrated on the strategic aspects, as well as putting some costs on the current arrangements; however, the underlying thrust in regard to technical concept remained unchanged. As part of assessing the costs of communication, it was found that at least £15,000 per annum was being spent on courier services for remote job submission and retrieval where installing a leased line could not be justified. One aspect of this was to emphasise the simplicity and effective data rate of transporting a vanful of the highest available recording media (6250 bit per inch tape, DAT tape, CD, depending on decade): the latency might be poor but the data rate has always been substantial.

Much political capital has always been made of the apparently irrational organisation of the lines leased from the Post Office in use at this time in academia. There was indeed a plethora of lines, of a variety of speeds, mostly interconnecting universities with regional and national centres. The SRC's laboratories, Daresbury and Rutherford, served departments in universities nationwide and, consequently, the star networks centred on these two laboratories were of national extent – as indeed were some of the links to the Computer Board national centres at ULCC and UMRCC.

A major component of the recommendations of the second Wells Report was to focus on encouraging the Computer Board and the Research Councils – primarily the SRC at that time – to collaborate on a national network infrastructure. Although the specific recommendations focused on potential economies such as would follow from rationalisation of lines, the real benefits would derive from the unified infrastructure: and to this end the Computer Board was recommended to allocate some of its budget for communications hardware 'towards the provision of coherent facilities serving the universities and Research Councils'. Furthermore, they were recommended to enter into negotiations with the Post Office together, with a view to provision of an academic network based on EPSS developments and to be operated for the sector by the Post Office.

The key enabling recommendation was 'that the Board and Research Councils set up a small, full-time unit to co-ordinate, guide and rationalise network development involving those bodies receiving their principal funding from the Department of Education and Science'. This was the recommendation which presaged the creation of the Joint Network Team three years later: the immediate effect was for the Computer Board to set up the Network Unit just over a year later to take the next steps.

Although in many ways the second Report had not really changed the essential thrust contained in the first Report, the idea of a network was now more familiar both to the Computer Board and the community generally – indeed various parts of the community were already feeling the need for such an infrastructure, demonstrated in part by the beginnings of regional networks and the SRC network, both of which had been taking place in the background during this time (see sections 2.2 and 2.3). The combination of demonstrable need and a Report which focused more on the strategic requirement now convinced the Board to act.

2.2 Regional Networks

Before looking at the activities of the Network Unit in the latter part of the 1970s, it will be helpful to have an appreciation of what had been happening 'on the ground' up to and during this time. In this section, we look at remote access developments funded by the Computer Board and in the next section we will look at the SRC sector.

We have seen that for the purposes of computer provision, the Computer Board tended to split the country into regions. In a number of cases these regional groupings followed those which were already established before the Flowers Report. Typically, there would be at one university in the region a system distinguished by being either large or providing specialist facilities, possibly both. By the end of the 1960s, the manufacturers of large systems had developed RJE terminals and regional star networks of these grew up to provide access to the regional centres. In the case of London and Manchester, the CDC-based facilities at each were designated to provide a national service as well as a regional one. Each developed links to regions outside its own, with UMRCC serving the northern part of the country and ULCC the southern.

While CDC 6600 or 7600 systems might be particularly appropriate for large numerical calculations, there were other types of large-scale data processing which could be better handled by IBM 360 or 370 systems. The regional systems at Cambridge and Newcastle were of these types and were designated to offer IBM-based services on a national basis. For example, the original NUMAC IBM 360/67 at Newcastle was for use by Durham and Newcastle Universities;[6] in the 1970s, when a larger system (an IBM 370/168) was funded, a proportion of the system was designated for use beyond NUMAC, and universities in the RCOnet region in Scotland also made use of it.[7]

The late 1960s and the whole of the 1970s was a time of growth for operating systems which could offer interactive terminal services. Examples included Titan and Phoenix at Cambridge, EMAS at Edinburgh, MTS (from the University of Michigan) used at NUMAC, Multics (originally the result of collaboration amongst MIT, GE[8] and Bell Labs) used at AUCC in Bristol, and ICL's George which was in use widely in the UK on the 1900 range.[9] Accordingly, a number of the regional networks also developed a capability for remote interactive terminal access, in addition to RJE access. Brief illustrative descriptions follow of some of the regional networks in operation in the 1970s (listed in Appendix B, together with those in operation in 2008).

---

6   Historically, in Newcastle, Armstrong College (an engineering college) and a College of Medicine merged to form King's College of the federal University of Durham, before King's College became a separate university, Newcastle University, in 1963.

7   Linda McCormick, private communication, September 2008.

8   Originally developed for the GE645; by 1973, GE had sold its computer business to Honeywell. (Peter Kirstein, private communication, January 2008.)

9   It was also the time of development of UNIX – which emerged from Bell Labs in 1969 – but it would be another decade before this saw widespread adoption and deployment in the UK, primarily, though not exclusively, on single-user workstations.

SWUCN

In the 1970s, the constituents of the South West Universities Computer Network consisted of the Computer Board grouping of the Universities of Bristol, Bath and Exeter, University College Cardiff (UCC), and University of Wales Institute of Science and Technology (UWIST).[10] Initially, the participants wanted to be able to share facilities across a set of identical systems in the region, as well as being able to submit jobs to the national service at ULCC. Subsequently, the South West Universities Regional Computer Centre (SWURCC) was formed to operate an ICL 2900 system situated in Bath to provide a regional service: it was connected to the network and began service in 1977. The following year, Bath and Bristol pooled resources to replace both university mainframes with a single, shared Honeywell Multics system, run by the Avon Universities Computer Centre (AUCC), a consortium of the two universities situated in Bristol. The Multics system was delivered and commissioned in 1979 and began providing a service over the network in 1980.

The South West Universities first began consideration of a network for the region in 1967 (Powell, 1980) and, under the auspices of the Department of Trade and Industry, SWUCN began design and implementation of its initial network with the Post Office and ICL in 1969. The network came into operation in 1974 and was in the form of a star, centred on a CDC 1700 minicomputer at Bristol acting as a message switch, interconnecting four ICL System 4s (UCC and UWIST shared a system) using 48Kbps lines. The systems were not only of the same type but also used the same operating system. By using what were for the time high-speed links, the systems were to a degree able to provide a single system image. The consortium was able to apply a measure of load balancing amongst the systems. Moreover, some facilities were only provided by one or other of the systems but were made available to all users by means of the network: an example of this was the interactive BASIC service available on the larger Bristol System 4-75, which was made available to the whole user community over the network.

The degree of sharing achieved amongst the four systems required that from an interactive terminal on any of the systems users could log into any of the others; files could be transferred amongst all four system filestores; jobs could be submitted from any system to any other, with the output being returned to the originator, all under operator control; and users could exchange stored messages with each other regardless of which system they were using. A user could also submit jobs to ULCC by constructing a CDC job and then using a SWUCN subsystem to perform the CDC UT200 RJE function.

The rich set of network facilities had been achieved by virtue of all the systems being identical, avoiding all issues of heterogeneity (other than RJE to ULCC) and exploiting native system facilities of the operating system and communications subsystems. By 1977, when the SWURCC 2900 had to be integrated into SWUCN and components of the SWUCN network were also in need of replacement, it was time to begin to rethink the network strategy. At the time, the need for a full rethink was avoided because the 2980 was defined to provide only a batch job service, so it was possible to implement a variation on the mechanism used for submission to the CDC service at ULCC: users created a 2900 job and then used another SWUCN subsystem which performed ICL-style (7020) RJE to the 2980.

---

10 UCC and UWIST merged in 1988 to form the University of Wales College, Cardiff; in 1996 it became the University of Wales, Cardiff. Then, in 2004, it merged with the University of Wales College of Medicine, separated from the collegiate University of Wales, and became Cardiff University.

However, it was recognised that such an approach was only possible because full function network access was not required at that time.

With the coming of the Multics system to AUCC, it was time to replace the SWUCN network components and embark on the process of moving to non-proprietary, more standard protocols (Thomas, 1978; and Powell, 1980). The new network was based around X.25 and the emerging set of 'Coloured Book' UK interim standard protocols (see Appendix E). The new Multics system was interfaced to these directly; the old systems were interfaced by means of protocol translation in a 'Network Interface Processor' system implemented on a GEC 4000-series minicomputer system which already had X.25 software available.

METRONET

The London Metropolitan Network was essentially a star network, based on ULCC, which grew out of the initial need to make the CDC 6600 batch job service available to colleges of the University of London in 1969. In 1970 a terminal service was offered on the companion CDC 6400 system. By 1972 the main service was based on a CDC 7600, and as one of the national centres the service was being offered outside London to the rest of the South East region as well as to the South West region. The initial network was based on use of native CDC protocols: RJE, for example, being provided by the CDC UT200 terminal. In the late 1970s investigations began into replacing the existing arrangements and moving to an X.25[11]-based network (METRONET II); however, a decision on this was delayed by the same issues as would affect the establishment of a national network and, although there were discussions with the Post Office about the possibility of basing METRONET II on an X.25 exchange operated by the Post Office, in the event ULCC would begin its own X.25 network construction to serve the University of London in 1981 (Brandon, 1978).[12]

RCOnet

The Regional Computing Organisation had its origin in a joint submission by the Universities of Edinburgh and Glasgow to the Computer Board in 1970 to support computing services at these universities and the University of Strathclyde, in particular by the development of a linking network, RCOnet.[13] Development of RCOnet began in 1971 and it came into operation in 1973, with the formal establishment of RCO. Interactive terminal access was provided as well as RJE, the service being provided by an ICL 4/75 and a rented IBM 370/155. By 1978 (see Davies, 1978), there were ICL 2900 systems at Edinburgh and Glasgow and an ICL 1904 at Strathclyde. There was no longer an IBM system in the region: IBM service was by then provided by the NUMAC 370/168 at Newcastle, with MTS terminal access available via NUNET[14] using ITP (Interactive Terminal Protocol,

---

11   *See* Glossary.

12   Rolly Trice and Danny Bramman, private communications, April 2008.

13   Linda McCormick, *ibid.*

14   *See* Glossary.

an interim protocol in use in parts of the UK before the Coloured Book protocols became available – *see* Chapter 3).

RCOnet was connected to EPSS when the latter came into operation and this provided the connection to NUMAC. RCOnet was sufficiently early in operation that replacement nodes were needed by 1979, before commercial X.25 switches were available. Thus although by then it was known that X.25 would be the protocol used in the academic network and external connections using X.25 were in use, the change to use of X.25 within RCOnet began later, in 1984, around the time when JANET became officially established, and was completed in 1987.

CYGNET

In common with other regional groupings at the time, the Yorkshire region began forming a network in the early 1970s in order to share access to computing resources. The region consisted of the Universities of Bradford, Hull, Leeds, Sheffield and York. Apart from York, which had a DEC10, each university had an ICL 1900 series system with the largest in 1973, a 1906A, being at Leeds. In order to access the major regional batch facilities a star network formed, centred on Leeds, based on ICL's proprietary 7020 RJE terminal protocol. Subsequently, when the larger, later model 1906S arrived in Sheffield, the network was expanded to enable access to the new centre for regional batch processing.[15] Again, this was an example of a network stemming from the need for shared access to processing facilities and, because the majority of systems were from the same range and used the same operating system, load balancing was also achieved.

GANNET[16]

The North-West Universities network had its origin, as others, in a common interest in enabling access to the Computer Board regional (and national) computer centre, UMRCC. In the early 1970s, each university in the region – Lancaster, Liverpool, Keele and Salford – had a 4.8Kbps line connecting an RJE terminal to the UMRCC service, by then provided by a CDC 7600. Members of the incipient consortium began to collaborate on a regular basis from 1972 and in 1974 approval was given for installation at Liverpool of an ICL 1906S (the largest in the 1900 range). This was also designated to augment the regional service via RJE links from Keele, Lancaster and Salford. The prospect of imminent proliferation of Post Office lines, with the concomitant cost, stimulated the beginning of discussion about forming a network. Apart from reducing the number of lines necessary to achieve the necessary RJE links in the region, it also offered the prospect of UMRCC providing a gateway for the region to EPSS and hence services in other regions. In common with other regions, the possibilities of other potential advantages were also identified, among them load sharing and access to specialist facilities at one or other of the sites in the consortium (Lindley, 1978; Rice, 1978).

Generally, the requirements identified for the North-West Universities Network included RJE terminal capability at each site capable of both local and remote

---

15 Private communication from Mike Wells, December 2008.

16 The name is an acronym stemming from the ICL project which developed the technology.

access; local and remote interactive terminal access; and file transfer amongst all of the hosts. In 1974 all sites, except Keele but including UMRCC, had ICL equipment consisting of a 1900 mainframe and a 7905, ICL's standard front-end system at the time. The system at Keele was replaced during the implementation of the network and, in order to simplify construction, an ICL 7905 front-end was installed together with a GEC 4000 system. The timeframe in which the network was required dictated purchase of existing technology. The only real contender, in view of the need to utilise existing ICL equipment, was to base the system on an R&D project called 'General Administrative Network' (GANNET)[17] which ICL had been pursuing since 1972 in conjunction with the DTI and the Central Computer and Telecommunications Agency (CCTA). The project was demonstrated, with favourable results, to the North-West Universities in 1975. In 1976, as the original GANNET project ended, implementation of the North-West Network began as a collaboration amongst the North-West universities and ICL, with funding from the Computer Board. The eventual topology of the network was a ring of 7905s, augmented by additional links so that every site had direct connections to the regional service sites at Liverpool and UMRCC. Experimental service began in the summer of 1977 with full service following in 1978.

MidNET

The Midlands group of universities in the mid-1970s was comprised of Birmingham, Aston, Warwick, Nottingham, Loughborough and Leicester. As reported by McConachie (1978), the network at that time was used only to provide access to the national centre at UMRCC. Each university had its own system, mostly ICL 1900s, though there were Burroughs and CDC systems at Warwick and Leicester, respectively, and Loughborough had twin Prime 400s in addition to a 1900. With an eye to the future, there were plans to develop the network over the next few years to provide more general interconnection facilities, using PDP11/34s to provide switching in addition to their other roles. The strategy was to use X.25 as the network protocol so that the network would be well placed to take advantage of national facilities when these became available.

By the autumn of 1979, progress had been made in both design and implementation (Harrison, 1979). The initial services would provide for all-to-all interactive terminal services and file transfer. Job transfer services, including those to UMRCC, would evolve to use of a non-proprietary standard protocol when available. By the end of 1979 the basic X.25 network was complete, using the 11/34 nodes in a ring topology to provide resilience, and application protocol implementation was well under way. Because design had begun in the latter half of the 1970s, the UK interim standard protocol strategy was already clear and MidNET implemented what would become the PSS and JANET core protocols from the start.

---

17 Robin Tasker and Jack Houldsworth private communication, September 2008.

What we see is regional networks all over the country moving in similar directions, albeit with varying priorities and timescales, influenced both by procurement cycles and differing emphasis regarding the nature of the services to be provided, so that by the time JANET begins, most of the regions are either already using a common core set of protocols or are close to doing so.

In retrospect, one can observe just how much the uncertainty surrounding the Post Office situation, both in respect of the monopoly position (see below) and the nature of the service which would be provided, introduced delay. Had the original Post Office timescales for a service not slipped – both EPSS and PSS operation were delayed by two years – it is possible that several regions might either have made use of the service directly or a spin-off service, public or private. Nevertheless, the presence of the national groups defining interim national standards, in which all regional networks participated, ensured that there was protocol convergence in all regions during the late 1970s and early 1980s.

Of course, networking activity was not confined to the research and higher education sector: industry was also busy. Just as in the 1960s, when the airline and banking industries had developed transactional networks to support the core of their businesses, so in the 1970s other sectors began to develop networks tailored to business requirements. For example, British Steel began consideration of how it could improve its business operations, particularly the role of computers, in 1972/3 (Dewis, 1978). This led to a ten-year plan to implement what would later be termed an enterprise network. The overall remit of the support required of the corporate information technology was substantially wider than what had commonly been undertaken in the past. Support was required not only for management information, administrative functions and order processing, but also for production, planning and control within the manufacturing part of the business at works level. By introducing a network, not only could money be saved by rationalising the number of lines rented from the Post Office – an aspect completely familiar in the academic sector also – but use of computing facilities could also be optimised in terms of performance and cost for the various functions and loads presented by the business. In this latter respect, of course, the design of the network could be specified much more tightly than is possible in any sort of open environment: one of the enduring differences between networking in the research and education sector when compared with industry.

By the beginning of 1978 the data network had reached the stage where all existing data applications were operating over the new network and the strategic flexibility which had been a major reason for its introduction had been achieved. The network was bespoke and used proprietary protocols, as could be expected from both the date of its design and its optimisation for a closely defined set of applications. As a demonstration of what could be achieved by the transition to a networked environment it was exemplary; and later, JANET would benefit indirectly from the experience gained in networking at British Steel.

2.3 Research Council activities

A number of references have been made to 'Research Council activities'. Actually, most of these were SRC activities. The Atlas, Daresbury and Rutherford High Energy Laboratories all needed to make available their mainframe computing facilities to universities nationwide. Each had developed a star network supporting RJE facilities and each had begun to appreciate the requirement for a terminal, whether it be for interactive or RJE use, which could provide access dynamically to more than one mainframe host. This is now familiar territory as we have already seen the same requirements in the regions; however, it should be pointed out that at this time – the first half of the 1970s – all the regions and the SRC were operating almost entirely independently of each other.

From a user perspective, the capability just described is an essential aspect of networked communication. From a network technology perspective it implies that lines can be shared amongst a number of individual end-to-end communications and that, given that there are not communication links connecting all parties to all others, there is some underlying mesh infrastructure of links and switches capable of switching data transmission streams so as to provide the equivalent of all-to-all connectivity.

Because the laboratories of the SRC served university science departments throughout the UK, it was quite common for a department or university to make use of several SRC facilities. Because the computing facilities could only support operation of dedicated terminals, this meant that such a department would need a separate terminal dedicated to access for each. By 1974, those separately responsible for remote access to each facility had come to the conclusion that development of dynamically sharable terminals was not only needed but becoming essential. It was this notion and its subsequent realisation which formed one of the strands of development culminating in SRCnet, a network of national extent to which all SRC computing facilities were connected, as well as access terminals and university departmental science computers operating under funds from SRC. On 22 March 1974, the first of a long-running series of meetings took place to realise the suggestion made by Trevor Daniels (Daresbury Laboratory), John Burren (RHEL) and Paul Bryant (Atlas Laboratory) that the three laboratories should create a network connecting the facilities and enabling more efficient remote access connections to these facilities. As suggested by Paul Bryant (1996) in an article in the final issue of ECN, the Engineering Computing Newsletter, this signalled the start of what became SRCnet.

Engineering Computing Requirements: The Rosenbrock Report

Since the early 1970s the Engineering Board of the SRC had been receiving a steady stream of requests for interactive computing facilities to support engineering research and design of every description. Initially a working group was convened by the Board in June 1973 to clarify the engineering community's requirements. The group reported in December 1974. The principle issue was that engineering needed more interactive facilities, coupled with substantial processing, than were currently available. The upshot was that in March 1975 the Engineering Board set up a Technical Group chaired by Professor Rosenbrock[18] to examine the whole situation and report on what was needed over the next five years to support engineering. In anticipation that interactive facilities were needed, it was tasked with specifying how these should be provided. The Technical Group had participation from universities, industry, SRC, Computer Board and Department of Industry, in respect of engineering, computing and networking expertise.

The main issue was not whether interactive facilities were needed – amongst those concerned with engineering there was no doubt about this – but more a

---

18   Professor H.H. Rosenbrock, University of Manchester Institute of Science and Technology.

question of finding a balance between what could be afforded and what was judged essential. All were agreed that local facilities were the ideal, primarily because of the ability to support advanced graphics. For major computation, however, there was no alternative to use of one of the major facilities provided by SRC, the most powerful being batch services. The interesting option just appearing was the small computer, affordable at the level of a department or even a group, dubbed a minicomputer or 'mini'. While a mini might even be contemplated as a single-user machine, as of 1975 it was really too expensive (and large) for this role: the time for the personal computer or workstation was not yet. As a multi-user system, the price was coming down but the multi-access operating system support was still immature. The overall view was that while these would shortly be available to support work where local graphics capabilities were essential, they were not sufficiently affordable nor quite mature enough for more general deployment.

The Rosenbrock Technical Group reported at the end of 1975. The essence of its recommendations was that an interactive computing facility should be set up, composed of some 'multi-user minis' or 'MUMs' as they came to be known, together with enhanced DEC10 facilities at Edinburgh and Manchester, and integration with national central facilities for substantial processing support. Following the Board's acceptance of the recommendations it was agreed that an Interactive Computing Facility (ICF) should be set up, based on a set of MUMs as local facilities for graphical processing, supported remotely by the major facilities.

In the context of networking, it was clear that whatever the balance between local and remote facilities, whether for access to processing or data, the development of a generally available network was rapidly becoming a necessity. It is also pertinent to observe that almost simultaneously, at the end of 1975, the Engineering Board of the SRC and the Computer Board had both reached the firm conclusion that a computer network of national extent was needed. The major result of the Rosenbrock report was the creation in 1976 of the ICF,[19] development and support for which was situated in the Atlas Centre of the Rutherford Laboratory (RL).[20] The combination of the various system development groups for the ICF MUMs and the mainframe facilities at Daresbury and Rutherford ultimately provided the technical driving force for the creation of SRCnet.

In the latter part of the 1970s SRCnet became an essential part of SRC's computing service provision. One notable aspect of this was that the ICF MUM systems were maintained almost entirely via the network, including the installation of new system software and rebooting the system. This not only saved substantially on the effort which would otherwise have been necessary but also enabled system faults to be rectified and development to take place much more rapidly.

As a postscript to this, it is perhaps worth remarking, for completeness, on a subsequent development which occurred in the astronomy community. In 1978 the

---

19 *See* Glossary.

20 RHEL changed its name to Rutherford Laboratory in 1975, signifying diversification to include support for other areas of physics beyond particle physics; the Atlas Laboratory merged with RL in the same year. *See* Glossary for details of the evolution of the names of RHEL, RL, etc.

Astronomy, Space and Radio Board of SRC (ASRB) became sufficiently concerned that computing support for astronomy research was inadequate, particularly in respect of interactive graphical facilities, that it convened a panel to advise on what should be done. The resulting report in early 1979 recommended an interactive facility on a model very similar to the ICF, though funded by a different part of SRC and organised and supported by a different part of RAL.[21] The hardware chosen for this facility, called Starlink, was different from either of the systems adopted for the ICF, in part to be compatible with the international community; Starlink was built several years later than ICF, becoming operational in 1980; and it initially used different, proprietary (DECnet), network communication protocols. For all these reasons, Starlink did not contribute to the development of SRCnet, though it later became part of JANET.[22]

Although most of the Research Council networking was within SRC, by the end of the 1970s networking was also becoming important for NERC. It had a number of small research laboratory establishments and by the beginning of the 1980s NERC also had a network which included its headquarters in Swindon. And as the Research Councils all gradually congregated at Polaris House in Swindon, so NERC took on the additional role of providing network access for all the offices there.

2.4 The Network Unit

As we have seen, following consideration of the second Wells Report, the Computer Board was persuaded in 1975 that a national network should be created. By now several regions had well-established communications infrastructures for delivering service and SRC had already begun its own network development (Sections 2.2 and 2.3). It was recognised that the next step towards a national network should be taken by the Computer Board in concert with the Research Councils. And there was another development: the Post Office, which had announced in 1971 its intention to establish EPSS, had begun development and invited organisations to collaborate in 1974, and a number of universities as well as the SRC laboratories had already joined this collaboration.

With the various technical network developments occurring in universities, SRC, the Post Office and in some parts of industry, it was clear that there were likely to be several technical options for provision. What was missing was co-ordinated organisation and assessment of the options for the academic sector as a whole, including how a national academic network should be managed. And there was now an additional factor, arising from the possibility that, given the EPSS trial turned out successfully, the Post Office would offer a service and, because of its monopoly position, the education and research community would perhaps be obliged to use this rather than constructing its own network.

The major result of the second Wells Report was that the Computer Board set up a Network Unit for two years, starting 1 November 1976. Its initial members were Mervyn ('MB') Williams, Roland Rosner and Chris Morris,

---

21 Rutherford Appleton Laboratory. RL changed its name again in 1979 as a result of merging with the Appleton Laboratory, which was responsible for atmospheric and astrophysical research and formerly situated near Slough in Berkshire. *See* Glossary for details.

22 For further information the reader is referred to the article by Disney and Wallace (1982).

with Janet Charles as secretary. The unit was accommodated in RAL.[23] 'MB', who was to lead the Unit (see Appendix C), had just retired as Head of the Post Office's Telecommunications Development, where he had been instrumental in establishing EPSS; Roland Rosner and Janet Charles were members of staff at RAL; and Chris Morris was from the University of Bristol. The Network Unit had the following terms of reference:[24]

> (a) to consider how the short term requirements of the Research Councils and groups of universities to interconnect could best be met, and how such interconnections could be related, in the longer term, to a national standard.
>
> (b) to act as the focus of liaison between the universities, the Research Councils and the Board on one hand, and the Post Office and Department of Industry on the other.
>
> (c) to initiate discussions, and propose solutions, concerning the management aspects of network developments.
>
> (d) to review the need for continuation of the Unit beyond 31 October 1978.
>
> (e) to make recommendations to the Computer Board and the Research Councils.

By the time of the start of the two-year lifetime of the Network Unit, the general utility of having a national network infrastructure had become both widely accepted and more clearly articulated. Terminals, both interactive and RJE, were proliferating rapidly, with those deployed by some of the regional centres and SRC being capable of accessing more than one computer by virtue of being connected to a network instead of a dedicated line: economy and convenience both militated in favour of adopting this model more generally. Specialised facilities, providing either a particular type of processing or access to unique collections of data, were appearing: connecting such facilities to an ubiquitous network was clearly the most effective way of making these available to the whole community. This latter aspect was a particular example of the value of resource sharing on a national scale. As we have seen, several regions were also developing network-based resource sharing of a different sort, where the idea was to have several mainframe processors available and for a user to be directed to whichever was appropriate at the time or for the particular type of processing required. (It is interesting to observe that, at the time, the latter concept was also being pursued quite independently in the research community as one aspect of distributed processing. In the decades to come, some of these concepts would eventually reappear in the context of processor farms and Grid computing – see Part C.) A significant aspect to emerge from these developments was confirmation of the argument made in the Wells Reports that a national network infrastructure would provide the flexibility to accommodate evolution in the balance between local and centralised provision as future circumstances, including economic and technological, dictated.

Of course, the issue of cost remained, although the focus had altered. It was now evident that a national network would be not merely useful but probably essential. Nevertheless it had to be paid for. Some potential economies were

---

23 At the east end of the top floor of R1, for those who know the Laboratory.

24 *See* the Introduction to the Final Report of the Network Unit (1979).

apparent from the sharing aspects already mentioned but it evidently could not be funded on that basis. Although the need for a national infrastructure was no longer in doubt, the question of funding this now emerged as a central and open issue, which would raise its head for discussion regularly in the decades to come. The Network Unit recognised that even to form a basis for any discussion, communications network costs would need to be established, and to do so accordingly formed one of the Unit's recommendations. However, in the short term, it was clear that the central funding model adopted by both the Computer Board and SRC had worked well for computing provision for a decade and extending this to supporting network infrastructure was natural. In the event, it would turn out that this established a *de facto* model for 'top-sliced' funding for academic network infrastructure which has survived to the time of writing with only minor perturbations.

How the organisation, operation and management of the network was to be arranged had been recognised as central issues from the very beginning. That some degree of national level organisation was needed was clear to all. However, existing academic expertise in both technical development and operation resided within individual universities, existing regional co-operatives and SRC's laboratories. But if the Post Office offered a national service, by now predicted to be available in November 1979, there would be the difficult question of how to organise co-operative management, organisation, operation and development. If the overall picture of how to organise and run a national network was still somewhat hazy, one aspect of future organisation was already becoming clear. Stemming from the Computer Board's decision to build upon the emerging co-operatives already apparent in 1966, regional networks were now in operation all over the country and organisationally it was clear that they had much to contribute, not just technically but organisationally as well.

Another important conclusion of the Network Unit was that provision of network infrastructure and computing (or any other) facilities should be separated in principle. This would allow facility provision to be optimised independently of the location of the supplier; in particular, it would free a particular institution from being bound to obtain some or all of its computing support from its local regional provider. Whether provision of network infrastructure and support would continue to be best provided on a geographic regional basis could then be decided on its own merit. A regionally devolved approach to such management was favoured by the Computer Board since it offered a scalable approach to the funding and resourcing of that management. On the other hand, further elaboration of this issue could not sensibly be pursued without resolving the issue of the Post Office monopoly and further details of what might be offered by PSS. Accordingly, further examination of how best to organise the network, including opening negotiations with the Post Office on behalf of the whole of academia, formed another significant part of the Unit's recommendations.

During the first half of the 1970s, the focus of network developments had been on achieving inter-institutional access to major facilities over the wide area. By the mid-1970s, the cost of access terminals began to fall; moreover, they began to be capable of providing shared access to several facilities. If ubiquitous network infrastructure could be provided within an institution then these two developments offered the prospect of providing access facilities distributed around a campus, all of which could be used to access a range of both local and remote facilities, the latter with the aid of a gateway system interconnecting the campus network and the national wide area network. The Network Unit identified the need to develop affordable support for such campus networks and initiated a number of studies in this area, as well as recommending that campus network infrastructure should qualify for Computer Board funds. Recognising that there were a number of competing technologies, none of which were yet mature, the Unit recommended that a programme of assessment and prototype trials should be undertaken, with the objective of funding university computer centres to undertake provision and operation of campus network infrastructure. Once the future JNT was in place with a budget, this thread was taken up and became part of the programme (*see* Chapter 5).

It will be seen that the Network Unit had completely endorsed and elaborated the original networking concepts articulated by the Wells Reports. In fulfilment of its remit, it had identified the beginnings of a substantial programme of organisational and development work. In order to realise a national network infrastructure for education and research two further ingredients were necessary:

- development of a network protocol strategy, including defining what network functionality should be provided; and

- creation of a permanently funded national network team, with a budget and remit to undertake all of the foregoing.

Although the Network Unit had initiated some of the activities described, it was the Final Report of the Unit, published in March 1979, which finally began to set in motion the creation of the network. Although much of the report was couched in terms of mainframes, remote job entry and interactive terminals for accessing mainframes, large or small, the majority of the ideas and strategy were really quite independent of the technology of the era. Although the Report did not dwell on the more technical aspects of network architecture, it did articulate the anticipated hierarchical structure of the forthcoming network, as shown in Figure 2.4. The Report was accepted by the Computer Board and endorsed by the Research Councils, and in November 1979, the Joint Network Team of the Computer Board and Research Councils, or JNT, was set up to carry out essentially the remit defined in the Report – the 'Joint' part of the title emphasising that the JNT was acting for both the Computer Board and Research Councils.

Before looking more at the strategy which was adopted for creating the network, it is worth commenting on one other thing originating at this time: the beginnings of the community which to this day continues to 'own', champion, develop, operate and manage the network. The very fact that there was a small group talking to people, organising meetings to focus on various aspects of how to create a network, collecting information and promulgating it, and generally acting as a focus for network activity caused the community to begin to coalesce. The major series of workshops called 'Networkshop' (*see* Chapter 4) began during the time of the Network Unit and continue at the time of writing; and it was during the time of the Network Unit that *Network News* began: the first issue was produced by Roland Rosner, dated 25 May 1977. The introduction reads as follows:

> 'We thought that one of the functions which the Network Unit could usefully perform might be the assembly of information of topics concerned with networking that we happen to hear about either formally or on the grapevine. The dissemination of news in these areas would help to keep people up-to-date and, if necessary, to pursue in more detail, issues which particularly interest them.'

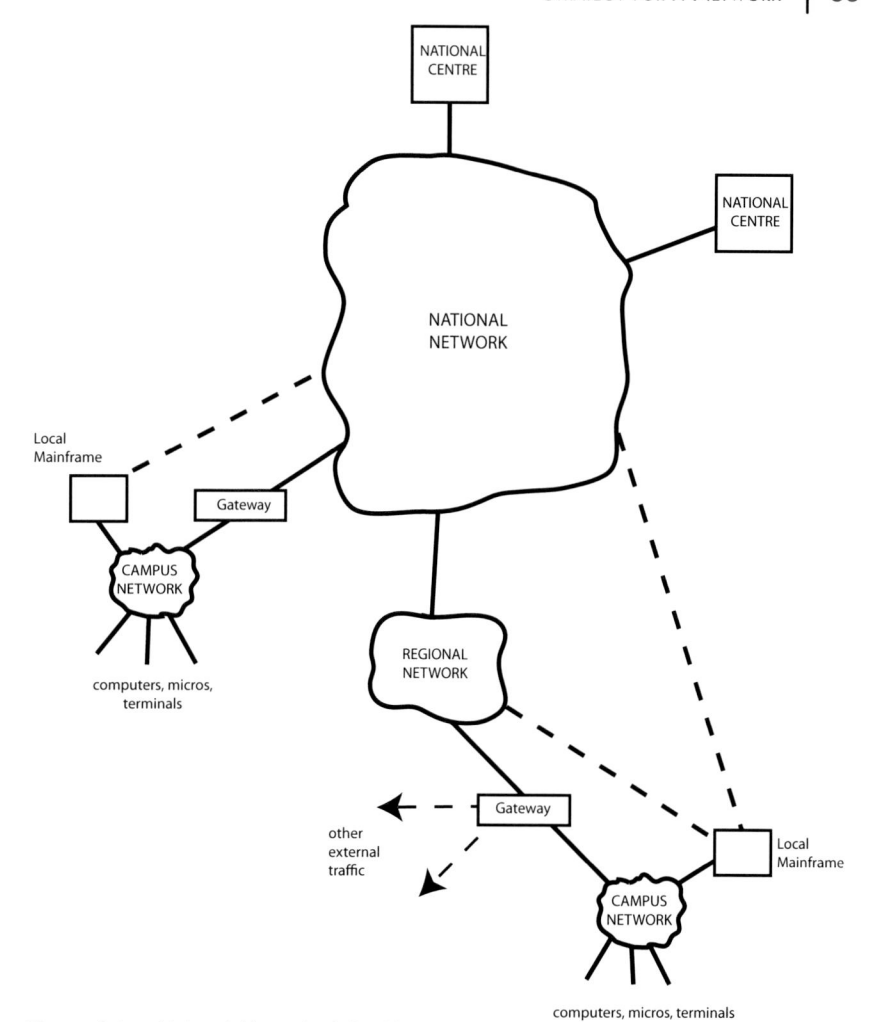

Figure 2.4. Network hierarchy (after Network Unit, 1979).

There are articles about various university projects, as well as network protocol and hardware development; and it included a final paragraph drawing attention to a recently published article by Metcalf and Boggs (1976) about some work done at Xerox PARC in California developing a new sort of fast network infrastructure which used distributed switching, usable in a restricted area such as a company or university campus and named an 'Ethernet' (*see* Chapter 5).

2.5 Strategy for a network

So, by November 1979, two major elements – the JNT and a budget – were in place to begin creating a network. Most of the strategic elements needed have been mentioned already and now formed the elements of the remit of the JNT. Two areas remained: whether the network would be based upon the prospective Post

Office PSS network service, now anticipated to start in 1980; and the protocols and services to be offered by the network.

Historically, in the 19th century, telecommunications services in Britain were initially provided by a number of companies. In the early 20th century (1912), public telecommunication services became one of the services provided by the Post Office and, as such, the Post Office gained an effective monopoly.[25] The Post Office was a public utility controlled by the government. The essential point at issue was that only the Post Office had the right to carry out 'third party switching', that is, to switch packets or messages on behalf of external legal entities. So, if a university or Research Council (being independent legal entities) switched traffic on behalf of other universities, it could find itself in breach of the monopoly. In the late 1970s, since the Post Office could not provide any such service, it had taken no action in respect of incipient academic network activity, in any case conducted over communications lines for which it received revenue, and no doubt in part because the activity was clearly not commercial. Without delving into the legal niceties, suffice to say that the position was not in the least clear-cut because licences for networks had already been granted to commercial computer bureau operations and all the existing star network operations supporting regional and national centres had already been accepted as being similar. Pro tem, the position was that the Post Office was willing to accept or license existing or planned operations but for any future network activities it reserved its position to refuse a licence and insist on use of its own service. As events transpired, it would only be about 18 months before the government of the day semi-privatised the telecommunications part of the Post Office to create British Telecom (BT) in 1980, initially as a part of the Post Office, before partially selling it into private ownership in 1981, though retaining a government majority holding.[26] With hindsight, with the discussions leading to this having been in progress for at least three years before 1980, and the eventual position regarding the monopoly being in some doubt,[27] the Post Office was in practice probably in a weak position to enforce it, particularly in the unique, government-funded education sector. While this engendered considerable delay, as well as a degree of uncertainty about the exact nature of the national network implementation, it also had the effect of providing strong guidance about some of the network protocols to be adopted.

The concept of a network protocol derives essentially from that of a diplomatic protocol, which is an agreed set of rules and procedures about how two autonomous entities (sovereign states) should conduct bilateral activities, be they business or anything else. A network protocol defines how a particular function, such as transferring a file perhaps, should be carried out, down to the details of when various messages should be exchanged, their semantics and composition. As has been mentioned previously, the protocols used in the early 1970s for constructing the RJE and interactive terminal star networks based on regional and national centres were all proprietary to the manufacturer of the associated

---

25 There were three exceptions to this, the only one still remaining being in the Kingston-upon-Hull area of South Yorkshire, where the telephone service continues to be offered by Kingston Communications, originally wholly owned by Hull Corporation, now independent and still in operation. *See* http://www.ofcom.org.uk/static/archive/oftel/publications/news/on61/5min0903.htm (accessed May 2010) for a brief chronology.

26 Privatisation of BT was completed in 1984, when the remaining government holding was sold off under the Telecommunications Act.

27 As indeed transpired: new legislation accompanying the creation of BT gave the Minister for Trade and Industry overall discretion in respect of telecommunications licensing.

mainframe computer. Apart from the need to develop the principles by which terminals could provide shared access to more than a single system, which meant replacing the master-slave relationship then subsisting between mainframe host and terminal with a peer-to-peer relationship, placing hosts and terminals on an equal footing, it was also necessary that such protocols should be widely accepted and non-proprietary: that is, standardised.

Interest in standardised protocols became widespread in the 1970s. Within the UK, the Post Office needed to establish protocols to enable subscribers to access its prospective network, beginning with EPSS. It set up the EPSS User Forum (later to become the PSS User Forum when PSS was announced) to coordinate this activity, which in turn set up several study groups to develop particular protocols. Then, in 1978, the DTI set up the Data Communications Protocol Unit (DCPU) under the auspices of NPL to coordinate UK protocol development, taking into account wider interests than just those of the Post Office. The members of the DCPU were Keith Bartlett, formerly of Derek Barber's group at NPL, and Peter Linington,[28] at that time seconded from the University of Cambridge.

As a result of the growing international interest in packet switching, in 1975 the CCITT[29] began discussions on a standardised protocol by which public packet-switched data networks could be accessed. In this context it was recognised that defining a widely acceptable and non-proprietary protocol was key to enabling the combination of a large market and equipment interoperability to ensure equipment availability at an economic cost through the economies of scale and commercial competition. The result was a protocol referred to as X.25, initiated in 1976 and first published in provisional form in 1978.[30] There followed fairly rapidly related protocols for terminal access and also a close variant for use in interconnecting networks. X.25 would turn out to be very influential in the development of packet networking, both nationally and internationally: in particular it would form the basis of JANET for its first decade. However, at the time, because the viability of packet switching was still not universally accepted, the author can recall headlines of the form 'X.25 [will be] dead in 25 years'. Since the telephone system was by then a century old, this was meant to indicate an ephemerally short life and the comment was intended as a devastating denouncement of packet switching! Though not in the sense intended, it turned out to be remarkably prescient: by the turn of the century, there was indeed not much use of X.25 (and JANET had ceased any use of it a few years previously), but in terms of longevity, quarter of a century is a very long time indeed in the computer communications industry, and nobody at the time of writing is now predicting the demise of the packet any time soon – though it is no longer an X.25-shaped packet!

From the perspective of a strategy for developing protocols by which JANET would operate, it was clear that (a) standardised protocols would be essential to ensure that all network equipment, hosts, and terminals could be obtained from a variety of manufacturers and interwork; and (b) the set of protocols should be consistent with those being defined for use with (E)PSS so that JANET would be in a position to capitalise on use of PSS if it wished or was obliged to do so. As we shall see in Chapter 3, the original protocols for JANET included those from all of the above sources: CCITT, PSS User Forum study groups and the DCPU.

---

28 Subsequently, Head of the JNT; later, Professor of Computer Communication at the University of Kent.

29 International telecommunications standardisation organisation. *See* Glossary.

30 *See also* Glossary.

3

Original JANET Protocols

During the 1970s network protocol standardisation was beginning both nationally in the UK and internationally. The importance of having such standards has been emphasised in the foregoing chapters and by the mid-1970s was universally accepted. However, it was also evident that the process was in general going to be too slow for a community of early adopters already in need of a network. The problem was not just one of inertia in standards making: good standards need to be based on experience, and at the time what experience there was (nationally or internationally) resided either in experimental networks or proprietary networks built to provide specific functionality or serve particular industries. In consequence, a number of groups formed in the UK to define interim standards. There was liaison with the British Standards Institution (BSI) which by early 1978 had set up a number of working groups to take responsibility for networking standards: for example, DPS/6, which was initially responsible for lower level communications, and DPS/20, which was concerned with architecture and higher level protocols. BSI also coordinated UK participation in international standards creation.

Mention has already been made of the activities at international level of the CCITT, the organisation responsible for international standardisation of telephony, which naturally began to take on packet-switching standardisation since it was generally assumed at the time that it would be the various national PTTs which provided public packet-switched services, especially as at least in Europe they generally already had the monopoly on provision of the (circuit-switched) transmission facilities which would be needed to connect packet switches to each other and to provide access links for terminals and hosts.

The CCITT was not the only international organisation to take an interest in computer networking.[1] By the latter part of the 1970s ISO, the International Organisation for Standardisation, was formulating its model framework for network standards, known as the Open Systems Interconnection Basic Reference Model (or OSI Model, for short), which enshrined the concept of 'layering' for protocols (see, for example, Linington, 1983). Those who had implemented networks and the protocols for them had found it useful to separate, for example, the low level details of how to control transmission of bits on various forms of cable and other transmission media from the higher level details of a remote terminal session or managing the transfer of a file, email or a job. And there are many levels of detail in between these two extremes which could also be amenable to layering. There were some advantages to the approach. Low-level detail specific to a given type of transmission could be hidden from higher levels; indeed, with care, one could even replace one type of transmission with another without

---

1   Also sometimes referred to as data networking to distinguish it from voice telephony and in recognition that devices other than computers might use it.

disturbing much of the higher level machinery. There was, from the earliest days of ARPANET, overall agreement about the general usefulness of this: the debate starts when one tries to agree just how many layers there might be and what functions are in each.[2] The OSI Model suggests there are seven layers; but the Internet, the protocols for which are defined by the Internet Engineering Task Force (IETF), is at the time of writing generally considered to have four layers (see, for example, Peterson and Davie, 2003). However, while the principles are useful, some of the layers in either model are in practice often composed of several sublayers, so that attempting to count the layers may degenerate into mere semantics – and insisting on a particular number is evidently futile!

Within the UK, by the mid-1970s the major protocol definition activities of future relevance to JANET were primarily under the aegis of the Post Office, with the aim of agreeing the protocols for use in EPSS. To do this a number of study groups were created, effectively within a club of prospective participants in EPSS called the EPSS User Forum. Since a major option – possibly the only option in view of the apparent monopoly situation at the time – for the future provision of a national academic network was expected to be use of PSS in some form, many of those concerned with the development of the early regional networks[3] and SRCnet participated. In terms of protocol architecture, the backdrop to this community effort included ARPANET, CCITT and ISO. Network and terminal access were both emerging from the CCITT and were expected to be adopted by PSS. ARPANET experience came through Peter Kirstein's group at UCL, and also John Laws' group at RSRE,[4] Malvern; ARPANET network layer protocols were evolving into those subsequently used in the Internet; file transfer and interactive terminal access were in regular use, and mail was becoming standardised and spreading. ISO OSI high-level protocols were anticipated to support end-user applications and be adopted worldwide – and were being promoted by governments seeking a basis for developing national information and communication technology into the international market.

The EPSS User Forum and its study groups began forming in 1974 at the time when the Post Office called for participation in EPSS. Study Group 3 had within its remit transport, network and terminal protocols. Other application-oriented protocols, such as RJE and file transfer, came within the purview of Study Group 2, later renamed the High Level Protocol Group, though it also pursued some early work on remote procedure call. The DCPU[5] had been set up[6] in 1978 with the mandate to identify and promote requirements for OSI. Its support for UK interim standards creation stemmed from this, with the corollary that their development should be oriented towards future adoption of OSI standards. In consequence, the definition of high level protocols increasingly came under DCPU organisation, a development which perhaps also suited the Post Office since the latter was primarily focused on providing a packet-switched bearer service, together with terminal access, both of which were seen as PTT services (and consequently subjects of standardisation by the CCITT). The DCPU ceased operation early in 1982, by which time the JNT had been in existence for just over

---

2   *See* Appendix D for a light-hearted take on early layering debates.

3   *See* Appendix B and Chapter 2.

4   Royal Signals and Radar Establishment: *see* Glossary.

5   *See* Section 2.5.

6   *See* Section 2.5.

two years. Subsequently, the DTI set up the Information Technology Standards Unit, ITSU, under Keith Bartlett, with a particular remit to promulgate and support ISO standards, particularly OSI. Within the academic community, the revision of existing protocols, together with the definition or adoption of new ones, devolved *ipso facto* to the JNT.

The rest of this chapter briefly surveys the original set of 'Rainbow' or 'Coloured Book' protocols, so-called as a result of the documents having distinctively coloured covers. A complete list appears in Appendix E; here we shall look at what might be described as the 'core' set of protocols which would be used for JANET for roughly its first decade of operation. Also included here, for completeness, in the section on terminal access, are a couple of protocols neither of which saw wide deployment: one contributed to the formative development of several early networks which later became part of JANET and the other was a part of the early JANET development programme. The other major group of JANET protocols were those for local area networks which were incorporated into JANET as part of the growth and development of campus networks, a subject in its own right which is dealt with in Chapter 5. An informal description of what a network protocol is has been given in Section 2.5 together with the general importance of standardisation. An early review of what would become the JANET protocols is given by Larmouth and Rosner (1982).

3.1 Architecture of a network

In describing the operation of a network, one can identify the high-level functions which users make use of and which involve end-user systems: email is an enduring and familiar example of such high-level functionality. At the other, low level, end of the technology are the techniques for transmitting bits, bytes and packets over various forms of wired or wireless media which link individual components such as terminals, hosts, switches, routers, etc. to form a network. In between these high and low level extremes there is the matter of how to arrange that a packet originating in one end-user system can find its way to its designated destination system. The ISO OSI term for this layer is the network layer. The two essentials are a scheme for addressing end systems, so that source and destination can be identified, and a system of determining from the addressing information what route a packet should take through the network and, in particular, which output link a system (end-user or internal network node) should use to forward a packet it has for transmission. There are a number of ways in which this latter function, commonly called routing, may be accomplished, depending on the type of network.

In traversing a network packets may be damaged or lost. The former occurs as a result of transmission errors. A packet may be robust against damage through redundant encoding or it may be beyond repair and abandoned: equivalent to being lost. Apart from being irreparably damaged, packets may be dropped by a network node because of congestion. In order to achieve reliable transmission, the flow of packets needs to be controlled ('flow control') and lost packets may need to be recovered, typically by retransmission. To what extent the task of providing either of these functions is shared between the network and the end systems is a matter of design, and either extreme is possible in both cases.

In defining and implementing the high-level, end-user services and protocols to support them, an economy of design can be achieved by ensuring they are independent of the underlying network technology. Moreover it is sometimes considered desirable to be able to construct such services using protocols which operate assuming the network provides a reliable service consisting of a two-way pair of ordered streams of bytes or records[7] (termed a 'full-duplex' channel). Such a network independent and reliable transport service can be provided by implementing suitable procedures above the packet-oriented network layer. The higher-level protocol architecture of the UK academic network was based on the use of just such a transport service. In defining the higher-level protocol architecture, there was awareness of the ARPANET architecture, which itself was evolving from use of NCP[8] and already had interactive terminal and file transfer protocols in regular use.[9]

Network and link layer protocols

Mention has already been made of X.25 (Section 2.5), the provisional version of which was published by the CCITT in 1978.[10] This was a set of protocols which defined how an end-user system could access a packet-switched network. Strictly, it did not specify how the network worked internally; however, an obvious mechanism for the internal protocols was to use a minor variation on the access protocol. Since the monopoly position of the Post Office meant that it might be the only legal provider of the network and it had confirmed that PSS would use (a version of) X.25, the academic community defined its network protocol architecture on the basis of use of X.25 in the wide area.

In terms of layers, X.25 consists of several, of which the uppermost is a network layer. Hosts are given numeric addresses similar to telephone numbers. When a host wishes to contact another, it establishes a call or connection somewhat analogous to a telephone call, with the difference that dialling and use of circuit-switched exchanges are replaced by an initial exchange of packets through packet-switches. This initial exchange of packets establishes an end-to-end route or path through the packet switches which is remembered by the switches for the duration of the connection. Subsequent packets do not have to be routed afresh: each packet contains an identifier indicating which call it belongs to and the switches just look up which link to use. It is this telephone-call-like style of operation which gives rise to the term 'connection-oriented' to describe this type of network layer operation. When the hosts have finished communicating, one of them clears the call – the analogue of putting down the telephone – by sending a disconnect packet.

The lower, link layer of X.25 defines how data frames containing packets are sent on the transmission links between network nodes. Two of the functions of any link layer protocol are: to indicate where packets begin and end; and to detect whether a packet has been damaged in transit. In the X.25 link layer, both of these functions require bit-by-bit processing of a type which is fairly easy (especially the error detection) in purpose-built hardware but which is very laborious and

---

7   A record here can be considered just an ordered set of bytes, with some means of delimiting the boundaries between successive records.

8   *See* Glossary and Chapter 6.

9   There was also some early work on a non-proprietary RJE protocol but it never saw widespread use.

10 *See also* Glossary.

slow using a computer. It took time – in practice, several years: hardware support for the link layer only began to appear in 1977 – for suitable hardware and supporting software to become available and, in consequence, several networks – SRCnet in particular – initially used an alternative link layer (called binary synchronous control or BSC) which only required byte-level processing, capable of being accomplished fairly easily and quickly in a conventional processor, and for which software support was already available.

Transport service

In the mid-1970s, packet networking was generally viewed as the emerging technology to support data networking among computers. As such, delivery needed to be reliable since the data in question was typically text, numbers or programs, none of which could tolerate errors. The transport service defined by Study Group 3 of the EPSS User Forum (*see* Appendix E) was of the network independent, reliable, full-duplex, record-stream variety described above, suitable to support all the functions then required of a general purpose data network. A significant consequence of the transport service being network independent was that it was the natural level at which to engineer the join between networks based on differing lower level technologies. This was no accident: historically, the transport service used over the academic network owed much of its origins to work on 'bridging' in EPSS – this being the term in that community at the time (1975) for the interconnecting of networks.[11] Essentially, a system termed a 'gateway' in the UK at that time,[12] placed at the junction of two or more networks, could implement the transport service for each and, given suitable addressing conventions, create end-to-end transport service connections over which higher-level protocols could operate independently of the intervening network technologies.

The specification was published – in yellow covers – initially in draft in April 1979 and in its first version in February 1980. Two revisions followed, the second in July 1982, by which time PSS had come into service (by then run by BT) and EPSS had closed. It was entitled 'A Network Independent Transport Service', and was known variously by its acronym NITS, as 'Yellow Book Transport Service', or by its alternative acronym, YBTS.

Included with the transport service specification were several descriptions of how to realise it over various underlying networks. Since X.25 was anticipated as the network protocol for the future network, this was specified in some detail. However, this was still in the future and evolution to its use would take time; meanwhile, there were other practical alternatives, such as using leased lines, the public switched telephone network (PSTN) or, possibly, X.21 circuit-switched networks. The latter included automated circuit switching and was still considered a potential candidate by some PTTs outside the UK. While use of X.25 could have been extended to cover these cases, in the absence of X.25

---

11   As an aside, it is of technical interest to note that this work also included study of transmission rate-based flow control algorithms, a topic which has recurred in subsequent decades, typically in the context of multiservice networking.

12   Terminology has varied between communities and over time. Such an object would have been termed a 'transport service relay' in ISO OSI. The term was also in use in the ARPANET community at the time, to denote an object which joined networks at packet level; it gradually gave way to the term 'router' as the current inter-networking architecture evolved.

implementations simpler procedures were possible and these were included. Subsequently, in 1983 a procedure was published which defined how YBTS could be realised over asynchronous lines, a simpler option than synchronous lines (as used by PTTs), with or without X.25, and suitable for use with microprocessor-based systems by then in widespread use. The other significant extension to the collection of YBTS specifications appeared in 1981/2, as campus networking was being developed and we shall return to this topic in Chapter 5.

Like all protocols, issues arose after YBTS had been defined, either as a result of experience or because of later external events, and these led to revisions. We mention just one here, partly because it had financial implications and partly because exactly the same issue reappeared a decade later in the more international context of broadband ISDN. As EPSS evolved into the commercial service PSS, so it was revealed how BT would charge for the service. Communication service charges can typically involve a number of elements: a per-call element, distance, duration, line speed and volume of traffic, for example. The charge for a fixed-line telephone call has traditionally depended on the first three. For PSS, BT tariffs, like those for ordinary telephone calls, included a per-call charge. For the common case of a computer with a number of users making use of the same remote service, such as a regional or national service, it would be common for a number of simultaneous calls to be in use between the two. If the transport service had the ability to support (multiplex) a number of transport connections over a single X.25 call then the per call element could be reduced to one for any given destination. This extension was added but, although significant, it is not clear how much it was eventually deployed because, as it would turn out, neither PSS nor any other commercial network would after all form the basis of the academic network.

Addressing

The transport service was the layer at which network interconnection took place and so it was intrinsic that it needed to handle addresses. However, it did not need to interpret them: that was a matter for each individual network and, in particular, other networks would use different addressing schemes from those for PSS. So, in order to maintain network independence, YBTS took the pragmatic approach that an address was just a string of characters whose interpretation was up to the underlying network: since the transport service had no role in interpreting such strings, they could also be names which would then be resolved to addresses by some additional underlying network service. If a transport service connection crossed several networks, each with its own naming and addressing, then parts of the string would in general need interpretation by different networks.[13]

It was mentioned above that a host might have several transport connections in simultaneous operation, either because of several uses of the same service or, if the host offered several services, because several services were in use. The latter case illustrates that although we have loosely considered an address as designating a host (strictly, a particular interface on a host) – which is sufficient when considering routing through the network – in fact one also needs to specify

---

13 A particular form of the problem is illustrated by telephone numbers. In the set of numbers dialled for an international call, the international prefix is interpreted in the caller's country while the rest of the number is interpreted in the destination country. *See also* Chapters 8 and 13.

which service the connection should be made to inside the host. There are many ways of arranging the details of this but they all amount logically to adding a bit more to the address, and so are often termed subaddressing. In any packet carrying an address, an additional field is introduced to carry the subaddress information and in many networks this field goes by the alternative name of 'port', by analogy with a host having many logical ports which one might choose logically to 'plug' into. Initially, when naming was introduced, it was applied to hosts and the subaddress or port number was just added as an extra by the application. However, it was soon recognised that it was useful to name the services also, which would effectively incorporate the full addressing scheme, including subaddresses, into the naming scheme. Within a particular network the interpretation of a service or host name was a matter for that network, just as was the interpretation of addresses. It should be noted that service names might or might not be related to the name of the host providing the service.

So what of naming conventions in the UK? Each regional network and SRCnet had over a period of years simply chosen names for its hosts according to its own conventions. By about 1980 or a little earlier, it had been generally agreed that it would be useful, if only to avoid choosing duplicate names for systems which might one day be connected on the same network, if the names included the host institution. It had also been recognised that a naming authority was needed; and use of a separator (most commonly '.') had been generally adopted between the institutional part of a name and the specific host part. In some cases an institution had effectively devolved parts of its name space to departments by adding another, departmental, component to the name. Furthermore, it was quite common to derive the names of the services offered by a host by adding a further name component to the host name.

And what of the ordering of the components of a name? Informally, host names were typically structured as 'institution.host' or 'institution.department.host' at the time. To name a service on that host one could add a component 'service' to get 'institution.department.host.service'. If it became necessary to specify which network this was on, 'net' could be added, in this case – it seeming most natural to prepend, to get 'net.institution.department.host.service', where 'net' might be PSS, ARPANET etc. This was approximately what had begun to come into use in the UK by 1980. Soon it would be formalised, including one small detail: the order described here of components in a name. It should be remembered that at this time naming schemes were only just being developed and names, like addresses, were considered peculiar to a given network, with no obligation for global uniformity.

The remaining two sections of this chapter describe briefly the other protocols of the original core set, with the addition of email. Recalling Chapter 2, by the early 1970s the main end-user functions requiring support were: remote terminal (keyboard and display or typewriter-like) access; file transfer; and job submission and retrieval, the generalised RJE function.

What about email? The position with this was different.[14] By 1970 a number of interactive operating systems had local mail systems, so that a user logged on

---

14   *See* Partridge (2008), for example, for an historical account of the development of email.

to the system could send mail to another user of the same system, who would either receive it immediately or when next logged on. By 1973, email[15] was in common use on ARPANET, particularly among Tenex[16] systems, but it was not entirely system independent and work on developing a standard model and the protocols and email headers to support it were only just beginning. Although the first RFC relating to email dates from 1971 (RFC196), it was not until 1977 that RFC733 appeared, defining what in many ways is still the email header in current use. (To date, it has only been revised twice.) In the UK, those who had access to email, including at UCL to ARPANET from 1973 onwards, were convinced of its usefulness, particularly as a collaborative tool, and used it heavily. However, in the absence of a national network this experience was not universal. Indeed, since computers and communication were expensive in equipment and effort, there was a view in some quarters that the whole infrastructure should be dedicated strictly to 'work', which was not considered to include exchanging messages via email. Moreover, the very success of email on the ARPANET meant that a substantial proportion of its traffic was email, and this reinforced a UK funding view not to see substantial resources expended on what it perceived as an expensive electronic toy! As a consequence, email never appeared in the 1970s or early 1980s as a required element of network service in the context of any request for funding, despite the fact that by 1980 a substantial part of the community was using it productively and regularly on SRCnet, the regional networks and via UCL to ARPANET; and, by 1982, there would be a UK interim mail protocol (see Section 3.3). Wider recognition of its importance for collaboration and its establishment as part of the general workplace infrastructure would take another decade.

3.2 Terminal access

A variety of terms have been used over the years to signify interactive computer access from a keyboard and some sort of display. In this section this is just referred to as 'terminal access'; elsewhere, the expression 'interactive terminal access' is used where necessary to avoid confusion, for example, with RJE terminals. Initially such interactive terminals were somewhat like old typewriters, except that printing on paper was produced by some sort of electromechanically driven head with characters on it. The earliest version was the Teletype, with origins in the telegraph industry. It operated at 10 characters per second and used an asynchronous coding that needed 11 bits per character,[17] so the overall linespeed was 110 bps. Printing onto a paper roll by the print head was either from the user typing in on the keys, the characters also being transmitted via the line to the computer, or from characters arriving on the line from the computer. Of necessity, user and computer effectively took turns. It was also possible to have all printing under the control of the computer. When the user typed in, the computer echoed the characters to the print head. The first mode of use was termed using local echo; the second form, remote echo.

---

15 Throughout the 1970s and early 1980s, the term 'mail' was typically used in connection with the programs and protocols used to support this application; 'email' became the common term during the 1980s.

16 *See* Glossary.

17 1 start bit, 8 data bits and 2 stop bits.

The next terminal[18] to come into general use consisted of a CRT[19] display and a keyboard: the whole was commonly referred to a VDU (visual display unit) terminal. This eliminated the electromechanical element and might operate at around 100 characters per second. The display might typically be 20 lines of 70 to 80 characters; typing beyond the last line would cause the screen to scroll, the top line disappearing. As with the Teletype, it could operate with local or remote echo. It initially operated at 300 bps (30 characters per second). While some computer systems expected the user to type a line ending with some form of newline character, after which the computer would reply with a line of its own, others assumed character-at-a-time interaction. The former mode of operation was typically used with local echo whilst character-at-a-time used remote echo – and with some systems, if you typed a letter or two and the computer recognised this as one of its commands, it would echo with the completed command. Used locally, this latter form of operation could be effective; used over a long-distance slow line it could be off-putting or even inconvenient: fast typists could get several lines ahead of what was being echoed back and displayed. Character-at-a-time working with remote echo required the line to support full-duplex (i.e., simultaneous two-way) transmission of characters. Line-at-a-time operation naturally operated with half-duplex (i.e., two-way, one-way at a time) transmission.

When it came to devising a protocol to handle terminals, these differing modes of operation gave rise to the need for a network protocol which could cater for a range of modes, as well as various conventions for newline. In fact, the variations did not stop there. Tektronix had introduced a graphics terminal based on use of a storage-tube display, which was a considerable step forward for scientific work at the time. The display could now handle either line (vector) graphics or characters, with commands to change mode between the two, and an extra command to clear the screen. And whereas a VDU scrolled, the storage display did not: it was necessary to clear the screen and start again at the top.

A simple protocol to support such terminals was one of the first to be devised for potential use over EPSS. It was called Interactive Terminal Protocol or ITP (*see* Appendix E) and defined in 1975. Several early networks used it, including SWUCN, NUNET, RCOnet and SRCnet. However, within a couple of years it became clear that it would have no future in the standards arena because, in conjunction with X.25, the CCITT was defining a packet-mode protocol which would support this type of terminal.

X.3, X.28 and X.29

This set of protocols was published alongside X.25 and between them specified how interactive terminal access to X.25 networks could be provided. The initial definition was published in 1978;[20] however, the more widely adopted version came two years later in 1980. The types of terminal described above all worked on the basis of a bi-directional stream of characters and were referred

---

18 There was also a terminal called the Silent 700, which used heat to print on special paper, operated at 30 characters per second, and used an asynchronous line with an encoding which needed 10 bits per character (only 1 stop bit, as compared with the Teletype), so the overall linespeed was 300 bps.

19 *See* Glossary.

20 *See also* Glossary.

to in the standard as character-mode terminals. Because the terminals operated in terms of characters and the network in terms of packets, some form device was needed to handle the interworking or conversion between the two: this was called a PAD, short for packet assembler disassembler. Of course, the concept was not new: the Terminal IMP introduced for ARPANET in 1971 had precisely this additional functionality.[21] Because of the various modes and speeds associated with terminals, there were a number of parameters associated with PAD operation and the CCITT standards were defined in X.3. Communication between a PAD and the local terminal was specified in X.28. The way in which X.25 packets were to be used for such terminal traffic in communication between the PAD and a remote host was specified in X.29. The set of three specifications was referred to as 'triple X' or XXX.

As with any standard, there were a number of options relating to the way in which triple X could be used, and the specification for use over PSS was published in its latest form in 1981 by Study Group 3 of the PSS User Forum, a revision of an earlier version in 1980. The specification was published in green covers and accordingly became known as the Green Book, and because it was still expected that the academic network would have to be based on use of PSS in some way, it was adopted for use in that context –encouraged also by its being an international standard, so that manufacturer support could be expected.

There was one awkward point about this otherwise apparently complete solution to handling interactive terminals: what if the underlying network did not use X.25, and in consequence X.29 was unavailable? By 1980 there had already been several years' development of alternative technologies for use in restricted areas like buildings and campuses, and universities were implementing campus networks (*see* Chapter 5) to act as the local access networks to one or more wide-area networks, including the eventual academic network. And the community had already developed the Yellow Book Transport Service to enable networks of diverse technology to be interconnected so as to allow high-level protocols to operate end-to-end over the concatenated set of networks.

The approach taken to this problem was to define a terminal protocol which had all the same features as X.29 but which operated over YBTS: this was known as TS29. By design it could operate over a concatenation of networks supporting YBTS. Since the features and capabilities of TS29 and X.29 were the same, an exact mapping between the two was possible. By defining the parameters and encodings in TS29 to be the same as X.29, the details of the mapping between the two were made simple. And, as a consequence, for a host which already supported X.29, adding support for TS29 was not too expensive. The Green Book in its 1981 version not only defined the details of triple X to be used over PSS but also contained annexes defining TS29 and providing guidance on implementing PADs.

Screen-mode terminals

Although the character-mode terminal described so far did enable people to gain remote access to computer services, the service it provided was in the nature of a lowest common denominator. Most manufacturers had additional

---

21 And UK users of ARPANET had been using the interactive terminal access facilities of the UCL TIP since its installation in 1973, initially by dial-in, subsequently via local and remote terminal access to the RAL 360/195.

features in their terminals, many of which were useful, most of which varied from one manufacturer to the next, and all of which were proprietary. Perhaps the commonest enhancement related to VDU-based terminals were those in which the user could interact with the host through the whole screen, including being able to split the screen in various ways and use each part to interact with different services or programs on the host. Alternatively, a block of text could be edited on the screen and then submitted as a block to the host.

Throughout the late 1970s and early 1980s, there were various efforts directed towards devising support for such screen-mode terminals both in the UK and internationally. The major difficulty was in defining a sufficiently general abstract model of a terminal (sometimes referred to as a 'virtual terminal') which would be both implementable and configurable to support a wide enough collection of real terminals. Development of such a 'virtual terminal protocol' (VTP) was not made any easier by its being a moving target. Despite this, largely due to efforts at Newcastle University, in 1985, the Fawn Book was published which defined a simple screen management protocol. Over the next few years, during the latter part of the 1980s, there were various developments based on SSMP. Implementations appeared for a number of single-user workstations and personal computers. There were also emulation packages for several well-known screen-mode terminals of the time, perhaps most notably the IBM 3270 which was used with all System 360 and 370 mainframes (and derivative architectures). A Z80[22] microprocessor-based 'box' was also developed which enabled a screen-mode terminal to operate over the network using SSMP by placing the box between the terminal and a PAD.

However, by the early 1980s bit-mapped, windows-based display handling, originally pioneered a decade earlier at Xerox PARC, was being deployed along with the introduction of the mouse. Although screen-mode terminal use based on the Fawn Book was deployed quite widely, by the mid-1980s graphical user interfaces based on a mouse and windows had arrived. Moreover, in 1984 MIT had developed the first version of X-Windows, a windowed virtual terminal protocol and system which was manufacturer-independent. Although it would be some years before X-Windows saw widespread deployment, the general move to bit-mapped displays and a windowed environment was by the mid-1980s becoming rapidly established.

3.3 High-level protocols

The remaining application protocols used in the academic network were all published, including revisions, during the period 1978 to 1982. Implementation was, of course, a much longer process and it was generally not until well into the middle of the 1980s that most of the operating systems represented in the community had implementations.

File transfer

The ability to access remote storage or move files around among systems is generally regarded as fundamental in any system of networked computer

---

22 A very successful and popular 8-bit microprocessor introduced by Zilog in 1976.

facilities, and the networks already in operation in the community in the 1970s were no exception. Among the more basic requirements were assembling data and programs on a system preparatory to running jobs. The file transfer protocol defined initially by the High Level Protocol Group of the PSS User Forum in 1978[23] was to enable a single file or document, binary or text, to be transferred from a process in one machine to another, over a reliable transport service such as YBTS. It was published between blue covers and entitled *A Network Independent File Transfer Protocol*, NIFTP for short. In the following year it was submitted to DPS/20 of BSI and it subsequently formed part of the BSI's input to ISO.

Some care was taken to ensure reliable delivery, particularly in terms of ensuring synchronisation regarding the state of the transfer. Because fairly large files were sometimes transferred over slow lines and connection failures were not uncommon, the protocol had a capability for resuming a transfer from defined synchronisation points. Systems did not all use the same conventions regarding end-of-line or newline, nor indeed the same set of character codes; to cope with this, as part of setting up the transfer, simple negotiation took place so that for text files basic conversion was possible between the conventions used by the two end systems. During the previous decade, the major purpose of accessing a remote system was to be able to submit jobs and retrieve the output; in an effort to avoid *ad hoc* conventions developing for filenames indicating jobs, it was possible to indicate during initialisation that a file was destined for submission to a system's job queue or that the requested file was job output, to be retrieved from the output queue.

After a couple of years' implementation experience, a revision was published early in 1981 by the File Transfer Protocol Implementers Group organised by the DCPU, correcting some aspects of the protocol, simplifying the options available during the negotiation phase, and clarifying use of some of the control exchanges used for state synchronisation between the parties to the transfer. The revised protocol also contained tables defining the ASCII and EBCDIC character codes[24] assumed by the protocol and the translation to be used in converting between them when performing text file transfers. (As an aside, these tables represented the outcome of substantial technical debate and negotiation over several years, particularly amongst those connected to SRCnet at the time, and had somewhat wider significance than solely file transfer – see panel 'SRCnet character code transition'.)

Job transfer

As will be apparent from the description of the way in which regional networks and the SRC network had developed, the principal *raison d'être* for their existence was the submission and retrieval of jobs. It was always appreciated that defining how this should be done in a machine independent and network independent way was a major task, particularly as the overall model of just how it should operate and what facilities should be offered were not properly understood. However, one thing was clear: however it operated and whatever facilities it offered, it would need to incorporate the transfer of files for both submission and

---

23 The 1978 version of NIFTP acknowledges earlier work which was published in 1975 as 'Basic File Transfer Protocol', HLP/CP(75)3.

24 *See* Glossary for expansion of these acronyms.

SRCnet CHARACTER CODE TRANSITION

Character codes refer to the encoding of graphical text symbols into the binary codes by which the symbols are represented in computer memory or during transmission over a digital communication link. These codes turn up in two contexts: representation of text documents to be read by people; and programming languages and command languages by which computers may be programmed or controlled. Computers do not all use the same codes: many use EBCDIC (an 8-bit code) or ASCII (a 7-bit code).[1] Unfortunately, both of these include minor variations in both symbols and encoding: classic areas of confusion being between {} and [], amongst £/$/#, and amongst the symbols used for logical OR, AND and NOT, which are of significance in some programming languages. For documents, it is important that the graphical symbols in a document are maintained as the document moves from one system to another. For a program, it is desirable that the graphical symbols are maintained so that the textual representation of the program is maintained; however, the primary purpose of a program is to communicate with a computer, for which purpose the code is significant, not the graphical symbol.

In the 1970s, there was no universal version of either ASCII or EBCDIC in use, still less a standard definition of how to map from one to the other. After a good deal of debate and negotiation amongst those familiar with a variety of machines in the community, a slight restriction of the printable codes in use enabled a definition of both ASCII and EBCDIC, together with translations in both directions, which were acceptable for all the machines on SRCnet, as well as some which were not but which would later be on JANET. Then one day in 1980[2] the tables in all the machines on SRCnet were changed: DEC10s, IBM systems, ICL systems, RJE systems, ICF systems, GEC2050s, GEC4000 systems, the PDP9 at UCL which connected SRCnet to ARPANET, Prime systems and DEC VAX systems. The number of systems was probably around 100 and involved co-ordinated action all over the UK: co-ordinated because until all the tables had been changed and became consistent, terminal, RJE and file transfer traffic were all at risk of character corruption. The author knows of no other similar simultaneous network-wide synchronised host system modification ever having taken place on this scale in a few hours.

---

1 Actually, it was worse than this, since not all systems even used 7 or 8 bits: e.g., the CDC 6600 used a 6-bit code – essentially a 6-bit subset of one of the versions of ASCII.

2 There is some uncertainty in the date.

---

retrieval of jobs. Accordingly, work on a job transfer protocol did not begin until work on the initial draft of NIFTP was nearing completion.

While the goal of a system independent scheme might be hard and still in the future, a start was made in the early 1970s on one aspect. It will be recalled that even for a single manufacturer's machine range, an RJE station could only serve a single remote host at a time, essentially because once signed on to a mainframe, that system took over control of the RJE station. This was a serious practical nuisance and, by the mid-1970s, several regions and the SRC laboratories had successfully tackled this problem in a variety of ways; but that still left the system manufacturer dependent, requiring a site either to have several systems or to run a number of emulators.

Once an initial version of NIFTP had been published, attention turned to job transfer. Although the file transfer protocol had included facilities to indicate that a file was either job input or job output, this was recognised as essentially a stop-gap, though a number of networks made use of it, including, for example, SWUCN. And while there still remained no model of what job transfer was supposed to encompass, it was certainly necessary that it should include mechanisms to find out how a job was progressing and possibly to alter some aspects, perhaps rerouting its output, changing its priority or cancelling the whole job. At one time there was the question of whether job transfer should embrace the idea of creating a system-independent job control language,[25] but this notion was never really seriously considered, largely because the range of what was possible on any given system was so diverse as almost to defy the abstraction of any useful commonality.

Following publication early in 1978 of the initial version of NIFTP, a group was formed by the DCPU to define a network independent job transfer and manipulation protocol (JTMP), the manipulation aspect to address the combined functions of enquiring about status and changing attributes of a job. Development of the model of job execution to be supported and specification of the protocol to support it took nearly three years: a draft was circulated in April 1980 and the protocol was published – this time in red covers – in September 1981.

By the late 1970s, several regions had progressed well beyond the simple model of RJE developed by manufacturers in the 1960s, most notably the early resource-sharing networks, GANNET in the North West and SWUCN in the South West. The eventual features of the model which JTMP supported included the possibility of quite a complex job which had a number of execution steps, each of which could happen on a different machine on the network. Individual output documents could be routed independently or collectively to network hosts. Only files included with the job or local to the system executing a particular step could be accessed, but by including a step whose only purpose was to access a local file and include it in the job, this actually meant that input could be assembled from a set of systems by visiting each in turn. Apart from the particular model of network job execution, the protocol had a couple of other features of interest at the time: the encoding was entirely character-based and was akin to a declarative programming language (with a BNF-like syntax definition); and considerable attention was devoted to concurrency, commitment and control regarding the various transfers and states of a network job. The data file transfers associated with JTMP were accomplished through use of NIFTP; the synchronisation issues during transfers and between steps were analyzed and the way in which the features of NIFTP were to be used were specified, so that if a transfer failed, a known state could be regained and suitable action taken. Parts of this presaged later work in network distributed systems.

During the next few years, JTMP was implemented on quite a wide range of machines in the UK and a portable, machine-independent implementation sponsored by the JNT was produced by Salford University, under the direction of John Larmouth, who also took a major part in shaping the ISO JTM standard, based substantially on the UK JTMP. However, with the arrival of the workstation and personal computer in the early 1980s, the role of major mainframes receded

---

25 There was work in this area in ISO at the time on Operating System Command and Response Language (OSCRL), which had UK participation.

somewhat as use of computers became more affordable and interactive. Certain types of large scientific or engineering calculations would remain more tractable with the aid of a large system, but many calculations would be capable of being tackled by a single dedicated machine simply left to do little else: it might take a lot longer but the machine was cheap! So, for a decade or two, managing the remote execution of jobs on a set of machines in a network environment became less important, only to resurface in the context of Grid computing at the turn of the millennium.

Email

The somewhat ambiguous position in which email found itself has been related above. However, by mid-1981, pressure for some sort of interim email standard to be agreed for use in the academic community had become irresistible. By then the importance of email as a tool to aid collaboration had come to be widely recognised. Moreover, big science was international, so a solution which had the potential to work internationally was needed. To address the email requirement, in August 1981 the JNT sponsored Chris Bennett of UCL to lead a working group to define a mail protocol for the academic research and education community. This resulted in publication of the Grey Book in January 1982.

There are typically two aspects to any mail protocol: the format of the mail itself and its transfer. The mail itself forms a file, so transferring it requires the use of a file transfer protocol. As for the mail format, it can be thought of as having a (header) part which needs to be not only readable by the correspondents but also processable by the mail systems responsible for end-to-end delivery. It contains items like who the mail is for and who it is from. Subject, date and time of origin, and a log of staging by intermediate systems also appear, along with a number of other items of more significance when things go wrong than of immediate importance to the correspondents.

Actually, there was not a lot to be done! By this time, ARPANET mail headers had been standardised in RFC733 in 1977 and use of ARPANET email was spreading internationally. It is worth noting that there were several international networks in operation with origins in the USA, notably BITNET/EARN and SPAN:[26] both, however, were proprietary – the former used IBM job transfer protocols generalised to include file transfer and mail and the latter used DECnet, which included DEC mail. There was also UUCP[27] networking, sometimes called UUCPnet, which was in the nature of a widespread informal club, mostly of Unix hosts, which would, amongst other things, forward mail. However, the only non-proprietary, non-operating-system-dependent mail protocol already fairly widely deployed was ARPANET mail protocol – and it did not take the working party long to decide to base UK mail on the ARPANET mail header format.

Transfer was a different matter. ARPANET email used its own simple mail transfer protocol – entitled just that, commonly referred to as SMTP. Technically, mail could use any file transfer protocol. Since the whole enterprise was regarded as interim until the promulgation of international standards, and NIFTP was already in widespread use in the UK, there seemed little point in importing

---

26   *See* Glossary and Chapter 7.

27   *See* Glossary.

another file transfer protocol when most of the systems at the time already had NIFTP and did not have SMTP.[28]

That just left the matter of email addressing. ARPANET mail addresses had the form 'name@service', where 'name' was the name of a 'mailbox'. This could be the name of a person (possibly a login id); a role name, like postmaster, which the destination system would resolve to the name of a person; or a mailing list, which again the destination system would expand to a list of names and then forward.[29] So far so good: the syntax of the mail address and the semantics of 'name' were adopted. What of the semantics and form of 'service'?

The naming of hosts and services as it was developing in the UK by the time of the drafting of the UK mail protocol in 1981 has been outlined in Section 3.1. Each of the networks already mentioned, including the incipient UK academic network, had its own naming conventions. In the absence of any standardised, universal scheme, naming was regarded as network specific. In this situation, it would be necessary to implement application or transport service gateways, if only to process addresses in passing from one network to another. In the case of UK naming, as already described, the construction of names followed a 'big-endian'[30] convention regarding order of components: the initial component was of largest scope and successive components of successively lesser scope, as in 'net.institution.dept.host.servicename' when read in the normal left-to-right convention. Although it was known that ARPANET and UUCP/Usenet[31] were adopting the opposite, 'little-endian'[32] convention, this was not then regarded as sufficiently significant to cause the UK to change, since ARPANET (and UUCP) protocols had no particular standing at the time in a standards context. Moreover, to have mail service names use the opposite convention to transport service names seemed both confusing and inconsistent. As we shall see, the consequences of this apparently innocuous, but ultimately contentious,[33] decision were to haunt JANET for nearly 15 years!

Reference has been made a number of times to 'international standardisation' during the 1970s and early 1980s. Standards can emerge and gain national or international acceptance in a number of ways; some of the national and

---

28 The major exception to this would become UNIX which, in a year or two, would be distributed widely with single-user workstations and would include the early Internet protocols.

29 Originally, mail could also be sent to a program or process which would interpret and act on 'instructions' in the mail: a feature long ago abandoned (in its original form) because of its open invitation for system abuse.

30 That is, largest scope or most significant component first: see Glossary for origin of the terminology.

31 *See* Chapter 5.

32 *See* Glossary.

33 Representations were made to adopt little-endian ARPANET and UUCP order, not least by UCL which operated the mail relay to ARPANET and the University of Kent which operated a Usenet gateway (part of what would become UKnet in 1987 – *see* Chapter 5). The decision to retain the big-endian order for the academic network was confirmed by Mike Wells as Director of Networking in July 1983, in the context of the form of names to be used in the NRS. (*See*, for example, presentation by Holder (2007), which contains a facsimile of a letter from Mike Wells to Brian Spratt, University of Kent.)

international standards organisations of relevance to the UK were mentioned at the beginning of this chapter. Since it was widely recognised that it was essential for networking standards to be international, not just in the UK, during the 1980s much of networking was regarded as interim in this sense. In order to encourage development and adoption of early international standards, governments in general encouraged the promulgation of whatever standards began to emerge and this happened in the UK, Europe and the USA. In the UK during the 1980s the Central Computing and Telecommunications Agency, the CCTA, had both an advisory and central procurement role. In the late 1980s it developed the UK Government Open Systems Interconnection Profile (GOSIP),[34] version 3.0 of which was published in February 1988, coincident with the EEC making it mandatory for equipment purchases over 100,000 ECU to be OSI based – although in practice this never materially affected purchases in the UK education and research sector. There was subsequently a US GOSIP, first published by the US National NIST in 1990 as one[35] of the Federal Information Processing Standards (FIPS). The standards in the GOSIPs were generally ISO OSI ones; however, they also included X.25 and X.400, the CCITT email standard which we shall meet in Chapter 5. As we shall see, the period 1982–84 was the time when the ARPANET officially adopted the new protocols which became those of the Internet; nevertheless, in the USA, US GOSIP protocols were mandatory for US Federal Government purchases from 1990 until 1995.

As we have seen, the topic of naming was already an issue by the end of the 1970s. Many schemes were being studied around this time: Grapevine (Birrell *et al*, 1982) and the Clearinghouse (Oppen and Dalal, 1983) were both at Xerox PARC and both incorporating a naming scheme; the ARPANET community was developing the scheme that would become the Domain Name Service (DNS); Sun was developing its 'Yellow Pages';[36] and later CCITT would develop its X.500 standard. The implementation of any naming scheme rests on several components. There is the set of conventions regarding the construction of names in the scheme; a supporting system for creating, maintaining and querying the database of names is needed; and a single point of authority is necessary to take responsibility for registering the creation of names within the community in which they will be used. And it could be argued that it was perhaps from the latter perspective of authorisation and registration that the UK Name Registration Scheme (NRS, *see* Chapter 5) had its origins.

---

34   *See* Article by Mark Gladwyn, (UK) GOSIP project manager, in *Network News*, No.26, July 1988.

35   FIPS 146-1, first published in 1990.

36   Later renamed the Network Information Service (NIS) to avoid infringing BT's rights in respect of the name of its classified telephone directory product.

4

JANET

By 1979 there was a substantial amount of computer networking activity going on throughout the UK, but there was still no national network service either in academia or for more general public use – domestic or business. The public telephone network was still essentially analogue and transmission was based on copper cable. Although it had been possible since the 1960s to rent from the Post Office lines of sufficient quality to transmit digital data at rates up to 40 or 50Kbps, these were very expensive and most lines in the 1970s were still operated at from 1.2 to 9.6Kbps. And because discrete components were still in use in such equipment, a modem[1] provided by the Post Office to attach a computer to the telephone line was the size of a drawer in a kitchen cabinet and weighed even more. An IBM modem for 40.8Kbps was even bigger at half the size (width) of a two-drawer filing cabinet – and comparable in weight!

The Department of Industry had convened a National Committee on Computer Networks (NCCN) in 1976 to advise on the need for a national network. The Department needed to know the nature of the commercial requirement for such a service, partly to inform its own view on the need for the Post Office (for which it was responsible) to provide such a service, and partly to inform UK computing and communications manufacturing industry of the nature of the commercial opportunity. The NCCN reported in October 1978, by which time there was substantial interest amongst the informed community and it was considered both important and urgent that such a utility should be made available (Howlett, 1978).

Notwithstanding all the general enthusiasm generated by now, it would be another three years before a public service became available. Meanwhile, uncertainty about the monopoly position supposedly enjoyed by the Post Office – and subsequently British Telecom – continued to impose on the academic community the equivalent of planning blight: should the community go ahead and build its own network – which it was clearly in a position to do – or would it find itself in court in breach of the monopoly?[2]

4.1 THE JOINT NETWORK TEAM

So, in November 1979, seven years after the initial Computer Board visit to ARPANET, the JNT – or, to give it its full title, The Joint Network Team of the Computer Board and Research Councils – was established, with a remit to co-

---

1 *See* Glossary.

2 A more detailed overview of the situation at the time from the academic network perspective may be found in Rosner (1978).

ordinate the development of academic computer networking with a view to creating a national network infrastructure to support research and education in the UK.[3] At the time, this more or less meant tertiary education and research: specifically, the university and Research Council community, both of which were funded by the Department of Education and Science (DES). Though polytechnics, at least in England, were funded by DES, this funding was through local authorities and for that reason, although there was representation on the Computer Board, the position with respect to prospective network support was at this point ill-defined. (It would not be until 1988 that the polytechnic sector would achieve more co-ordinated funding through the PCFC.[4])

The JNT had a budget sufficient to provide a full-time team of six plus a secretary, to fund network development activities and to support community building. The first head of the JNT[5] was Roland Rosner, who had headed the Network Unit following the departure of Mervyn Williams after its first year. He was joined by Janet Charles, secretary to the erstwhile Network Unit, who became the first secretary to the JNT. By the end of 1979, the team had swelled to four: Barrie Charles, who had previously worked in Roland Rosner's group at RAL in online physics data acquisition systems; Ken Heard, from AERE Harwell, where he had already participated in computer networking projects; and Bob Cooper, who brought from British Steel experience in management and control of its enterprise network (see Section 2.2). Initially accommodation, as for the Network Unit, continued to be provided at RAL as part of SRC's contribution, but the following year, in 1980, the JNT moved into new offices in the recently constructed wing (R31) of the Atlas Centre,[6] which would continue to be the administrative headquarters of the UK academic network for the next quarter of a century. 1980 also saw Les Clyne join the JNT from Logica, the London-based software and systems consultancy. Subsequently Jim Craigie joined, so that by 1982 the JNT had finally achieved its full complement of staff.

Notwithstanding the potential position of legalistic limbo in which the JNT might find itself in respect of actually constructing the network, there was plenty of network construction going on. All the Computer Board national and regional centres, as well as the SRC centres were admitted as being akin to commercial computer bureaux for the purpose of licensing under the monopoly (see Section 2.5). And since Daresbury Laboratory and RAL were both part of SRC, the fact that they had created a sort of distributed bureau was also deemed acceptable.

A substantive conclusion of the Network Unit was that whatever network was constructed, it should conform to network protocol standards, preferably international. Apart from its being peculiarly necessary for a network to agree the 'language' of communication, the more widely agreed the protocol standards, the greater the likelihood of affordable products. In view of the additional necessity to prepare for a position from which to capitalise on a national service predicated on standards, the goal of standards-based network components became a clear objective identified and initiated by the Network Unit, and it subsequently became a major plank of Computer Board and JNT policy. The previous chapter

---

3   *See* Chapter 2.

4   Polytechnics and Colleges Funding Council; *see* Glossary.

5   *See* Appendix C for a list of the Heads of the JNT.

6   The name by which the erstwhile Atlas Laboratory building complex became known following its merger with the Rutherford Laboratory in 1975.

has already given an overview of the interim standards developed in the UK at this time to meet the identified needs of the community.

Following the report and initiatives of the Network Unit, the Computer Board had recognised that it was important to maintain momentum toward the creation of a network, in spite of the planning vacuum created by the uncertainty over the monopoly position. So the JNT was created, with remit and budget, and a community technically and operationally poised, but hamstrung by the lack of ability to plan in the presence of an unresolved monopoly position. The JNT would become the point of intersection of politics, funding and technology and, for its first five years, the orchestrator of components in imminent expectation of a network.

Standards and implementation

What would become policy with respect to network standards for JANET was formulated by the Network Unit in 1977. Most computer manufacturers at that time could only offer proprietary communications, although the recent X.25 international standard was being taken seriously. There were perhaps around a dozen manufacturers supplying at least a score of machine and operating system ranges to the academic community. Contractually supported network software and hardware was essential for the eventual network service, particularly now that it was universally appreciated that it would become a cornerstone of academic research and education infrastructure. The Computer Board endorsed the Network Unit's recommendation that from 1979 a mandatory requirement for all computer systems procured with Computer Board funds would be support for UK interim network standard protocols, as approved by the Computer Board for academic networking and promulgated in a published list. The Computer Board authorised a letter to be sent to this effect early in 1978. The initial list included X.25 and NIFTP, with YBTS, for example, being added in 1979.

Once the JNT came into being in late in 1979, with a budget, it became possible to pursue this policy more proactively. In the context of computer procurement, the Computer Board already made use of the CCTA to assist with the technology aspects of computer procurement. Now, the JNT took on a similar role for Computer Board procurements in respect of the technical detail of the network capability and standards conformance to be provided. On occasion, this was of course seen as a 'policing' role and an unwelcome intrusion amounting to inteference in a university's own procurement business – a situation exacerbated in those cases where there was fundamental disagreement about the protocol standards policy: to the pressure to 'buy British' and satisfy CCTA technical requirements, there was now an additional hurdle of support for UK interim protocols.

However, with a budget, the JNT could now place contracts with university groups and third-party software contractors to assist in developing support for the protocols. By 1980, there were a number of implementer groups for the protocols. Experience could be shared and protocol details clarified, and, later, several of the protocols underwent revision in the light of experience, sometimes by reconstituted definition groups, sometimes by the implementer groups themselves. An important outcome of this experience was the gradual elaboration of the specific choices made within the academic community with respect to options and parameter usage in the protocols, and also, particularly for JTMP, the definition of functional subsets chosen to provide a coherent set of services to be offered by the national and regional centres.

A consequence of this community activity, which now involved several hundred people, was the substantial corpus of knowledge and experience which accrued; this not only informed the later development of the network but also influenced the development of subsequent international standards in ISO. Specifically, YBTS contributed to (connection-oriented) transport; NIFTP formed an initial contribution to file transfer and access;[7] and JTMP formed the major foundation of job transfer.[8]

As we shall see, remaining compatible with PSS continued to be politically essential until 1982 because of the expectation that PSS would constitute the major part of the eventual network. This ensured that whether network development was in pursuit of a PSS connection or part of constructing a private regional network or SRCnet, all began to evolve toward a common set of protocols from the late 1970s. As it became apparent that protocols or functionality need no longer be bound to those of PSS, the JNT took over direct responsibility for defining and achieving community endorsement of these for academia, following the example of electronic mail.

Community

In the late 1960s and early 1970s, there was no networking community of national extent. There were perhaps as many as ten regional and national groupings, mainly around the early centres. Each of these groups was initially focused primarily on installing the manufacturer's RJE solution to the relevant centre and the centre provided any necessary assistance in doing this. In a few cases, perhaps because access to more than one centre could be provided on a shared basis or because some other more specialised resources were to be shared, more specific early networking collaborations came into being, examples being GANNET, RCOnet and SWUCN.

With the announcement of plans for EPSS and the formation of the protocol working groups of the EPSS User Forum came the beginnings of a wider exchange of ideas and experience. Participation in these groups included representatives from industry, university national and regional centres, NPL, the Post Office, RSRE and UCL; the latter had become the portal to use of the ARPANET and, together with RSRE and other users, provided access to the experience being gained by that community. Initially, contact was all through early UK protocol definition; but then implementation began in the late 1970s and now the community began to swell rapidly as various groups took on the job of implementing protocols on a range of machines and operating systems and testing the implementations against each other. Throughout, there was liaison and information dissemination provided through the Network Unit, the DCPU and then finally the JNT. By the time of the latter, in 1979, there was already a considerable corpus of experience; some of the regional networks, as well as SRCnet, were already running X.25, and machine range and operating system groups were coming into existence to pool experience. In the early 1980s there was a further strand of activity bringing

---

7   ISO FTAM: *see* Glossary.

8   ISO JTM: *see* Glossary.

people together as local area networks and campus networking arrived on the scene (*see* Chapter 5).

A notable community activity also began at this time. The Network Unit got wind of a workshop that Peter Barry of Glasgow University, one of the participating institutions in RCOnet, was organising. Although initially planned as an RCOnet event, representatives of those involved in university regional and national networking throughout the community were invited.[9] The workshop was, held over two days in Glasgow, 29–30 September 1977, in the Queen Margaret Hall,[10] one of the halls of residence of the University. There were 70 delegates at this meeting, indicating how significant networking had already become. Following the success of this initial meeting, there was immediate interest in ensuring it became a regular occurrence. Liverpool hosted the second one, in April 1978, and so Networkshop was born. It was to become (and remains) the longest running series of workshop conferences in the subject in the UK.[11] For the first five years it was held biannually, in spring and autumn; thereafter it became an annual spring event. Significantly, from the second event onwards, numbers had to be restricted, such was its usefulness and general popularity. Within three years the event had doubled in size, and in more recent years it has been a three-day event with typically 400 to 500 delegates, the limitation being the capacity of the venue.

So just why was Networkshop so popular and important? In the early days, one aspect was the building of a community-wide collaboration. Since the late 1980s, both the importance and the overhead of building and maintaining successful collaboration has become more widely understood, perhaps particularly from the experience of international collaboration in Europe in both service-oriented and research and development projects. Initially, Networkshop enabled all those involved to meet, whether that involvement was as funder, administrator, manager, operator, developer, implementer or researcher. Of course, the majority of participants were from higher education but from very early on there was also participation from industry, starting with the Post Office but soon broadening to include manufacturers, initially associated with host protocol implementation but then expanding to include third party suppliers and specialists in network equipment. Early participants in the latter category included Camtec (the manufacturer of the JNT PAD, see below) but there were soon others too, as campus and local area networking grew and as X.25 switches were needed for the wide-area network. Much of the workshop was conducted in plenary, which enabled opportunities for people to take time out to develop appreciation of the bigger picture. However, the main focus of the workshop was undoubtedly technical. As well as progress reports of protocol definition and implementation, there were also opportunities for floating ideas for new services in 'birds-of-feather' sessions, often informally later in the day. Of course, as the community grew and expanded and with the coming of an actual network, the workshop naturally took on a more service and operationally focused orientation: and yet, with the widening of the range of services and technologies as well as with the broadening of the community to include the whole of education since the beginning of the millennium, there has been a steady increase in the number

---

9   *See* Proceedings of the RCO Networks Workshop held in Queen Margaret Hall, Glasgow, on 29th and 30th September, 1977.

10 The building in Bellshaugh Road, Kelvindale, in which the first Networkshop was held was demolished in 2000 and a new building opened on the same site in 2002.

11 *See* Appendix F.

of topics and strands of interest so that much of the conference is currently conducted in parallel sessions.

In the first decade of JANET, education, practical experience and training in respect of network operation and support were something provided from within the community. In 1982 it had become clear that networking in general and X.25 in particular were about to become part of service operation throughout the UK, no longer confined to those who had been participants in its development and operation thus far. In April of that year, at Networkshop 10, in Nottingham, the JNT PAD was demonstrated; and in September, the University of Exeter ran a course entitled 'Practical X.25 Networking' (subsequently repeated several times), which encompassed all the protocols in use in the UK at the time as well as hands-on sessions. Other opportunities soon appeared, particularly those associated with some of the new technology to appear for local campus networking: for example, the University of Kent began a regular course on the Cambridge Ring local area network in April 1984. In more recent times, stemming from the appearance of the Internet as ubiquitous infrastructure in the early 1990s, courses including practical demonstrations of this sort have become a commercial proposition and are offered world-wide by a number of companies. With the expansion of the community served and the services offered, however, the need for focused courses within the JANET context became increasingly important from the mid-1990s; in consequence, providing this type of support would come to be recognised as an intrinsic part of managing and operating JANET and its services.

4.2 COMPONENTS OF A NETWORK

From 1980 onwards the technical push towards making a national X.25 network a reality gathered momentum. There was still no PSS, nor any indication as to whether it or a private version of it would form the basis of the network; and, consequently, nor was it possible to say who would operate the network or, indeed, whether different parts would be operated by a variety of different managements. Nevertheless, technical planning went on as if PSS would be available and it would be used. In the interim, the combination of national access to centres and continued development of both regional networks and SRCnet meant that all the technical development needed could continue in spite of the vacuum affecting the planning of management and operation of a national network.

From a technical perspective there was a recent arrival on the scene at this time: the micro, or more properly the microprocessor, a computer processor on a single chip. Micros had been under development since the late 1960s and in the mid-1970s they became a reality with the release of a number of 8-bit processors: the Intel 8080 and Motorola 6800 (1974); the MOS Technology 6502 (1975, a derivative architecture based on the 6800); the Zilog Z80 (1976); and the Motorola 6809 (1978). They began to appear in several roles: domestic computers; peripheral controllers in larger computers; and special-purpose devices performing dedicated functions. In a networking context, they enabled developments in both the latter roles, as we shall see below.

By this time, all those concerned with providing computer services, whether national, regional or as part of the Research Councils, were committed to X.25 and Coloured Books. Some had already begun, others planned phased introduction in concert with the regular cycle of computer replacement. The technical work to be done could be split broadly into two categories: host attachment, along with implementation of user application services based on end-to-end use of higher-level protocols; and network components. One component, the terminal concentrator, which enabled a number of interactive terminals to connect to a network, was really a miniature host; however, because derivatives of the device developed for JANET had many roles, it is described separately below.

Host support

Connecting a host computer to a network requires both hardware and software. In terms of the layered network architecture, hardware is necessary for the lowest layer and it may assist some of the other lower layers. We have already observed (Section 3.1) that some parts of the X.25 link layer are best performed in hardware. Performing all of the link layer procedures in hardware, however, is neither necessary nor particularly desirable because of complexity and maintenance. The arrival of the microprocessor made it possible to design an intelligent peripheral controller at reasonable cost which would not only provide the hardware connection to the network but also process the X.25 link layer at a satisfactory speed. The earliest such controllers began to be available in the UK in 1977, and several groups developed support and gained early experience of their use in experimental prototype host connections. It turned out that achieving a correct and reliable implementation of the X.25 link layer was not so easy. However, by 1980 most manufacturers had incorporated such link layer support, as well as software support for the rest of X.25 within the operating system.

Of course, this was only one step toward achieving anything useful for end-use of the system. Typically, the next step was to provide facilities for interactive terminals. Evidently there are two ends to a terminal session: an interactive user may initiate a (client) session to a remote system; and, correspondingly, a (server) system may offer a service for terminal sessions initiated remotely from terminals elsewhere on the network. Most systems offered both client and server facilities, which were implemented simultaneously – since the first step in testing was usually to test client-to-server on the same system. Thereafter there followed a long period of testing with other implementations over the network. File transfer followed a similar pattern of developing and testing.

With file transfer available it was then possible to implement job transfer and email. Email was the simpler of the two and, encouraged by its being so useful for collaboration – not least for further network development! – implementations began appearing fairly rapidly after the appearance of the Grey Book. Job transfer was intrinsically more complicated because a multiplicity of functions could be supported and agreed subsets were needed before coherent services could be implemented.

When coupled with the number of operating systems and machine architectures to be supported, it will be apparent that all this represented a considerable amount of effort spread through the community: implementation, testing, deployment and support took time. In spite of the absence of a nationwide infrastructure, there was considerable infrastructure available in the regional networks and SRCnet. Access lines could be provided to support implementation and testing. Of course, all this took time. Obtaining lines was slow. Testing between different systems situated remotely from each other was slow and painstaking, often involving several people at each site. Nevertheless, as the number and variety of implementations increased and more experience was gained, this became easier.

Network elements

The fundamental element of any network is a set of switches situated at the junctions of the links making up the mesh constituting the network. Each switch takes in bits, bytes or packets off incoming links and forwards them onto outgoing links towards their destinations. In the case of circuit or 'virtual' circuit networks (like an X.25 network), the output link to be taken at any switch by packets belonging to a particular call is determined as part of the routing for that call which takes place at the time the call is initiated.

Once it became clear that PSS would be based on X.25, a number of prototype X.25 switches were developed by the community. Most of these were based on mini-computers, a number of which could provide multiple interfaces for communication links. As part of planning for the eventual network, the Network Unit in collaboration with NPL drew up the specification of an X.25 switch in autumn 1977.

Packet switches were needed in a variety of sizes. The switches for the wide area network needed to be able to handle higher speed lines – 48–64Kbps in 1980 – and to switch between perhaps 10–20 links, though probably only 2–3 would be at the highest speed. To serve a campus and connect it to the wide-area network, a switch of intermediate size might be needed; and to provide access for small computers and terminal concentrators, a relatively large number of small, cheap switches would be needed. So, ideally, a switch design that was configurable was required.

In the late 1970s, several sites were working with GEC, which made the GEC4000 range of mini-computers; systems from this range were used as the switches in SRCnet and SWURCC, and also for some of the MUMs in the ICF. Eventually GEC would provide the initial wide-area and campus network switches for the national network. The small campus switches would come as part of the JNT PAD developments (*see* below).

One other thing was needed to make the network practical: the equivalent of the telephone book, so that people could look up the address of the service they wished to use. Actually, some sort of unique authority for issuing addresses in the first place is needed to prevent duplicates. For PSS addresses, the Post Office issued an address as part of host connection. Initially, York kept a list of PSS addresses which were made available over the network: this reduced the problem to remembering the address of the service at York and the service being available. For addresses on the national network, the JNT was the natural authority; the scheme adopted was a simple two-level numerical hierarchy, based on numbering the packet switches in the network and the hosts on each switch. And, of course, each switch needs to know which link to use to get to any host in the network, something which was achieved by loading a complete table into every switch.

Transport and application service naming has been described in Chapter 3 and since names are more memorable than numeric addresses, most people used these. Again, an authority was needed, which for the academic network was the JNT. There were a set of conventions for deriving the names of well-known services from host addresses or names, by appending well-known components (like .FTP for file transfer, .FTP.MAIL for mail, and so on). Of course, now this meant that hosts all had to have tables to enable the translation between names and addresses. Initially that was exactly how it was done, though even then there was a need for system administrators to be able to access an up-to-date master table. The construction of a service to cope with this requirement was one of the early developments of the network and was called the Name Registration Scheme (NRS)), the story of which is told in the next chapter.

JNT PAD

In the late 1970s, when most campuses had an increasingly large number of computers on site, together with a large number of interactive terminals situated all over site, mostly in terminal rooms and laboratories but also beginning to make their appearance in offices, enabling these terminals to connect at will to any of the computers on site, instead of being connected permanently only to one, became an urgent priority. What was needed was some sort of circuit switch which enabled a user to 'dial-up' the system of choice. This sounded like something a telephone system might be able to accomplish and indeed it was: private automated branch exchange (PABX) systems had been in use for some years in offices to handle internal calls between offices and to route external calls to individual office phones. A variation on this, called a private automated computer exchange (PACX) became available in the 1970s – a particularly popular model was made by Gandalf, a Canadian company – and many universities and laboratories in the UK installed these in the late 1970s and early 1980s to meet a pressing need.

With the advent of a national packet network on the horizon it was evident that the equivalent facility would be needed to enable terminals to connect to computers anywhere on the academic network. This was exactly what the PAD described briefly in Section 3.2 was intended for. So one of the first things the JNT set in motion was a programme to create PAD facilities and in April 1980, at the Durham Networkshop, the initial specification produced by the JNT Terminal Handling Group was published. Shortly afterwards, the specification was circulated to potential suppliers to begin exploring the possibility of a commercial product. The initial outcome of this was, in effect, that it was too early, particularly as PSS had been delayed and the date for service was still unknown; a few products were on offer but at prices far in excess of those acceptable to the academic community. In addition, the working group responsible for the Green Book on terminal handling was in the midst of a revision to be completed by the end of the year.

However, there was considerable activity throughout the community programming PAD facilities on a variety of hosts, so that although a terminal might be attached to a particular machine, it could nevertheless be used to initiate a terminal session on another machine on the network. With this experience, several groups began investigating the possibility of using microcomputers as the basis of small network devices, including a PAD. In particular, Chris Cheney and Mike Guy at the University of Cambridge Computing Service (UCCS) had already undertaken some preliminary work in this area based on a Z80 and, later in 1980, UCCS, ULCC and the JNT collaborated in further investigation, this time looking for a commercial source for the hardware with the software to be supplied by Cambridge under contract to the JNT. The target price was £2000 for a device which could handle around 16 terminals.

At the beginning of 1981 the JNT announced that it was sponsoring development of a Z80 PAD at Cambridge, based on a system by then already in operation. What Cambridge had developed was actually more general than just a PAD. Known in-house as 'the transport service box', it was capable of switching X.25 or YBTS connections. Interestingly, even at this early stage, it was envisaged that it might, eventually, serve as a small switch, providing the PAD function on a Cambridge Ring local network (*see* Chapter 5) and host printers, creating an X.25 network printer server. In August 1981 it was announced that a contract for commercial development, together with an initial order for just over 50 JNT PADs,[12] had been placed with Camtec Electronics Ltd of Leicester. With development proceeding satisfactorily a further 85 units were ordered in December 1981 and the first two prototype units were assembled in February 1982, followed by delivery of a few pre-production models in April 1982, in time for demonstration at Networkshop at Nottingham.

And so began the introduction of one of the more ubiquitous components of UK academic networking, illustrated in Figure 4.1. It is interesting to observe that it was selling successfully before it was actually known how the eventual academic network would be built. Over time, the device was further developed to support all the functions noted above and originally envisaged for it. As line speeds increased, so the capacity of the Z80 proved insufficient and in later versions it was replaced by a 32-bit micro, the Motorola 68000, which had been introduced in 1979 and by the early 1980s was in widespread use in small, single-user computers. By November 1982, the JNT PAD had become a commercial success, and negotiations had resulted in support being transferred from Cambridge University to Camtec.

4.3 Where to get a backbone

In retrospect the period 1980–1982 can be seen as a turning point. Certainly by 1982, technically, the community had within its grasp all the knowledge and technology necessary to create a national network, much of it realised over national and regional centre networks and SERCnet.[13] However, the community was by no means convinced that it wished to take on the responsibility of operating such a service. In the autumn of 1980, a number of academic sites conducted successful pre-service tests over PSS; and in the spring of 1981, after delays amounting to about 18 months,[14] BT finally announced PSS was open for service – although full service did not formally begin until 20 August 1981, once the final details of charging had been announced. So, in theory, BT was by then in a position to offer a national network service: perhaps a national academic network was in sight at last.

Following earlier recognition of the common interest that the Computer Board and Research Councils had in computer communication, the Computer Board and the SRC Facilities Committee for Computing (FCC) had by 1980 begun holding regular joint meetings. Although a committee of the SRC, the FCC had

---

12 Camtec is still in business as of 2008: see http://www.camtec.uk.com/ (accessed March 2010).

13 The Science Research Council (SRC) was renamed the Science and Engineering Research Council (SERC) in 1981, as part of a restructuring of funding for research in the physical sciences which included increased recognition and emphasis of the role of engineering. Correspondingly, SRCnet began to be referred to as SERCnet.

14 In February 1978 *Network News* 3 announced that the Post Office was aiming to offer PSS service by mid-1979. Subsequent issues of *Network News* document four slippages before the eventual service began.

WHERE TO GET A BACKBONE | 67

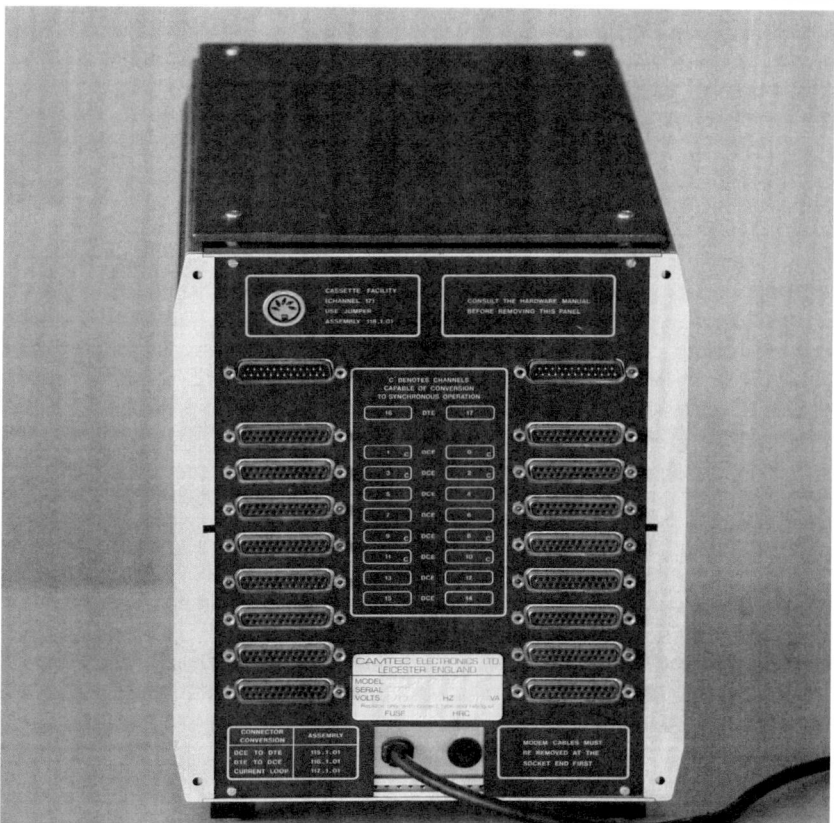

Figure 4.1. The JNT PAD: front (top) and back (bottom). *(PAD provided by Rolly Trice. Photos: Sue Srawley)*

representation from the other Research Councils of the time (NERC and SSRC[15]) and, between them, the two bodies were responsible for all major computer facilities in the country. First the Network Unit and subsequently the JNT had conducted periodic discussions with the Post Office regarding both its attitude to licences under the monopoly and the possibility of its providing network services tailored to the academic communities needs. By October 1980, when there were still no answers to either of these questions, the Computer Board and the FCC set up a Joint Working Party on Networks 'to examine issues associated with the integration of communications in the academic community'.

The working party reported in January 1981. The two major issues upon which it concentrated were the options for a national network and how to fund and manage it. It was estimated at this time that between them the Computer Board and the Research Councils were spending in excess of £1.4m per annum on communications costs. Since rationalising communications lines was expected to reduce these costs, this provided initial guidance on the budgetary estimate for any national network to serve the community. Several options were explored but BT[16] was not able to offer anything other than the standard PSS service at its published rates. This was unacceptable on two counts: it removed all control over the nature and future development of the network service from the community; and the cost was both unacceptable and unmanageable. Charges for PSS included usage-related elements, one part of which depended on traffic volume. This had two problems: budgeting would be hard because traffic patterns and volumes were generally unpredictable; and furthermore, it was already known that, even with some of the known traffic volumes, it would be unaffordable.

That left some form of private network. A possibility mooted and immediately rejected was to contract some third party to build and operate it. This met with no support anywhere, largely because it seemed likely to combine the disadvantages of being more costly with less control than if the community were to do it itself. The approach which emerged as most attractive was quite simply for the existing networks gradually to interconnect and evolve towards a single network. At least operationally and technically, this seemed quite a reasonable approach, with no obvious pitfalls in respect of service disruption. Of course, that raised the spectre of the Post Office monopoly again.

However, the community was now in a stronger position – in practice, if not strictly formally. BT had demonstrated that it could not offer a service equivalent to a private network in terms of either price or meeting requirements. This would have been likely to make any formal refusal to grant a suitable licence embarrassing, especially since the education community was, and remains, self-evidently non-commercial, still less in competition with BT. This aspect in itself might not have been sufficient but as part of the deregulation represented first by the creation of BT and then the preparations for its subsequent flotation on the stock market,[17] the Government had introduced legislation in the form of the Telecommunication Bill in November 1980 which included provisions to alter the position in respect of granting a communications network licence. This was going through Parliament at the time of the working party report, with the expectation that it might become law by July 1981. The effect would be that overall discretion

---

15   *See* Glossary.

16   Post Office Telecommunications had been renamed BT in 1980 but was still part of the Post Office. This was the first step in the Government's implementing the recommendation of the Carter Report in 1977 that the telecommunications part of the Post Office should be split off.

17   This eventually happened in November 1984.

Overnight commuting for lack of a network

The level of frustration being experienced by 1980 at all levels with the lack of integrated communications within the academic community can perhaps be illustrated by one user's experience known to the author. At the time the BBC was launching its computer literacy project and had selected as the domestic computer to accompany its programmes a 6502-based microcomputer (which became known as the BBC Micro) being developed by Acorn Computers, a spin-off company from the Cambridge University Computer Laboratory. Part of the design depended upon a semi-custom chip[1] designed with the aid of GAELIC, an early specification language available on the ICF DEC10 at Edinburgh. Although available over SRCnet, there was no access from Cambridge at the time, in consequence of which Andy Hopper[2] (then a post-doctoral researcher at the Computer Laboratory) ended up commuting nightly by car the 160-mile round-trip between Cambridge and RAL in order to complete and debug the chip design in time for the BBC Micro launch at the end of 1981. And this at a time when around £1.5m was being spent each year on communications by the Computer Board and Research Councils.

---

1   Technically, the chip was an uncommitted logic array (ULA); the design related to the interconnection of the pre-formed logic elements.

2   Professor Andrew Hopper, CBE, FRS, Head of the Computer Laboratory at the University of Cambridge at the time of writing.

---

on the granting of such a licence would now be vested in the Secretary of State for the DTI. Accordingly, the report recommended that assurances should be sought from the Minister regarding an application from the community for a licence.

In respect of how to manage and fund the eventual network, the working party recommended that as a first step to unified management a Network Management Committee (NMC) should be created. This Committee would have overall responsibility for policy, approving planning and future development, and submitting annual bids for funding and annual reporting to funding bodies. The funding bodies would initially comprise the Computer Board, SERC and NERC but would include other Research Councils as and when this became appropriate. In order to carry out the day-to-day management of the network, and assist the NMC in estimating the budget, producing annual reports, and planning and implementing growth and enhancement, a Network Executive was recommended. The Network Executive would also have responsibility for operation of the network. If the network were operated by the community then the Network Executive would consist of a team sufficient to staff a network operations and management centre together with a staff member at each of the institutions where a main packet-switching exchange was sited.

Following endorsement of the Joint Working Party's report, the Computer Board and FCC established the NMC, chaired by David Hartley.[18] It met for the first time in June 1981. The immediate task was how to achieve a national network from the current position. The details of that position in terms of network equipment, lines installed, and an estimate of costs had been established the previous year by the JNT. The NMC commissioned Scicon Consultancy International to report on the design parameters of the community's requirements based on this information.

---

18   Dr David Hartley, then Head of the University of Cambridge Computing Service.

At the same time, it established a working party chaired by Tony Peatfield of ULCC to evaluate this report and draft an operational requirement. The idea was to proceed cautiously, keeping BT informed at every step of the way to ensure it had the maximum opportunity to respond. However, as well as serving as the basis for any tender should that become appropriate, the document would also serve as the initial planning document for a private network created, owned and operated by the academic community.

By the time of the next meeting of the NMC, in September 1981, the Telecommunications Bill had become law, PSS was in service and BT was on the way to becoming a separate company (initially wholly owned by the Government). The granting of any necessary network licence to the community could now, in principle, be made by the Secretary of State for the DTI. Furthermore, by its next meeting in November 1981, the preferred option of creating a national network out of parts already existing became the *de facto* plan. But still there remained the question of how to do this.

There was, of course, a ready-made candidate. David Hartley recalls[19] that at this point 'SERCnet was being viewed covetously from the Computer Board side of the fence.' He and Roland Rosner agreed that the thing to do was to use SERCnet to form the wide-area backbone of the future network. So, the only remaining issue was: how to persuade SERC of this? In the end, they telephoned Geoff Manning,[20] then Director of RAL, who was immediately supportive. Furthermore, he not only attended the Computer Board at that time but was also a member of the NMC and was SERC Computing Coordinator. At the next meeting of the Computer Board, during discussion, having sent Roland Rosner a note to the effect that there was duplication of effort, Geoff Manning suggested to the Board that SERCnet should be handed over to become the interconnecting wide area component of the academic network. And that was it really, 'bar the shouting'!

4.4 1984

So, by the spring of 1982, it had become clear what to do to create a joint national academic network for the Computer Board and the Research Councils: just interconnect all the regional networks by means of SERCnet, rationalising the leased lines along the way; set up a joint operations and management infrastructure; and put in place a joint structure acceptable to the Computer Board and the Research Councils to oversee policy and budget, and to approve planning and spending on the network. Oh, and by the way, do this carefully, that is, without any of the proverbial 'shouting' of the expression in the last section, so as to avoid disturbing the delicate and now rather murkily-defined monopoly position in any unproductive way.

The term adopted to refer to the agreed approach of simply linking up what was already there to create the network was 'evolutionary'. By March 1982, David Hartley was able to announce that this plan had been endorsed and its implementation was being planned. There were several advantages to the evolutionary approach. Firstly, it would offer minimal disruption to the

---

19   Conversation with Ben Jeapes in 2004.

20   Dr Geoff Manning, Director of the Rutherford Appleton Laboratory from 1979 until 1986. Ben Jeapes interviewed him for his memories of JANET in 2004. Sadly, he passed away on 21 December 2006.

community: people could continue working as before, except that connectivity would steadily improve as previously disconnected groups of sites became reachable from each other. Secondly, it offered a manageable and affordable approach. And finally, should there be any objection from BT at any point, the process could be paused, reviewed, and potentially altered at relatively little cost or disruption, especially since increasingly all the parts were becoming PSS compatible. Now that the approach had been chosen, it was time to set up the joint management infrastructure and, a decade after his first seminal contribution to the creation of a national network, Mike Wells again stepped forward, this time to chair a Common Management Study Group to elaborate the details of the joint management and funding arrangements recommended earlier by the Joint Working Party on Networks. Both Mike Wells' management group and Tony Peatfield's technical operational requirements group reported to NMC meeting in March 1982. The overall structure of a Network Executive responsible for network operation, managing a set of Network Operations Centres (NOCs) and a small Network Control Centre (NCC) was agreed, very much as originally recommended in the Joint Working party report of January 1981, though the smaller NCC had replaced the somewhat larger network management and operations centre suggested in the report. There remained a substantial amount of negotiation to be completed both in respect of the effort required to operate the network, budgetary provision and operational details.

In the autumn of 1982 it was announced that a temporary part-time post of Director of Networking would be created, thus ensuring senior participation from both university and Research Council sectors in organising the creation of the network. The plans were aired at a public meeting in January 1983 and received sufficient support to proceed. Now it was time to translate all the talk into action. Geoff Manning, now that it was agreed that SERCnet should form the basis of the academic network, formally intimated to the NMC that he wished to see the new management structure take over responsibility for SERCnet as soon as possible. A temporary executive group was formed, by secondment from SERCnet (DL and RAL), regional operations groups (ERCC and SWURCC) and NERC, to begin the technical evolution of SERCnet and the regional networks.

Now that the effort he had invested in the Network Unit and the JNT was finally coming to fruition, Roland Rosner decided that it was time to take up a new post as Head of the Communications Group at ULCC. Barrie Charles took over as acting head of the JNT; and shortly thereafter, in the summer of 1983, Mike Wells became the part-time Director of Networking, with the incipient Network Executive reporting to him. A little later in 1983, the first two staff were appointed to the Network Executive: Ian Smith,[21] who had been working on network development and operations at DL; and Keith Mainwaring from network operations group at RAL. The NMC was now reconstituted as a smaller advisory body, initially entitled the Network Advisory Committee (NAC). Interestingly, there was a small but significant change at this point. The NMC had reported to both the Computer Board and the SERC FCC; this new committee advised the Computer Board, which was the major stakeholder in the new network. What was evolving was an arrangement in which policy for the network would be decided by the Computer Board, which would be the major fund provider of the network. These funds would, in respect of the JNT and the Network Executive, be channelled through SERC to RAL, where both the JNT and the Executive would

---

21   Dr Ian L. Smith, from Network Operations, Daresbury Laboratory.

be based. Of course, the Research Councils would continue to have representation on both the Computer Board and the NAC.

The completion of the new management structure occurred shortly before the end of 1983. As had always been envisaged, the (part-time) post of Director of Networking was for a fixed term of 5 years, to oversee and guide the transition. The next step toward an integrated structure was to have a new post of Joint Head of the JNT and Network Executive: this was achieved in November 1983 with the appointment of Peter Linington who, since the DCPU ceased operation, had been managing a large collaborative network research project[22] for a couple of years. This appointment achieved what had been articulated earlier by Mike Wells:[23] it brought together overall responsibility for the network in a single organisation, with the JNT becoming the development arm and the Network Executive the operations and management arm.

Now the physical assembly of the national network began. SERCnet by this time already had packet switching exchanges (PSEs) at Cambridge, Daresbury, ERCC, RAL, SWURCC and ULCC. The NERC network, which was attached to SERCnet, also had PSEs at Swindon and Bidston. The SERCnet exchanges all used X.25 packet-switching software developed by Andrew Dunn at RAL for GEC4000-series computers (see Figure 4.2). Though originally designed as a small general-purpose machine, the system was based around message passing with micro-coded hardware assist. As such, the design was very suitable for packet switching. To gain performance, practically all of the standard operating system had been stripped out; while good for performance, it also meant that there was little in the way of logging or management facilities. At around this time, as a result of a tender exercise, GEC had been selected as the preferred supplier for the PSEs for the national network, but with manufacturer-supported software incorporating management facilities. Towards the end of 1983, two of these new exchanges were installed at UMRCC and Belfast; after a period of debugging and commissioning, they went into full service in the first quarter of 1984. There were now 10 PSEs, situated at as many NOCs, together with a small NCC situated at RAL, and for the first time there was a national, joint academic network – shown in Figures 4.3 and 4.4.

And what of the position *vis-à-vis* the emerging regulatory position of BT, as recently modified by the Telecommunications Act 1981?Telecommunications Act 1981? As it would transpire, the spectre posed by the monopoly in respect of either its possible contravention or the granting of a licence under its provisions gradually receded. There had been great care taken by the community, specifically Mike Wells, Roland Rosner and David Hartley, on behalf of the Computer Board and the Research Councils, to ensure that BT was kept informed and given every opportunity to make an appropriate offer to this unique, non-commercial sector. The final act in this saga took place in January 1983. At this point, the evolutionary approach to realising what would amount to a private network for the sector having been agreed, implementation was about to be initiated. It fell to David Hartley, as chair of the NMC, to write to BT. The gist of the letter was quite simple: PSS was too expensive; and BT had made no proposal regarding how it might provide the community with facilities equivalent to a private network. The evolutionary approach was explained as one which allowed for the use of

---

22 UNIVERSE: a three-year collaboration amongst three universities (Cambridge, Loughborough and UCL), three companies (BT, GEC and Logica) and a government laboratory (RAL), 1981-1983 (Burren and Cooper, 1989).

23 Article by Mike Wells in *Network News* 17, July 1983.

Figure 4.2. GEC 4000 PSE at University of York, 1986. *(With permission of Suky Thompson and University of York.)*

a BT-provided solution, should it be offered; and it was emphasised that a BT solution which met the community's requirements was one which would be preferred. The closing paragraph, in the politest possible diplomatic language, presented BT with an ultimatum which, in the vernacular, amounted to: 'put up or shut up'! No reply was ever received to this salvo. In part, this may well have been because by then the next stage of deregulation was beginning. Technically, BT still retained its monopoly position as a result of the 1981 Act. However, the Secretary of State had the power to license other operators and in 1982 Mercury Communications, a subsidiary of Cable and Wireless, was granted a licence to operate a public telecommunications network. In the same year, the Government formally announced its intention to privatise BT, which was eventually confirmed by the Telecommunications Act 1984. This also finally removed BT's monopoly position. In retrospect, it is hardly surprising that no response to the ultimatum was received – although there were a number of helpful discussions with the legal experts at BT. The outcome of the latter was essentially that BT would not

74 | 4. JANET

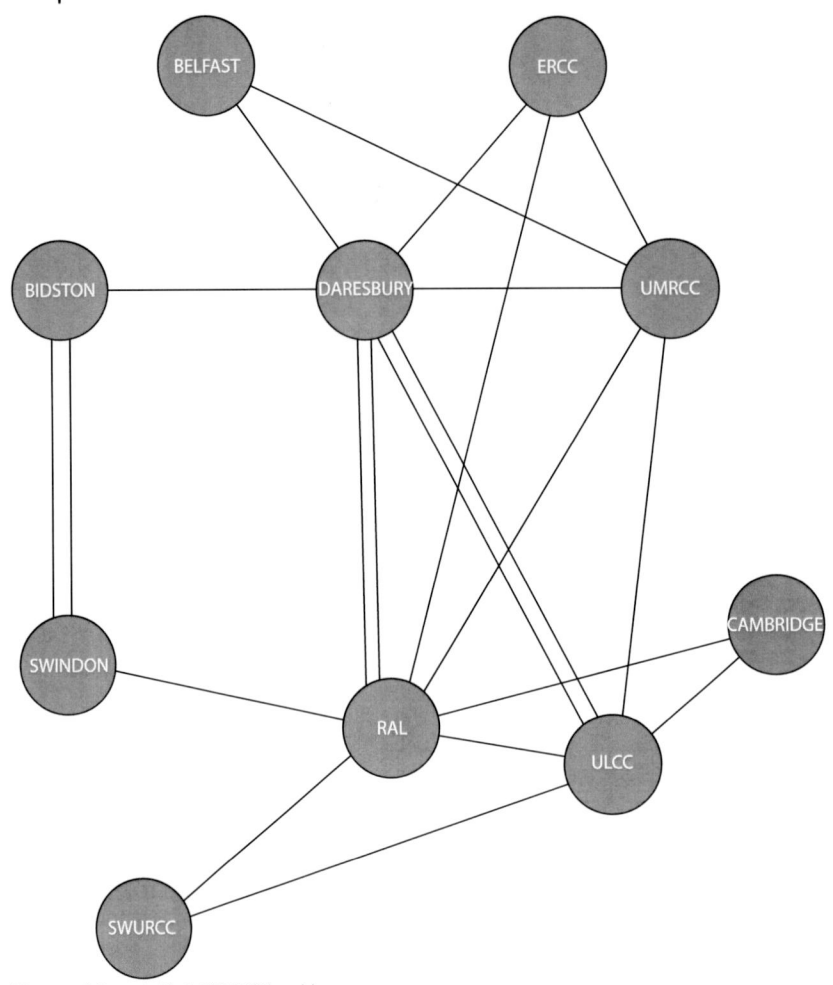

Figure 4.3. Initial JANET backbone.

interfere, provided the academic network refrained from encroaching on areas which could be interpreted as competing with BT. However, the position was anything but transparent at the time – and indeed there were various legal technical suggestions made subsequently about how the position of the national academic network might be regularised: these were mostly based on regarding the network as being 'private' and licensed to some legal entity of sufficiently adequate weight and scope – suggestions varied from the DES to the Crown – or being within the Crown prerogative as being in the public interest. A couple of years later, in 1986, one of these options was pursued further when Kenneth Baker, then Minister of Education, agreed that he formally 'owned' JANET and its existence was in the public interest: in effect, JANET was operated by the Crown and required no licence.[24]

And so, finally, after the addition of a couple of new exchanges, in Manchester and Belfast, to what had existed as SERCnet, the joint – i.e., Computer Board

---

24 See *Network News* 24, December 1986.

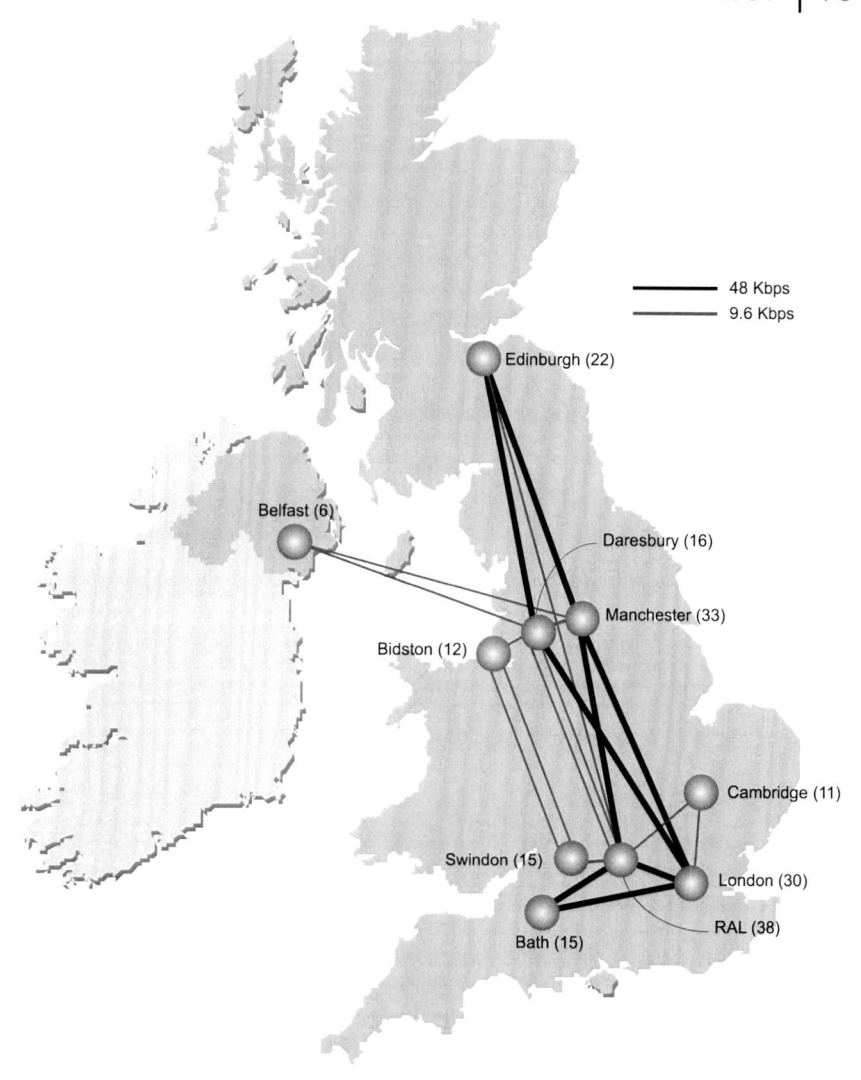

Figure 4.4. JANET backbone, showing geographical locations of and number of connections to each PSE (after Wells, 1986).

and Research Councils – academic network came into being. The first financial year for which there was a specific line for the joint academic network in the Computer Board budget was 1984/85. And what was it called? Joint Academic Network, apart from being heavy with its political origins, hardly tripped off the tongue. On the other hand its acronym, JANET, did and was in informal use by early 1983;[25] and by *Network News* 18, November 1983, the name appears in print, apparently in regular use. And, of course, the choice of name was sealed by the serendipity of Janet being the first name of the secretary to the Network Unit and

---

25 Conversation with Roland Rosner, September 2008.

the JNT during its first years.[26] So, by 1 April 1984, JANET not only had begun operation but it also now had a budget, and so this became the official date of the beginning of JANET, a little over a decade since the idea was first mooted.

In some respects the construction of JANET had proceeded on faith that a way to realise it officially would eventually be found. Almost all the early host software came out of the community. In the later stages there was some collaboration with manufacturers as the latter gradually committed to supply and support the protocols. The initial PSEs came out of SRC work, to be replaced during and just after the formation of JANET with commercially supported products. Likewise, the PAD came out of the initial hardware and software design at Cambridge. The academic community was an early adopter of the technology; and, at the time, this meant that there were no products on the market: there was really little option but to build the network 'in-house' in the community. This was a considerable achievement and provided a wealth of hard-earned knowledge and experience.

It came, of course, at considerable cost in effort. So who paid for it all in the first decade? Ultimately, the DES, primarily through the Computer Board and SRC. Equipment and lines appeared as specific purchases or rentals, each proposed by an individual institution to its immediate funding body. As to effort, this was less transparent. A large centre typically had a budget which included an element of development effort which was effectively part of the very substantial technical system administration support associated with such a service, because the only way to configure a system to specific needs in the 1970s and early 1980s was to make alterations to the code of the operating system – there was little in the way of configuration scripts or options. This meant that all centres had a pool of expertise, some of which gradually migrated to network development. With the coming of JANET came the beginning of a more transparent budgeting of the effort and equipment for the national education and research network infrastructure.

---

26 There was the obligatory naming competition and associated bottle of champagne (private communication, Peter Linington, May 2010) but the name was effectively already *un fait accompli*.

Part B

Development and Transition to Internet Protocols

Introduction

It is almost a truism that solving one problem promptly reveals several more to be addressed. The original JANET addressed the question of how to interconnect users and computing resources on a national scale. The computing resources were very expensive and physically large, and not very many of them could be afforded either nationally or institutionally. By 1984 JANET was on the verge of enabling access for most people to undertake substantial computation remotely, initially via remote terminals and shortly thereafter by means of the precursors of the modern workstation or personal computer. Even before JANET operation began, the question arose of to how to arrange local, campus-wide access and interconnection: indeed, campus access was needed by the end of the 1970s and was already under active investigation a few years before JANET was born. Much of the 1980s was taken up with this question, which happened to coincide with the deployment of several new, competing, shared network technologies that came to be referred to as local area network or LAN technology.

The LAN promised much, partly because the technology was intrinsically attractive in apparently offering flexible connection and partly because it enabled new distributed systems technology. The latter offered potential separation of the various components of a system, so enabling systems design to respond more flexibly to end-user requirements. But the LAN came with its own set of standardisation and market wars which, curiously, were effectively settled by the solution to the next challenge of how to make local networks scalable, now that everybody had a computer and all needed connecting to the network. Much of the early technical debate was taken up throughout the 1980s and early 1990s with whether a ring, bus or tree topology was best – a debate in which confusion about resilience, how each technology worked, and convenience of cabling management was compounded by warring vested interests in the various technologies.

The issue of which LAN technology and standards to adopt was paralleled by the corresponding world-wide debate about the wide area. Technically, this focused on whether a network should operate in a connection-like fashion, incorporating features of the century-old, successful telephony model, or whether it should embrace the brave new world of the LAN, where transmission was about a million times less prone to error and the seemingly simpler approach of just launching each individual packet towards its destination worked fine. Of course, it didn't happen quite like that. Buried inside the debate were a number of far-reaching architectural design issues: resilience (as with LANs); flexibility to interconnect and incorporate multiple network technologies; and what types of traffic could be carried. But such debates are only partly technical and in the wide area the network would need to span national boundaries. The one thing which governments and communications corporations (monopolistic or otherwise) could publicly subscribe to was adherence to standards – preferably international ones. Even here, though, there are *de facto* ones, which may originate in many different forums and gain international acceptance, and there are those created by overtly international bodies like ISO and ITU-T (or the CCITT as it was in the 1970s and most of the 1980s). In Europe, the latter form of standard was generally favoured, perhaps partly encouraged by the prospect of a larger market as well as the practical need to cross many borders. In the USA, the size of the home market was adequate for business and, although the US Government backed international network standards to some extent, *de facto* or US standards held sway: many either originated or were subsequently standardised under the aegis of US national standards bodies such as ANSI and IEEE and later ratified as international standards.

Against this international background, during the 1980s there emerged a plethora of more or less international network 'exploits'. Some were like clubs, several of which had institutional, community or corporate backing; others stemmed from the international telecommunications community; and in the background was the steady spread of the Internet. JANET participated in and was influenced by several of the exploits but, of course, only the Internet lasted. Part B describes how JANET evolved, much of it during the 1980s, to the stage where it became integral to the way education and research was conducted and, in the early 1990s, joined the Internet. During this same time, international communication grew steadily in importance: Chapter 7 provides an outline of how its development affected JANET, starting from its beginnings in the 1970s, during the 1980s and into the beginning of the 1990s, up to the time when development of TEN-34 began. And providing a recurring theme to the decade was the perpetual pursuit of increased capacity for the burgeoning national network.

5

Early Evolution of JANET

At various stages in JANET's evolution, growth and development, new technologies and ideas for new services have appeared. Those described here can be characterised as having occurred approximately during the 1980s – though in several cases they have their origins in the late 1970s or somewhat before JANET was officially born, and some lasted into the early 1990s.

A number of new applications and services came into being in this period. Possibly those of most lasting significance had to do with information services; though the technologies of these early examples passed away a couple of decades ago, what they presaged was the use of a network to purvey information and the need, as soon as the network became too big to search manually, for an automated aid to finding that information.

Local access has already been mentioned a number of times in a university or research laboratory campus context. Simply replicating on a local scale the technology used for the main network was an obvious option and was exploited. However, stemming from origins going back to the late 1960s and early 1970s, a new technology had appeared on the horizon by the second half of the 1970s: the local area network or LAN. This offered a number of possibilities, many of which were pursued in the UK by the research community and as part of the JANET effort. All LANs were based on sharing a cable transmission medium capable of what was for the time a very high transmission rate – roughly a thousand times the typical wide-area communications link in 1980 – over a restricted range of a kilometre or so, at a bit error rate about a million times better. It was almost as good as having a computer bus that covered a building and in the 1970s, in research laboratories around the world, together with the microcomputer, it sparked distributed systems research and development in which various parts of a computer system were split up over a LAN, enabling the expensive bits to be shared amongst an officeful of cheap, small personal systems. Then, as affordable versions of the technology began to be deployed at the beginning of the 1980s, it also offered an attractive solution to wiring a campus and for roughly the next 15 years, cable-based, shared-medium campus and office LANs spread worldwide.

A constant theme in many success stories is that of growth and JANET was no exception. Indeed, even at its moment of birth, that was its immediate challenge. SERCnet had been operating on line speeds no higher than 9.6Kbps, even interconnecting its switches in what might be described as its trunk network. Suddenly, this trunk was faced with handling a community perhaps tenfold larger, though estimates are hard to make at this time. What is certain is that from the moment it was known in 1982 that this was how the trunk would be formed, upgrades in transmission capacity were among the first priorities.

Of course, as network performance improved, so it enabled new applications – or sometimes just expanded versions of old applications. Once local area networks providing megabit per second user data rates appeared, the wide area was under even more pressure, constrained by expensive line tariffs which bore an ever decreasing relation to costs, and with the challenge to keep up not only with population growth but expectations increasingly influenced by what could be achieved with higher performance.

5.1 New services

As JANET connected up and became the national network, the first new services to appear were both informational and both related to the network: one for users of the network and the other, ultimately, to assist the network to function.

Name Registration Scheme

The early naming and addressing of network services in JANET has been briefly described in Sections 3.1, 3.3 and 4.2. Generally, a user of a network service needs to be able to look up its name or address in order to be able to contact it. If the services offered by hosts all follow the same scheme for deriving service names or addresses from the corresponding host names or addresses, it may be sufficient simply to publish the host names or addresses, and indeed a table of host addresses was what was provided initially for the PSS user community by York University in 1981.

In the early days of SRCnet, addresses were assigned by the committee which was overseeing its development, the effective unique address authority, thus avoiding duplicates. As JANET began to form, the JNT naturally became responsible for assigning addresses within the network. Attaching names to addresses typically originates as a convenience to users of the network since memorable or mnemonic naming of hosts or services may eliminate the need for repeated reference by users to a directory. Using names instead of addresses implies the need for a table listing the complete mapping of names to addresses which, at its simplest, can just be implemented as a table in each host. To ensure that a given name refers to the same service throughout the community, a naming authority is needed, as for addresses, and all the host name-to-address translation tables need to be kept synchronised or, alternatively, a name should always be looked up in some uniquely maintained system. The JNT naturally took on the additional role of naming authority for JANET; developing the supporting infrastructure and management for name registration and distribution was undertaken by a team at Salford University led by John Larmouth.

Assigning addresses for hosts in JANET was effectively achieved as a part of the operation of connecting a host, much as BT did for PSS. As has already been mentioned in Section 4.2, the actual scheme was a simple two-level hierarchical one of numbering each host relative to the PSE to which it was connected, and concatenating that with the switch number. Naming was also hierarchical (as implied in the description in Sections 3.1 and 3.3): names were composed of a number of concatenated components, the top-level component of which was institutional. The top-level component to be used by an institution would be agreed with the JNT and all the lower-level components were the responsibility of the institution. The institution might repeat the devolved management pattern for subsequent components by delegation to departments. Once chosen, the names needed to be registered and published to users and distributed to hosts on the network.

The need for registration was first discussed by a group composed of members from the Transport Service Implementers Group and the JNT in mid-1981,[1] essentially elaborating the requirements summarised above in anticipation of the formation of JANET. Once it was known by the beginning of 1982 that JANET would be constructed by using SERCnet to join up the regions, a detailed design study was commissioned by the JNT from Salford in 1982 and completed early in 1983. By May 1983 implementation of what by now was known as the Name Registration Scheme, or NRS, had begun and an initial system was undergoing trial by March 1984. Registration was still by paper because authoritative signatures were needed and also email was not yet pervasive enough for it to begin to replace established formal communication channels. At the end of 1984, development was sufficiently advanced for production service planning. During 1985 a prototype service was opened by Salford and a full service followed in 1986.

We shall return to the Name Registration Scheme in Chapter 6 but it is important to realise what sort of service it was. It was conceived at a time when the majority of service computing was still provided by mainframes of various sizes, though some were small and cheap enough to service groups or small departments, as well as the large, expensive ones which provided university or community-wide services. The number of such computers in the country was still measured in hundreds – comparable with ARPANET in 1982. The service which the Name Registration Scheme provided was that of managing a single global table for the network, so that it could handle new registrations from institutions and could make available the directory table for inspection and retrieval: in the latter case, in a form suitable to form the basis for constructing a host table. Although the ARPANET and Sun systems were already in deployment, the NRS did not build on either of these.

News, information and mail services

Disseminating information about academic networking was something that the Network Unit had begun in 1977 with *Network News*, long before the network existed. The first *Network News* was dated 25 May 1977. It was written by Roland Rosner and was simply a paper entitled 'Network News', produced by the Network Unit of the Computer Board and Research Councils, reference 'NUCBRC P24'. The next two followed suit and then, with issue 4 in 1979, it acquired not only its first banner header but also its first editor, Barrie Charles, who continued for five years until issue 20. Subsequent editors included Peter Linington, Bob Cooper and Shirley Wood. *Network News* acquired its first ISSN[2] in 1988 and ended with issue 44 in June 1995. The JNT had been superseded by UKERNA the year before and now, after an interval of a little over a year, the JANET newsletter was reborn as *UKERNA News*, which ran from October 1996 until issue 39 in June 2007. Following the adoption by The JNT Association of a

---

1   Reported in *Network News* 11, August 1981.

2   International Standard Serial Number; 0954-0636 for *Network News*. When a serial publication changes its title significantly a new ISSN is issued. *UKERNA News* was 1365-3636; *JANET News* is 1755-2397.

new trading name, JANET(UK) in June 2007, the newsletter was also rebranded as *JANET News*, starting with the September issue that year. *JANET News* is currently a quarterly publication on paper and in HTML, published in March, June, September and December.

One of the earliest means of disseminating news, eventually worldwide, was Usenet (mentioned briefly in Chapter 3). Based on UUCP, it was a means of passing on items of interest by means of emails categorised into groups, topics or threads of discussion. Originally developed in the US in 1979, by 1982 it was in quite widespread use, including in Europe (where it was known as EUnet). By 1983 the University of Kent was providing access to Usenet news, which in 1984 had to be put on a chargeable basis because there were real costs to be recovered. This was used quite widely in the UK, including by the academic community, to the extent that by 1987 a spin-off company, UKnet, had been formed to provide the service. The same year UKnet was contracted to provide a JANET-wide Usenet newsfeed. UKnet subsequently became GBnet and the feed continued, ultimately changing to use of Network News Transfer Protocol (NNTP) over IP in 1994.

By 1997, the JANET Usenet News service was being provided by PSINet, though it was still a service to which individual institutions subscribed directly. This resulted in a large number of replicated feeds into JANET, helping to swell the traffic on what was by then a 10Mbps SMDS[3] connection almost to saturation. To handle this traffic more efficiently, in 1997 six feed sites[4] were chosen to receive feeds from commercial ISPs under contract with UKERNA and these sites then undertook distribution to subscribing sites on JANET. By 1998, as this service became fully operational, it also became an early example of JANET applying filtering, in pursuance of that part of the JANET Acceptable Use Policy (AUP) forbidding use of JANET to convey obscene, offensive or illegal material – though the UKnet feed in 1990 already excluded some groups, the 'rec.arts.erotica' group being an example.[5] We shall return to the topic of regulation more generally in Chapter 13.

Popularity of the service continued to grow, so that in 1999 further developments were needed to cope with the volume of traffic – a result not just of the growth in number of users but also of the increased volume of traffic being generated in Usenet News, particularly by bitmap (pictorial) information. By 2001, use of readers accessing the JANET distribution sites directly was encouraged, so that only requested information was transferred across JANET; and a Usenet News cache service was being developed. The reader service also enabled access for people at organisations without a feed service. However, the next few years would see a gradual diminution in demand which continued until, in 2006, two of the feeder sites were removed; people were again encouraged to move to using the reader service in an effort to reduce the use of feeds even further, and residual feeds were now configured to provide only those news groups specifically requested. Already there was talk of phasing out the service – foreshadowing the announcement late in 2009, as use dwindled even further, of plans for closure of the JANET Usenet News service at the end of July 2010.

Of course, once JANET existed, making news about progress of the network and its facilities available over the network became an obvious need. The first

---

3   *See* Chapter 10.

4   Daresbury Laboratory, Imperial College London, RAL, University of Wales at Swansea, University of Strathclyde and University of Warwick.

5   *See* Houlder (2007).

step in this direction was taken early in 1985 by the JANET network management team, the Network Executive. A read-only account (id NEWS, password NEWS) was made available on a GEC 4000 system, where ASCII text files could be posted by JANET staff, including information submitted by the community: this was accessible as UK.AC.JANET.NEWS via X.29. Mailing lists covering a wide range of local and national interest groups also sprang up in the next few years but creating and maintaining such a list was a chore requiring detailed technical knowledge of some particular mail distribution system – and there was no way of finding out if a public list already existed relating to any particular topic.

By 1988, it was becoming increasingly urgent to tackle the issue of information dissemination over JANET and the topic was aired at Networkshop 16 (Reading) in a discussion group on 'Networked Tools for Information Services'. Since there were already a number of systems either available or in development world-wide, taking a variety of approaches and providing a range of facilities, the upshot was the formation of a group to investigate these, consult the community and suggest a plan to provide a community facility: so began the JANET Networked Information Services Project (NISP) at the University of Newcastle upon Tyne, funded by the Computer Board in November 1988 to develop the tools to enable users to set up and administer 'information servers, electronic journals, bulletin boards and of course electronic-mail-list servers as part of a Networked Information Services Infrastructure'.[6] The project included assessment of user needs, producing a functional specification and implementing a trial pilot system.

At almost the same time, a bulletin board system funded by the Computer Board was being mounted to meet the needs of the National Information on Software and Services (NISS) project, a project to catalogue and make available information about resources available to the academic community, including pertinent software. The bulletin board was based on the software already in use for the Humanities Bulletin Board (HUMBUL) at the time.[7] Funded in 1987 and in service since September 1988, access was by interactive terminal login to a menu-based system. By 1989 it was in regular use and by 1990 other databases of information were being added, coupled with a programme of development and upgrade. Of the several databases now supported, one was for jobs available in UK academia – foreshadowing the jobs.ac.uk service familiar since its launch in 1989 at the University of Warwick, currently run by Eduserv in Bath, the successor to NISS, founded 1999. By 1994 the need to manage user identities for access to a range of information had led to the development of a system called Athens (see Chapter 13), at this time part of the internal support for the NISS service.

Meanwhile, by the latter part of 1989, NISP had completed the first part of its work, producing a report summarising user needs. This could be retrieved by sending an email which contained a command 'send nisp <filename>' to a NISP identity at Newcastle, an example of the use of email as the front end to an information retrieval and file transfer system. There was an email list for discussion relating to the project (hosted on one of the IBM systems at RAL), subscription to which was also was managed through an automated email interface. Rapidly following the report on requirements, in December 1989 a functional specification for a networked information server was circulated; again, retrieval was through an email interface, this time to an identity 'mailbase@uk.ac.

---

6   *Network News*, No.27, p.2, 'JANET Networked Information Services Project' (Jill Foster, University of Newcastle upon Tyne).

7   USERBUL software created at Leicester University, which provided the HUMBUL service at the time: *see*, for example, the articles by Lee (1994) and Stephens (1997).

newcastle'.[8] This signalled the first use of the prototype system that the team at Newcastle had been building in parallel with the requirements and design phases – in view of the project being limited to two years duration. The system was called Mailbase and provided an interface to the prototype information server. The system provided for the management of discussion lists, including email, access to all files related to the discussion and an archive of contributions to the discussion list. The interface was by email – and to start things off the discussion list for NISP itself had now been moved from the system at RAL to Mailbase. The system proved popular: by August 1991, funds had been awarded by the Computer Board/ISC for a 3-year follow-on 'NISP II' from July 1991 – essentially to turn what had been a test service for 18 months into a production service with training and support. The next version of the system was available in the latter part of 1991 and now included interactive terminal access. So was born the service which became the UK education and research email list management system for the 1990s.

Of course, by the latter part of the 1980s many disciplines were in need of a general means of publishing information on a network as an aid to collaboration, as a result of which out of the particle physics community would come the World Wide Web. In 1994 a Web-based interface to Mailbase was introduced, effectively replacing the interactive terminal interface. In 2000, JISC put the service out to tender, as a result of which it moved to RAL as one of a portfolio of services provided under the JISC Assist umbrella. The service would now be called JISCmail and based on LISTSERV,[9] a system first developed in 1986 and by 2000 commercially available from its author's company, L-Soft International.

PSS

During the latter part of the 1970s a number of universities, regional networks and SRCnet had participated in use of EPSS, to the extent that it was providing an interim interconnect. Then, in 1980, the PSS trial service began. This ran in parallel with EPSS for a year, at the end of which time EPSS was withdrawn and PSS became a full service, the major aspect of which was that it was now to be paid for. People had begun to rely on the previous service – and it was still possible that some variant of PSS would form the basis of JANET – so SERCnet, the regional networks and some universities mplemented gateways to PSS: before the end of 1981 there were already 21 academic connections to PSS.

Now arose the issue of the mechanics of paying for PSS use. Setting up a PSS call from a gateway required an account identifier and password to be quoted for billing purposes. A variety of such accounts were set up at institutional, departmental, project and group level. Disregarding the issue of principle about network services being 'free at point of use', this presented some practical problems. An interactive terminal call could typically be handled by an application relay in the PSS gateway: a call would be made to a PAD process in the gateway, where another call could be initiated to the PSS destination. During this process, the terminal user could be prompted to supply the account and password.

---

[8] Mail systems at this time commonly provided the capability to deliver mail to a program, providing the basis for a variety of automated systems. As with the corresponding facility in Web server systems, such capabilities – at least in their original form – have generally long since been removed from most services because of the avenue they presented for abuse.

[9] See also Glossary.

However, for file transfer, including mail transfer, no user was present and a way needed to be found to include the account information for the onward PSS call. In the event a somewhat *ad hoc* (and insecure) addition to Yellow Book Transport Service addressing syntax and semantics was made to enable such information to be included in the addressing component relevant to transit across a particular network, enabling the onward call to be made by the gateway to complete the end-to-end transport service connection. This problem was not peculiar to PSS: it had existed for IPSS since the service began in 1978, though IPSS use had typically been for X.29 terminal access, since other networks generally did not support JANET protocols.

By contrast, the UK-US gateway at UCL had, since its inception, used IPSS to carry a mixture of application traffic to the US, as an alternative to the ARPA SATNET link. PSS access to the gateway was also provided. The use of IPSS continued until after the service moved to ULCC (*see* Chapter 7). However, once JANET came into operation, use of PSS within the JANET community generally dwindled to communication external to JANET, including access to Europe and elsewhere via IPSS (now accessible via PSS). The need for multiple gateways steadily diminished. The erstwhile SERCnet PSS gateway became the JANET PSS (and IPSS) gateway and funding for other gateways to PSS was withdrawn by the Computer Board in July 1985. Subsequently, to provide improved reliability through redundancy, a second PSS gateway was provided at ULCC. Then, in 1989, in a further effort to improve reliability, a third gateway was introduced in Manchester, with intent to provide automatic re-routing of calls in case the PSS service was unavailable at one of the gateways.

Automated re-routing of calls presented an additional challenge for accounting, since users typically used a particular gateway and accounting was by debiting usage against a regular allocation on a particular gateway: rerouting resulted in debiting a different allocation on another gateway. These sorts of issues provided an early example of both the additional mechanisms needed and the additional effort arising from the lack of a peering arrangement, in this case between JANET and PSS.

By 1994, with the continuing move away from X.25 to use of IP (Chapter 6), use of PSS would dwindle to terminal calls only, for which a conversion service for outward calls (from Internet terminal protocol, TELNET, to X.29) would be provided. This would eventually become the only connection to PSS once the JANET X.25 service closed in 1997.

5.2 Campus access

Much of what has been described so far has been from the gross, wide-area perspective of how to interconnect a set of computers dispersed across the UK, a few per university or research laboratory, with the major purpose of providing both RJE and interactive terminal access to run batch jobs and provide interactive services. The extension of the wide-area paradigm to a more local context has been described (Section 4.2); and how, for the original JANET, this was achieved using CPSEs and the mini-X.25 switches derivative from the JNT PAD. The capability for the latter had been in the original Cambridge design and had been exploited quite early in the PAD to enable PADs to be daisy-chained together without the need for a separate X.25 switch: it was but a short step to remove the PAD support and produce a small, inexpensive switch. Extension of the WAN X.25 technology into the campus worked satisfactorily in respect of both cost and extending the wide-area functionality to campus level. Moreover, the campus network almost immediately enabled 'federated access' off-site to the wide area, eliminating multiple lines which had been needed to support sometimes multiple access to SERC facilities and Computer Board centres.

However, by the late 1970s a new local technology for networking was emerging from research laboratories. Transmission rates some four orders of magnitude higher than commonly available in the wide area could be achieved over short distances of around a kilometre and at bit error rates of at least six orders of magnitude better; and all this over copper cable, some of it rather similar to that used for TV or VHF radio. Actually, apart from achieving the speed and impressively error-free transmission, there was another reason why the extent of the network needed to be restricted. The cable could be shared between a hundred or so stations and there were two ways of wiring all the stations together: either in a ring or on a 'bus' – a length of cable with the stations just 'tapped' into it anywhere. For either topology, one could just arrange for the stations to take turns; alternatively, for the bus, they could just transmit and hope for the best. The latter idea had been successfully used by the University of Hawaii in 1971 on a radio network called the ALOHA network; but there the propagation time was quite long compared to the transmission time for a packet. In a cable, the two times are much more comparable and the risk of collision correspondingly higher, so it was worth a station listening to find out if its transmission collided with another and hence knowing whether to resend. To do this reliably requires that the cable not be too long for the packet size and transmission rate. In the case of taking turns, if the extent of the network is too big, a station will have to wait so long for its turn that unless it transmits a huge packet – 'hogging' the network for an antisocial amount of time – the overall rate at which it can transmit data will be depressingly low.

LAN technologies

Historically, among the earliest LANs were Söderblom's ring, developed in 1967, and the Pierce loop *circa* 1971. Both used the idea of a circulating 'token' which a station had to 'claim' before it could transmit; after transmitting its packet, a station would return the token to the ring for another station to claim. It turns out to be quite complicated to organise the management of such a ring and it was not until 1981 that the first practical token ring appeared, marketed as the Apollo Token ring by Apollo Computers. Meanwhile, two other LAN developments had taken place.

The simplest is the one characterised above as 'transmit and hope' – known more formally as 'random access' or 'stochastic' – and with a couple of refinements it was developed at Xerox PARC during the early 1970s, resulting in Ethernet in 1973 (Metcalfe and Boggs, 1976). The refinements consisted of a station listening to the network and only transmitting if it was silent, and then listening to its own transmission to detect whether another station had started transmitting at about the same time, thus corrupting both transmissions. (Later, the whole mechanism would be referred to as 'carrier sense multiple access with collision detect' or CSMA/CD for short.) Xerox was already using this network in-house during the latter part of the 1970s but it only appeared as a commercial product in the mid-1980s. A potential *de facto* standard for Ethernet was published in 1980 by the consortium of DEC, Intel and Xerox (DIX): this was Version 1.0, generally referred to as the 'Blue Book';[10] it was later revised in 1982 to become DIX Version 2.0. At almost the same time there emerged from the IEEE the first version of the 802.3 standard for a 10Mbps CSMA/CD network. Although technically stable, it had slight differences in respect of low level operation and existing chips worked according to the DIX (2.0) specification. Also, some aspects of the framing structure were different, causing difficulties for interworking with existing equipment. The upshot was delay in production of widely adopted products and supporting software. In the UK, although some groups began gaining early experience around 1982, wider take-up did not really begin until around 1984.[11]

Meanwhile, at the Cambridge University Computer Laboratory, another form of ring LAN had been developed which was based on a fixed number of small, circulating 'slots', each with a flag to say whether it was full or empty. A station could claim a slot by setting the flag to indicate full, insert some data and send it on round the ring to its destination. In all ring networks, by virtue of the topology, there is the option for the receiver to set a flag to let the sender know the data arrived safely. The Cambridge Ring design chose to do this so it becomes the sender's obligation to mark the slot empty again. The choice is a trade-off between capacity and reliability. (A derivative design by BT known as the Orwell Ring[12] took the opposite decision – it was intended as the basis of a distributed PABX and reliability is at less of a premium when carrying voice, as a consequence of which its capacity could be higher than its transmission rate.) The Cambridge Ring design was initially published in 1975 and a working ring was commissioned in 1976–7 (Wilkes, 1975; Hopper, 1978).

The Computer Science Department at UCL adopted a copy of the Cambridge Ring for LAN-based experiments in the late 1970s, at the time when Cambridge was using the ring for constructing its distributed system[13] (Needham and Herbert, 1982). Soon after, commercially produced copies of the Cambridge design became available from SEEL[14] and, a little later, a re-engineered version from Logica VTS.[15] A number of universities installed rings for early experience and development, some for prototype service trial and others for research under the aegis of the SRC's Distributed Computing System research programme (1977–84); the University of Kent, which would contribute to the early efforts in standardisation of the Cambridge Ring, participated in both aspects of this.

There were two distinct threads in the efforts to commercialise the Cambridge Ring. On the one hand, to take advantage of mass-production techniques as well as reducing the component count, shrinking the electronics from a board to a

---

10 Not to be confused with the UK NIFTP specification, published in covers of a similar hue!

11 Deployment of LANs based on token-ring ideas suffered a different sort of setback when in 1981 Söderblom was granted a US patent on a token ring technique, which the US patent office subsequently broadened in 1984. Indeed, Professor Roger Needham, then Head of the Computer Laboratory at the University of Cambridge, received a letter at about this time suggesting that the Cambridge empty-slot ring was in violation of the Söderblom patent (private conversation, c.1984). Subsequently, the US patent office rescinded its 1984 ruling and a claim by Söderblom for patent infringement by Madge Networks was denied in the UK High Court. But all this took years and added complication to the choice of LAN technology.

12 Named for the river which flows in the vicinity of the former BT Research Laboratories at Martlesham.

13 Begun in 1978.

14 Scientific and Electronic Enterprises Ltd, near Edinburgh.

15 Marketed under the brand name Polynet.

chip is an essential step. At Cambridge the same Ferranti ULA technology as had been exploited in a part of the BBC Micro[16] was being used; simultaneously, SERC commissioned Swindon Silicon to design a custom chip. The other thread was standardisation. The Cambridge Ring had its own protocols: an intermediate-level datagram-like protocol, called basic block, and a stream protocol called byte stream protocol (BSP). Ian Dallas and co-workers at Kent developed a YBTS version of BSP which was published as Transport Service over BSP or TSBSP in 1981. The significance of transport service (TS) in the UK protocols was now becoming clearer. It was not just that transport was the layer at which network independence, that is, independence from low-level technology such as LAN or WAN, could be achieved: that same independence also meant that it could be used as the layer at which to interconnect networks. By implementing TS on each network technology, it could be used as the common layer above which any application protocol could operate end-to-end. Subsequently, in 1982, this specification, together with other parts of the Cambridge Ring specifications were reworked into what became known as CR82 (Cambridge Ring 1982), the Orange Book, which was in a form suitable for submission to standards bodies.

Of course, such undertakings took time and effort: in the event, not only did commercial versions of Ethernet begin to appear but other commercial LANs had also come on to the market. A subsequent, faster (50Mbps) version of the Cambridge Ring (late 1980s) – and, indeed, even later, a 500Mbps version called the Cambridge Backbone Ring (early 1990s) – were developed but these remained in the research laboratory.

As mentioned above, the token ring, having lain dormant for nearly a decade, appeared first as a commercial product from Apollo in 1981 and then, in 1985, as a product from IBM, as well as being standardised by IEEE as 802.5. At the time, the IBM product was 4Mbps, slower than the Apollo version four years earlier at 12Mbps. Subsequently IBM would market, and IEEE would standardise, a 16Mbps version of 802.5 in 1989; but by then substantially faster versions of token rings were available – as we shall see in Section 5.3.

However, independent of the precise technology, underlying all this development there was a continuing debate about the relative merits of bus and ring LAN architectures. In a bus network, if the cable is cut, a terminator is removed or a station develops a fault which affects the whole network, it is essentially impossible to determine where the problem is other than by disconnecting stations or physically checking the cable. In a ring (though not in all designs), it is possible to tell fairly precisely where the problem lies – but it comes at a price: it contributes to the complexity of rings referred to above.

The period from the mid-1970s to the mid-1980s saw LAN development emerge from the research laboratory into commercial deployment. A feature of this emergence was that it coincided with the emergence of commercial single-user workstations and personal computers, all of which along with office word processing facilities were in need of ubiquitous connection to networks: local for such things as printer and storage sharing, together with access to wide area networks for email and wider exchange of document files. That distributed systems technology was still a research topic caused a confusion between what was ready for product development and what was still subject to experiment: Ethernet, slotted ring and token ring were all vying for the position of leading

---

16 In the video driver chip; the ULA was primarily digital but it also had some uncommitted analogue circuitry, which was exploited for driving a TV display in the case of the BBC Micro and for driving the low-level analogue transmission in the case of the Cambridge Ring.

LAN; the PC was appearing, along with diskless workstations and fileservers; but the work patterns which would be associated with these developments had not yet emerged. By the second half of the 1980s Ethernet technology had gained the lead in the market place but it was not clear whether it would be capable of scaling to deal with the explosive need for greater capacity. Although rings were more complex, because they were deterministic rather than random they had more predictable performance and gave greater promise of scalability. We shall see what actually happened in Section 5.3.

Application of LANs to campus access

If the situation regarding LANs seems confused, this was correct in the first half of the 1980s. Research and development was proceeding rapidly, with new products emerging all the time – only a tiny selection have been mentioned here, essentially those which had an impact on JANET at the time – and there was no clear commercial winner until the second half of the 1980s. Nor indeed did any of the technologies initially have more than one supplier (a partial consequence of lack of standardisation) – with the exception of Ethernet and, in the UK, the Cambridge Ring. This left JANET and campus service providers with a serious quandary: not only was it not clear which LAN technology to install, but each was proprietary and used a different type of cable. Indeed, not only were the cables different for each LAN, they were also different from telephone cables and the type of cables which had been used at the end of the 1970s for PACX terminal systems (Section 4.2). The cables themselves were not particularly expensive but installation was very expensive, especially in older buildings where there were no ducts and building work was needed as well as pulling the cables. Having just gone through PACX installations, followed by X.25 campus switch and PAD installations, the notion that now the answer was the installation of yet another technology with yet more, different cables was greeted with some scepticism. Moreover, there was also beginning to be an entirely new sort of cable to contend with: optical fibre. Since there was actually no answer to all this at the time, campus service managers resigned themselves to focusing on how to make re-cabling easier in future: thus was the notion of structured cabling born.

Initially, LAN cabling tended to follow the topology of the LAN: so, for a ring – which is composed of directed point-to-point links – the cabling might take a circular tour of a number of rooms or a small building. Ethernet, although originally a single piece of cable (a segment), almost immediately developed repeaters which allowed segments to be joined together to a form a tree-like pattern in which there were no loops. This was quite convenient: one could in principle run a vertical segment as a 'riser' to link floors and then use repeaters on this riser to attach segments along each corridor. A ring began to look less attractive until it was realised that one can 'deform' a ring in various ways while preserving its essential (ring) topology. One possibility is a long, thin shape, like a bus with two cables interconnected at each end. Indeed, one can then 'pull' 'lobes' out of the cables and start making self-similar or fractal patterns of repeating stars like a snowflake, all the time maintaining the logical ring topology. Once manufacturers realised this, they made wiring centres and supplied single physical cables that incorporated 'out' and 'return' segments, so that all that was needed was to plug these and the wiring centres together and create a structured wiring pattern to choice. In fact, because a ring had active repeaters everywhere, signal-to-noise could be kept good everywhere, whereas only a limited number of repeaters were allowed in any path across an Ethernet, and every time a joint or station was inserted there was additional noise and signal loss, which could give rise to unexpected problems when a new station was added. Either arrangement was ideal for cabling a building and, accordingly, patterns of rising ducts connecting floors and distribution horizontally along corridors became the norm. The key here was that the components for achieving this using LAN technology became affordable because of the combination of simplicity and mass production.

In practice, of course, it was not so simple. If terminals were to be connected to a LAN, a suitable PAD was needed; and if a LAN was to be connected to the main X.25 network, a suitable gateway was needed. In 1984, both of these became available for Cambridge Ring from Camtec, based on evolutionary designs from the original PAD and X.25 switch. This enabled the use of Cambridge Ring LANs for terminal traffic to the X.25 network. What about connecting mainframe hosts to a Cambridge Ring? In 1982/3, a number of hosts (including PDP11, VAX, Prime and IBM System/370) had been interfaced to the Cambridge Ring as part of an experimental project called UNIVERSE which had linked some 200 systems (mostly micro-based) using a satellite and Cambridge Rings (Burren and Cooper, 1989). However, experimental development is one thing: a supported service product quite another. So, in 1982 the JNT had commissioned Acorn in conjunction with ICL to develop an adaptable interface which could be configured to accommodate a range of computers. The device was referred to as the MACE[17] and was designed around a 32-bit 68000 microprocessor. This was a substantial development at the time and it was not until late in 1984 that it began to come to fruition, with units being delivered in 1985.

This picture of early LAN technology exploitation is typical of what happens in the early adoption phase of a technology. Most of the basic research for LANs had been accomplished in the mid-1970s; but bringing the technology to market, with the accompanying standardisation, software development and overall product support, together with the eventual market shake-out which has to occur before market and product stability are achieved, took about a decade. In this case there were several other factors. The PC and the single-user workstation both appeared in the same timeframe. The PC was an almost immediate success in the office; however, disk storage and printers were expensive. For sharing information and expensive peripherals in this context, the Ethernet began to proliferate quite rapidly. Its success was cemented as workstation manufacturers, most notably Sun, also adopted it. A Sun workstation came complete with an Ethernet interface and, moreover, it came almost from the start with a derivative of BSD Unix which not only incorporated Ethernet support but TCP/IP software, as recently adopted by ARPANET. Other manufacturers followed suit, as for example DEC with Ultrix. We shall return to TCP/IP in Chapter 6; for now, it is enough to note that, from about 1985, Ethernet would become the LAN of choice, ousting almost all others including, in the UK, the Cambridge Ring; and this also included the IBM ring, standardised as IEEE 802.5. The latter had been a latecomer to the market but originally operated at 4Mbps. Although token ring performance is technically better at high load than Ethernet, this was not enough to overcome the disadvantage of the slower speed, especially as the popularity of Ethernet rapidly reduced its price. The 16Mbps version of 802.5 did not appear

---

17 After a 6809-based board known by that name originally developed at the Cambridge Computer Laboratory, which could provide a basic block interface to a 68000 system.

until 1989, by which time people were looking for a factor of 10 improvement on 10Mbps, not a factor of 4 on 4Mbps.

Of course, the question arose of how to integrate Ethernet LAN access into JANET. This had initially been discussed early in 1983, after the appearance of the initial version of 802.3 in December 1982. This effectively signalled the initial confrontation of the ISO standards with what would emerge as the IETF Internet standards. IEEE LAN standards would effectively be 'rubber-stamped' as ISO standards. Operating IP over IEEE LAN standards was straightforward so that support and exploitation appeared immediately; ISO standards were nowhere near this stage. This was not a question of which standards were 'best' – if there were any commonly held criteria upon which such a judgement might be arrived at – but much more issues of politics and which community people belonged to. The majority of users had no interest in esoteric debates about the relative merits of one technology or protocol over another: just so long as it worked, did more or less what was wanted, and was capable of support at not too exorbitant a cost.

Within the user community, one of the things that was wanted was international collaboration; however, in the case of the computer science community, collaboration meant much more than email and file transfer. At this time, by far the commonest platform for development was Unix, along with all the applications, including network ones, that were being developed in that environment. And, indeed, use of Unix and its rich set of applications was beginning to spread to other communities, encouraged by increasing deployment of workstations, all of which came with Unix and TCP/IP. However, government (including the US) and non-US PTTs subscribed, in principle, to international standards – which did not include TCP/IP – for at least the first half of the 1980s. This did not mean that TCP/IP was not catching on but it did mean that it was not yet available on many systems, particularly ones of UK origin, and that continuing support remained in doubt. Many manufacturers, especially of mainframes, were still keeping their options open; it was expected in many quarters that TCP/IP or a close variant would be accommodated in some form as part of the ISO suite but the outcome remained unclear.

This presented the JNT, along with many others, with something of a dilemma. The Computer Board had previously initiated a programme of funding campus networks, overseen by the JNT, and by 1983 this had included shared-media LANs. There seemed little alternative at this point to protecting the Computer Board's investment in its existing kit and strategy, particularly since, if the expected outcome in the form of an endorsement of ISO standards was realised, the incipient JANET ought to be well-placed. Accordingly, 1983 saw several meetings of those interested to devise a way of incorporating Ethernet – but using ISO-oriented standards rather than TCP/IP. Two options were available for providing a YBTS or equivalent service over Ethernet, which would then enable a transport service gateway to be constructed, as for Cambridge Ring – or, in principle, for any other network which came along. Here it is not the intention to delve further into the interests and remits of various 'standards' bodies, but it should be appreciated that apart from ISO, CCITT, and various US national bodies such as ANSI and IEEE, there were also manufacturer consortia and regional consortia with particular agendas. An example of the latter was the European Computer Manufacturers Association (ECMA), which had its own industry standards and strong influence in ISO standards. In any given context, there might be conflicting views amongst these various lobbies. As it happened, with respect to what connection-oriented transport service to implement over Ethernet, neither option was technically superior to the other and a participating watching brief was pursued for both, which included some groups implementing the upper layer of X.25 over Ethernet, one of the options for connection-oriented network service (CONS) over Ethernet. Of course, the real issue was whether to adopt a CONS or a connectionless network service (CLNS), datagram-oriented approach to network interconnection, but with the uncertainty about which way either the international standards or industry would go, it was premature to consider such a substantial change of direction at this point. In the event JANET decided to recommend the ISO-endorsed path of X.25 layer 3 over the link layer recommended in 802.3 (1984), also endorsed by ISO. Guidance on implementing this was published in 1985 in one of the last of the Coloured Books, in this instance pink. While this was a straightforward continuation of the JANET strategy of converging with international standards as they became available, in retrospect, it was also perhaps the first occasion when the JANET strategy came into overt conflict with the TCP/IP lobby.

The decision on Ethernet CONS could be considered the beginning of an explicit strategy to move JANET in the direction of ISO protocols, now that some of these were becoming available, at least in draft form. The very last Coloured Book (Peach), which was part of this strategy, was published in 1988 and contained guidance on use of OSI protocols over slotted rings commonly referred to as OSI CR. Actually, there had been an earlier addendum (jocularly known as the 'raspberry addendum') to the Pink Book giving guidance on this but, in the interim, CR82 had after some years become an ISO standard (8802/7) though it had suffered the same fate as Ethernet in standardisation: it wasn't quite the same as the original! The Peach[18] Book updated the guidance (based on using the existing Cambridge Ring, not the IS 8802-7 empty-slot standard). However, in the absence of either viable ring chips or earlier standardisation, the 10Mbps Cambridge Ring had by then long since passed its sell-by date as a service vehicle and this was hardly even of academic interest.

JANET protocols did gradually become available for Unix and on workstation systems using Ethernet during the latter part of the 1980s but, increasingly, what was wanted was TCP/IP because more and more people in research and education were using it. Then, in 1989, several events occurred which soon made the pressure irresistible – the story of which we shall return to in Chapter 6.

5.3 Coping with growth

In 1984, the trunk links connecting the main wide-area X.25 switches were being upgraded from 9.6Kbps to 48Kbps; correspondingly, those universities which were not already connected at 9.6Kbps were generally being upgraded to that. Of course, with the rapid expansion of the number of sites connected and the increase in access speed, it didn't take long before more capacity was needed in the trunk network. By 1986 the next upgrade was being planned;[19] a more complicated arrangement in which third-party multiplexers would be used to sub-divide 2.048Mbps links from BT to provide 256Kbps links interconnecting

---

18 A fruity colour possibly constructed from a palette of Yellow and Raspberry Pink.

19 This was the first occasion on which JANET expenditure reached a level where a bid at the Public Expenditure Survey (PES) stage from the DES was appropriate (*see also* Chapter 10). The bid was for £5m and was constructed at very short notice – in an hour – by Ted Herbert (E.J. Herbert, then Secretary to the Computer Board) and Peter Linington (then Head, JNT), who had been called out of a session at Networkshop for the purpose (private communication, Peter Linington, April 2010).

a central core of PSEs at RAL, Daresbury, London and Manchester. The other PSEs at Bath, Belfast, Cambridge and Edinburgh would be connected by multiple 48Kbps lines to at least two of the central core. (It will be noted that the erstwhile NERC PSEs at Bidston and Swindon are absent from this list: the lines were diverted to larger exchanges elsewhere and the number of PSEs reduced to 8 as the evolution of the SERC/NERC/CB networks into JANET proceeded in 1984.) All institutions would be connected at a minimum of 9.6Kbps; institutions with congested access would be upgraded to 48Kbps access.

In terms of capacity, this marks the first appearance in JANET of the notion of an explicit higher-speed trunk or backbone network in the wide area, to which other regional PSEs were connected in redundant fashion, at least in part for reliability. This was a trend which would continue in the future, though the substantial additional cost of the redundancy would always dictate just how widely its application could be deployed.

Now JANET was a demonstrable success and a process of continual expansion of both the community and the network was under way. Processing power had become substantially cheaper, finally allowing pent-up demand to begin to be satisfied, and giving rise to more systems and people to interconnect. Polytechnic funding was reorganised in 1988 with the formation of the Polytechnics and Colleges Funding Council (PCFC), and although it would not be until 1991 that a concerted effort to connect polytechnics would begin, it did serve to underline in 1988, when JANET was again in dire need of additional capacity, that this need could be expected to continue. And this was in both the wide area and on the campus. This led to three separate, if closely related initiatives.

What was needed was a way of extending capacity in a much more controllable fashion than merely ordering the next upgrade in linespeed from a PTT and paying charges related more to telephony than data. This latter aspect had affected the way in which PSS had been charged, making it unattractive; now, the telephony approach was making the price of lines unattractively high, largely because the pricing was based on how much could be charged for the number of telephone calls which could be carried, rather than a mark-up on the actual cost of the 'raw' transmission facility. Even with deregulation so far, prices were still too high for academia. And furthermore, at the time, the highest linespeed available to subscribers was 8Mbps; higher rates were used in the operators' trunk network but, as there was no published, regulated tariff for these, they were not typically available; and even had they been, they would have been unaffordable if priced pro rata with lower rates.

The case was made in April 1989 for a programme in three parts. The strategic objective was for a major new backbone, possibly constructed from fibre: this would become a major project, would need additional government funding, and was named SuperJANET. Since this would clearly take time to investigate, funds were not requested, but approval was given to proceed with investigation and planning in sufficient detail to be able to prepare a bid to government for the funding. Because of the anticipated amount, this would exceed not only the Computer Board's funding ability but that of the UFC[20] (Universities Funding Council) and even the DES (at least in part) and would therefore need to be put forward as part of the annual Public Expenditure Survey (PES),[21] the means by which the annual funding for Government departments was arrived at and a

---

20 *See* Glossary.

21 *See* Glossary.

part of the process leading to the Government's annual budget. We shall return to SuperJANET in Part C.

The immediate need was in the wide area and funding was obtained for a five-year programme of upgrades, referred to as JANET Mk II. At the same time a five-year pump-priming[22] initiative called the High Performance LAN Initiative was approved, with the goal of providing potentially every university with a campus backbone with a capacity of at least 100Mbps.

JANET Mk II

By 1988, many sites had 64Kbps access while the remainder had 9.6Kbps. Funding was agreed by March 1989 for an immediate overall upgrade, with the general plan being to upgrade some 64Kbps sites to 2Mbps access, some of those at 9.6Kbps to 64Kbps, and a 'commensurate' upgrade to the backbone – though it was not quite clear how this latter would be achieved in view of the cost of 8Mbps lines.

For the backbone, upgrading line speeds was not all that was now needed. The GPT[23] switches were based on computers adapted for use as switches. With the prospect of line speeds now a factor of 40 to 100 times faster than originally, it was time to consider replacing the switches. The switch chosen was the Netcomm 2000. Netcomm was another spin-off company from UK academia, in this case Imperial College London. The Netcomm 2000 had sufficient capacity to support 12 2Mbps lines, making it suitable for use in the wide area backbone. It was also configurable, so that a smaller one could be used on campus. SEEL (Edinburgh) also supplied campus switches (the SEEL Multipac) and support was available for 2Mbps lines, making it a suitable option for campus access.

The strategy for the backbone upgrade was effectively to install a new, parallel core network (Figure 5.1) using Netcomm 2000 switches and a separate set of 2Mbps lines amongst all the NOC sites (except Belfast, since capacity to Northern Ireland was both more difficult and more expensive to obtain). The two networks were interconnected at the NOCs. This arrangement was chosen primarily as a means for testing out the Netcomm switches and then enabling a gradual transition. However, it also had the incidental property of providing temporary additional upgrade capacity in the backbone, while options other than using multiple 2Mbps lines in the trunk were investigated. The link topology in the 'Netcomm network' was slightly different to that for the 'GPT network', which also gave the option, as the two networks were merged, to provide increased redundancy and capacity.

The plan for the Mk II upgrade was to divide up the programme into three initial phases, of approximately a year each. Upgrading the backbone was a prerequisite for upgrading any access links, so in Phase 1 (mid-1989 to mid-1990) the primary task was the backbone, though one or two access links were upgraded to 256Kbps, one of these being the first Mercury line to be incorporated into JANET. By the end of 1990, although the Netcomm Mk II trunk was not yet complete, sufficient progress had been made that Phase 2, the upgrading of access

---

22   Jargon in funding circles for a programme that initiates an activity which, if successful, is then expected to obtain funds for continuation through alternative means.

23   The original JANET switches, but GEC was reorganising, GPT had just been formed and it had taken over telecommunications products – see Glossary.

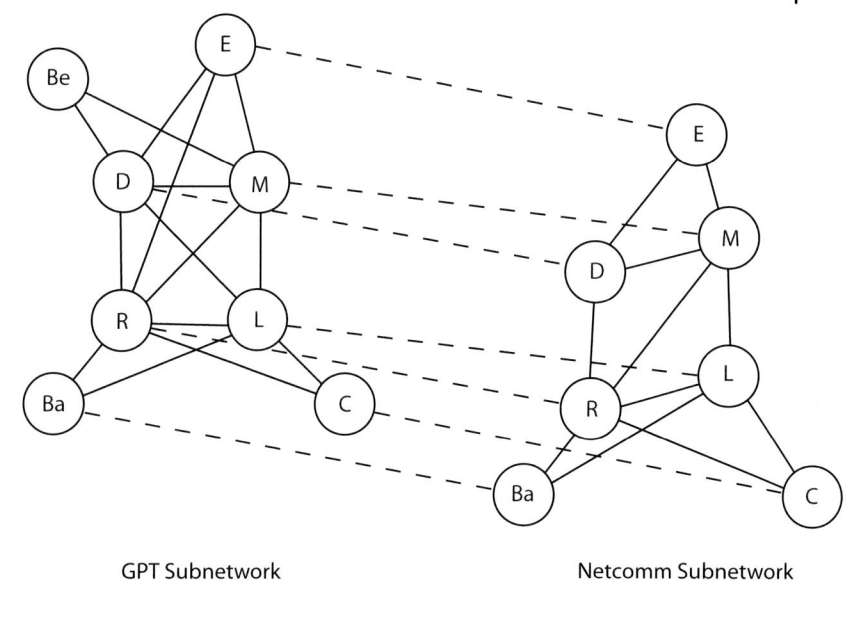

Figure 5.1. JANET Mk II transition topology.

Key
Ba Bath
Be Belfast
C Cambridge
D Daresbury
E Edinburgh
L London
M Manchester
R RAL

lines and moving them onto the new trunk had begun. The effective topology of JANET in 1990 is shown in Figure 5.2. As the new trunk began to take substantial traffic, so the inevitable teething problems for the new switches came to light. Nevertheless, by July 1991 the Netcomm trunk was in operation and Phase 3 had already begun, namely decommissioning the GPT switches, with some accompanying rationalisation of lines.

Now, however, there was change in the wind. As we shall see in Chapter 6, by this time an IP service had been introduced, operating over X.25. This had immediately proved very popular and, increasingly, the X.25 service was being used merely as a carrier, with rapid movement away from Coloured Book to TCP/IP applications. What was now needed was extra capacity for IP. Although at the end of 1991 SuperJANET was still some way off – and little could be said about it publicly, partly because of tendering being about to begin, partly because of a considerable amount of pre-tender information having been provided under non-disclosure, and partly because what technology would ultimately be the outcome of the SuperJANET procurement was unpredictable – funding for it had been approved and it was becoming clear that the substantial capacity upgrade for the data network, which would be one aspect of SuperJANET, should be used to boost the IP service. Given all these factors, and the fact that if the IP traffic were removed, then the X.25 network would be more than adequately provisioned, the JANET Mk Mk II upgrades took on a different complexion. Urgently needed upgrades in access speed continued to be attended to; likewise, the programme of removing GPT switches and rationalising lines continued; but upgrades were

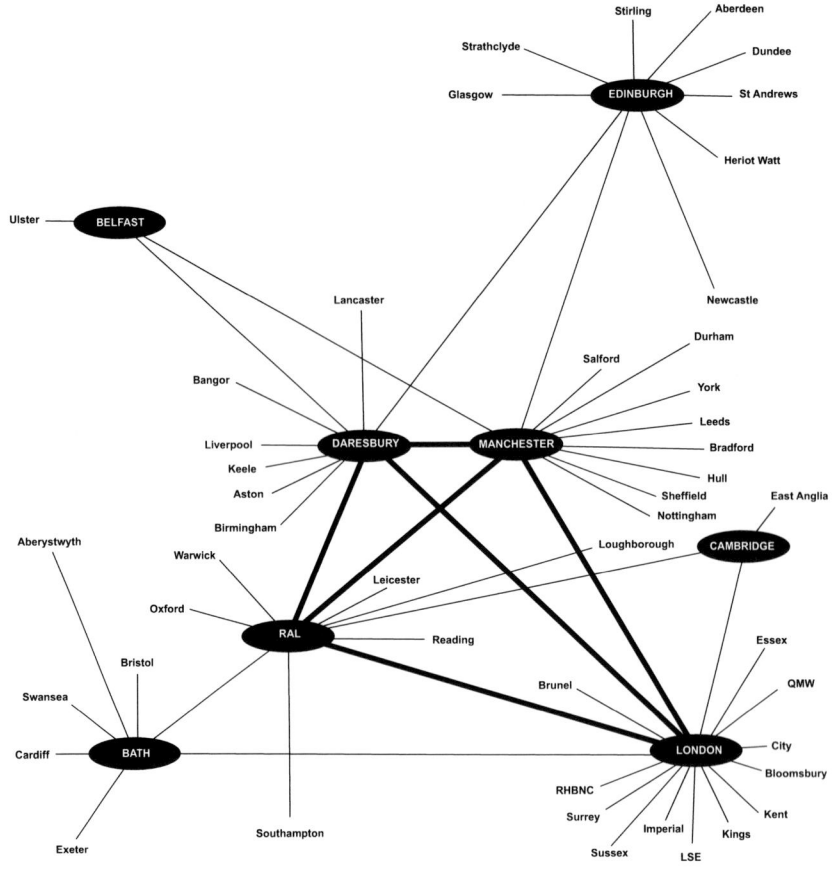

Figure 5.2. JANET in 1990.

now moderated by considerations of avoiding expenditure commitments which would have no longer-term benefit. In effect, the Mk II upgrades tailed off during 1992 and the beginning of 1993, in anticipation of SuperJANET.

High Performance LAN Initiative

As use of LANs took off in the mid-1980s, it became clear that much more capacity in the campus network would shortly be needed. The reasons were three-fold. The rapid increase in reliance on computers in all sectors of university operation produced explosive growth in the number of devices now demanding attachment to the network. And whereas before, a campus network had been used to provide relatively low-speed interconnection of a few computers, a number of interactive terminals and perhaps a few printers, now workstations and PCs were being installed and these expected access speeds of at least one to two orders of magnitude faster. The third reason was intrinsic to LAN technology as it was in the 1980s (and would remain until the mid-1990s): it might be very convenient to have a single cable everywhere to which attachment could be made but it was a technology based on a shared medium – that very cable. It didn't really take very long to fill up the capacity of a 10Mbps cable, no matter what technique was used to share it.

Technologies; FDDI

Ethernet, which had been so successful that it dominated the market, nevertheless had a problem when it came to making it go faster in its shared form. Owing to the need for reliable collision detection, if the transmission rate were increased by a factor of 10 then the packet became shorter by the same amount, leaving less time to detect any collision unless the overall length of the cable were made shorter by the same amount.[24] Of course, another possibility was to make the minimum packet size 10 times longer. This had two disadvantages: there were plenty of packets which were much shorter than that, so they would need to be artificially extended; and an Ethernet using short packets for its size and rate is inherently less efficient in its use of the cable capacity.

Two approaches were taken on what to do about this during the 1980s. The first was really a strategy for delaying the issue for a bit. Since LANs had a limit on the order of a 100 on how many stations could be attached, more than one was needed on most campuses and in many buildings, so the question arose of how to join them. It turned out there was a neat solution, which was dubbed a 'bridge', largely because initially it was used to join just two LANs. The neat aspect of how a bridge worked was that it didn't have to be told anything: it learned and automatically configured itself. Moreover, no further protocols were needed: hosts were unaware of the presence of a bridge. When first started, a bridge joining two LANs – which we can regard as being on its left and right sides – would forward all packets on its left to its right (and vice versa). However, it would observe the originating address in each packet and build up a table of LAN addresses on each side. Then, if it observed a packet on one side destined for an address in its table for that same side, it could avoid forwarding it to the other side. Gradually, it would learn all the active stations on each side – typically, it learns them quite fast for the active stations, helped by virtually all communication being two-way. The good thing about this from the point of view of capacity limitation on a LAN is that if LANs are spread around a campus in such a way that it is the norm for stations to communicate with other stations on the same LAN and more exceptional to communicate off-LAN, then the set of interconnecting bridges will automatically ensure that a LAN will carry as little traffic as possible which is of no interest to any of its stations.[25] This developed into the notion of a backbone LAN used to interconnect a collection of 'satellite' or 'peripheral' departmental or building LANs. However, at this point the need for a faster LAN reappears because in most scenarios, whether uniformly distributed traffic or provision of access to centralised high-speed servers or off-site services, the backbone LAN needs to have higher capacity than the peripheral LANs.

The more fundamental approach was to look for a LAN architecture that showed promise for operation at higher speed. Rings, being essentially deterministic, seemed to offer possibilities. In addition, it is quite easy for a ring to exploit fibre transmission, whereas it is hard for shared fibre Ethernet to serve

---

24 Technically, the relevant length is known as the 'diameter' of the LAN and is the maximum transmission length between any two nodes (including repeaters).

25 This property of bridges is generally referred to as 'filtering'.

more than a few stations (a practical form if this was never developed). And, indeed, there were several fast ring developments in the first half of the 1980s. Proteon, a spin-off company from MIT, had successfully marketed a 10Mbps token ring and subsequently developed and sold an 80Mbps version, though technically it turned out to be quite hard. The Computer Laboratory in Cambridge also developed a 50Mbps version of its ring (Temple, 1984), though it also encountered problems in doing so.[26] The source of one particular problem in both cases was that the rings were synchronous: all the stations transmitted and received at the same rate. Ensuring that all the stations maintained synchronisation, regardless of temperature, ring size, and stations being switched in and out, turned out to be hard, involving a non-linear process[27] which was ultimately limiting for synchronous rings. However, this was not the only possibility.

The early 1980s saw an alternative design for a token ring begun, called the Fiber Distributed Data Interface (FDDI), which was not synchronous. Originally it was conceived as a 'back-end' network, that is, one used in a large computer room for interconnecting peripherals such as disks and tapes in flexible fashion with large mainframe processors. It was designed to utilise fibre as the transmission medium. It was also designed to be much larger than the typical LAN: it had a maximum length of 100km. Reliability was an important criterion and it used two counter-rotating rings. Stations could be attached to both ('dual-attach') or just one of the rings. If part of the ring developed a fault, the repeater part of the dual-attach station access logic would reconfigure the ring into a single ring of twice the length (sometimes called 'folding' because the dual-attach stations on either side of the fault would 'fold' the ring back on itself by connecting the incoming ring on the fault-free side to the outgoing counter-rotating ring on the same side). In a further effort towards reliability, no monitor station (typically a single point of failure) was needed: there were distributed algorithms for establishing the token, detecting faults and reconfiguring the ring. But what led to its successfully achieving 100Mbps effective data transmission speed was that it was not synchronous. Each station had its own transmit clock which operated within well-defined limits but was not synchronised to any other station. The receive logic synchronised with the up-stream transmitter. To accommodate the small, but limited, difference in the transmit rate and the receive rate, so-called 'elastic' buffers were incorporated between receive and transmit,[28] large enough to absorb the difference that could build up while repeating a maximum size frame. FDDI had one other interesting property: its timed token algorithm for sharing access was amenable of complete mathematical analysis (Johnson, 1987 and 1988; Jain, 1990), so that it was possible to allocate shares in the capacity of the ring in a predictable way.

Although begun quite early in the era of shared LANs, FDDI was not simple, either to develop the standard or to develop a product, and it was not until 1987 that the MAC was first standardised (under ANSI). Because it provided the possibility of allocating defined access shares (as well as having a priority system), to be complete it also needed a management scheme, and this was not completed until 1992. To complete the story there was a subsequent elaboration

---

26 The later experimental Backbone Ring of the early 1990s, which operated at 500Mbps, incorporated elastic buffers for the same reason as FDDI (see below).

27 Phase-locked loop technology; the non-linearity meant that it was hard to analyse either stability or frequency properties, making it very difficult to provide stable, reliable synchronisation.

28 Interestingly, the (later) Cambridge Backbone Ring, which operated at 500Mbps, also used elastic buffers.

of FDDI, called FDDI-II, which provided isochronous (circuit-like) access as well as traditional asynchronous[29] data access; and then there was a follow-on concept called Fiber Follow-on LAN (FFOL), which targeted much higher rates.[30] FDDI-II reached partial standardisation but no product and FFOL never got past the concept stage: both were effectively overtaken by events. Although the token scheme for FDDI could provide shares for data traffic, in practice it was seldom used in other than the most basic mode, perhaps because the capability was beyond the typical needs of basic data networking at the time, but also – and probably more significantly – because there was no support for the necessary management.

FDDI Advisory Group

By 1988, following progress in ANSI, several universities had become interested in FDDI and at that year's Networkshop, in March, it was proposed that an FDDI group be set up to track the progress of the technology. The possibility of pilot studies was also mooted. Towards the end of the year an FDDI Advisory Group, chaired by Sue Weston of the JNT, was constituted, at the time when the case for the performance upgrades in both local and wide areas was being assembled. Although FDDI was now becoming an ANSI standard, it was still in the early stages in the international arena in ISO and products were still in development.

The following year, in April 1989, funding for the High Performance LAN Initiative[31] was approved, spread over five years, with actual spend to begin in financial year 1990/91. As usual with such Computer Board pump-priming initiatives, the funding would provide the initial equipment: installation, cabling, maintenance and future upgrades would come out of university funding, including that from the UFC[32] (Universities Funding Council). The FDDI Advisory Group's role now took on a more serious note since its advice would have a direct bearing on how the funds were spent. The objective was to provide all universities with campus networks of appropriate performance to match the forthcoming SuperJANET deployment. The FDDI Group was charged with elaborating the detail of the programme, beginning with assessing whether FDDI would be the right choice. There were other technologies appearing, including Broadband ISDN[33] (B-ISDN) and DQDB (*see* below). The former stemmed from developments happening in both the computer communications and the telecommunications industry. Standards for B-ISDN were still too immature in 1990 for it to be a contender as a campus network – there was little in the way of suitable products and software support at the time: however, it was certainly

---

29 The terms used here are not those used in FDDI, which was eccentric in its usage of timing terms.

30 A synopsis of the main issues may be found in Jain (1993), an article from his authoritative book (1994).

31 At various times during its progress, this initiative was referred to by several sobriquets, possibly reflecting aspects of the technology, its potential roles and phases of the programme: so 'High Performance Backbone LAN Initiative', 'High Speed LAN Initiative', 'Fibre LAN Initiative' and 'High Speed Multi-Vendor LAN Initiative' all refer to the same programme.

32 *See* Glossary.

33 Broadband ISDN: *see* Glossary. A note of caution: the term 'broadband' has been reused many times in the communications industry; in this case it distinguished a faster and more flexible successor to the already existing ISDN (*see* Glossary), which was a circuit-oriented PTT service limited to 2Mbps (more precisely, 2.048Mbps) and below.

a potential constituent of SuperJANET, and since its eventual role was there, it is described in Part C. DQDB, on the other hand, could be viewed as a direct competitor to both FDDI and FDDI-II; it had emerged in early product form and was undergoing standardisation in IEEE. Since, in addition to its being a candidate for a campus backbone, a derivative technology formulated by Bellcore[34] would form the initial major deployment in the SuperJANET, we shall look briefly at DQDB below.[35]

DQDB

Metropolitan Area Network, or MAN, standardisation goes back for about the same length of time as FDDI development. The IEEE 802.6 committee was formed in 1980. Unusually, its brief was not to standardise an existing technology but to come up with a standard for LAN interconnection and data networking suitable for use in a metropolitan context. Needless to say, during the subsequent few years this resulted in a plethora of proposals and considerable evolution of concepts – among which was the Orwell network (mentioned earlier in this chapter under LAN technologies, Section 5.2), a derivative of the Cambridge Ring developed by BT.

DQDB or, to give it its full name, Distributed Queue Dual Bus was another technology to emerge as a spin-off development from a university: in this case the University of Western Australia (UWA) in Perth in collaboration with Telecom Australia.[36] Originally termed QPSX (for Queued Packet and Synchronous circuit eXchange), it was invented at UWA during 1984–86 and then, in conjunction with Telecom Australia, proposed as the 802.6 MAN standard in 1986. Within a year it was the preferred choice for 802.6; and, in 1987, Telecom Australia and UWA launched a company, QPSX Ltd, to market QPSX, whereupon the 802.6 committee renamed the technology it was standardising as DQDB.

Like FDDI-II, DQDB was a combination of asynchronous transport over a synchronous bearer service, capable of providing both a synchronously multiplexed (isochronous) circuit service and a statistically multiplexed packet service. Unlike FDDI, it was based on use of the existing (and forthcoming) standard CCITT and ANSI transmission rates in use in the telephone industry, from 1.5 to 155Mbps. Technically it was based on use of two unidirectional slotted buses, to both of which each station was attached. The buses could be private transmission lines or provided by a public network operator. The buses operated in opposite directions, providing a full-duplex service between any two stations on the dual bus. The circuit service was based on pre-allocated slots (called pre-allocated or PA) and the packet service used a distributed queuing scheme (called queue-arbitrated or QA), in which stations observed requests on one bus by stations wanting to transmit on the opposite bus. By keeping count of these requests, and observing the passage of occupied slots, any station could tell the state of the combined queue up stream so that it would also know when it was its turn to transmit to satisfy an earlier request of its own. To achieve a measure of resilience against failure, the whole dual bus network could be configured as a loop, with one of the stations acting as both head and tail of the two buses; if

---

34   *See* Glossary.

35   A review of some of the technological developments at the time is given in Cooper (1991).

36   Renamed Telstra as part of divestiture and privatisation in the early 1990s.

there was a failure, the stations on either side of the break would then each take up the role of being head of one bus and tail of the other. A disadvantage of the scheme was that it was not necessarily fair to all stations as those nearer the ends of the dual bus received better service under some circumstances. Various fixes were developed but these were not complete remedies; they also complicated things further.

In practice the scheme was complicated – like FDDI-II. And, just as FDDI had success but FDDI-II never became a product, so DQDB had some initial success as a product with the QA mode of access to support asynchronous packet access but the PA circuit-mode access was never seriously taken up. There were other, perhaps more fundamental reasons why the 802.6 MAN standard never saw wide deployment, stemming from its being based on a shared medium. We've already seen how by the end of the 1980s there was demand for a backbone LAN network operating at around 100Mbps for a campus or large building. This was not confined to the university sector: it was a general requirement. The idea of a MAN was that the shared medium would be capable of serving an urban area. But if individual institutions were now demanding this sort of capacity, would a shared network of similar capacity be capable of serving a whole urban area with many such institutions? By the time the 802.6 standard was agreed, in late 1990, it was already apparent that there were going to be new forms of traffic in the wide area, video in particular, which meant that the answer to the question just posed was, No, there would need to be many such networks to provide the capacity.

There was also a quite different sort of observation regarding the use of any network for which the medium was shared in such a way that the data for one subscriber passed through the premises of another. By 1990, hacking and other malicious activities on networks had already begun: having one subscriber's data pass through another's premises did not seem a reassuring strategy! Moreover, by this time there were alternative, switched technologies becoming available to public network operators. Not only do switched technologies not suffer from this intrinsic disadvantage of a shared medium in a public network, they typically also scale better. One of these switched technologies – Switched Multi-megabit Data Service (SMDS) – was based on compatibility with DQDB and we shall meet it in Part C because it played an important part in the initial development of SuperJANET. However, in the context of choosing a campus backbone LAN, the future of DQDB was sufficiently uncertain in 1990 that it did not appear a serious option. And there was another problem: although equipment was available to interface DQDB to other network technologies, there was no widely supported range of interfaces available for computers, unlike what was beginning to appear for FDDI. Although DQDB was considered for inclusion in the pilot programme, it was evidently too immature in 1990; and subsequently it was overtaken in both the metropolitan and the campus context by switched technologies.

FDDI backbone deployment

By the end of the first year of the High Performance LAN Initiative, the FDDI Advisory Group was in a position to confirm that FDDI was the only suitable contender as a backbone LAN at the beginning of 1990, in time for initial equipment purchases to begin in 1990/91.[37] However, this was a new technology and experience was needed before it would be sensible to commit major spend on equipping universities. As had been anticipated from the beginning, a pilot study was mounted in 1990/91 in which five universities participated. Apart from gaining experience in installation and use of the technology, including interworking between different manufacturers' products, there were two areas where practical study was needed: host interfaces, including use with applications, and network interconnection. Bridging has already been described in the context of Ethernet, where it was first developed. Since then, two things had happened: there had been developments in the way bridges routed between a number of interconnected LANs; and, with the advent of several standard LAN technologies, the IEEE had undertaken the generalisation and standardisation of bridging. Bridging now offered one option for interconnecting a variety of LANs with each other, including now also to an FDDI campus backbone. For sites now supporting a variety of protocols on campus, this was an attractive option since it did not imply commitment to any particular higher layer protocol.

The other option for interconnecting LANs was to use Internet routers. Many institutions had already been using Internet protocols internally for some years because they were necessary for Unix workstations which came with IP support included; there was also support for it on PCs, enabling them to share the workstation infrastructure. By 1990 there was growing pressure for JANET to support the Internet protocols generally (and, more specifically, to be able to route the fundamental Internet Protocol, IP) and there was already discussion within the JNT and its advisory groups about mounting a pilot study for a JANET IP service (*see* Chapter 6). In the context of the FDDI pilot, the prospect of JANET supporting IP meant that gaining experience of router support for FDDI was a clear requirement.

There was also another, perhaps broader issue here. As in the USA, there had, during much of the 1980s, been considerable debate regarding the relative merits of bridges versus routers when constructing an extended network. There were various aspects of the debate. Bridge performance in terms of simple throughput measures like packets per second was consistently higher at this time than that of routers. Of course, routers did a more complex job, in particular dealing with choice of routes that might be available in a general mesh network, a capability that had application to resilience in the presence of component or link failure. The original 'learning' bridge could not cope with a network that had loops in its topology since it was liable to see the same address appearing to be on both of its sides. However, the solution to this problem had been developed at DEC (Perlman, 1985) in the form of a distributed algorithm which defined a tree[38] of links over the mesh network. This tree 'spanned' the network, meaning that all parts of the network could be reached solely by use of the links in the tree; the bridges then shut down the links which were not part of the tree and the normal learning algorithm could then operate. While this 'spanning tree algorithm' did not allow the additional links to contribute to the capacity of the network, it did enable their use for resilience in case of failure, since any alteration in the topology of the network – loss or insertion of a link or bridge – triggered the spanning tree algorithm to be run again. Routers had much more sophisticated features for selective traffic filtering than bridges, enabling them to be used in

---

37 In addition to chairing the FDDI Advisory Group, Sue Weston of the JNT also managed the FDDI deployment phase of the High Performance LAN Initiative.

38 A logical network topology without loops.

defence against malicious (or sometimes just erroneous) use of the network. But of course, bridged networks could carry a wide range of protocols – and there were plenty of those around at the time, such as those from Novell and Apple. The debate continued essentially throughout the 1980s but, as router performance in the 1990s began to match that of bridges, so the debate could focus more clearly on functionality.

In the wide area, routing was rapidly gaining in complexity caused not only by the expansion of the Internet but also by the increasing constraints imposed by the variety of stakeholders in different parts (subnets, *see* Chapter 6) of the network. As the general unmanageability of routing at two independent layers of the network, neither of which knew anything of the other, gained wider appreciation, so the practice diminished; by general consent, bridging (and, from 1993, switching) would become confined within an organisation, typically within a single management. Where an institutional network was dispersed over a wide area, extended bridging might be appropriate; equally, in a large organisation, it might also be appropriate to use routers internally for traffic filtering and access control. Gradual diminution of any requirement to support unroutable protocols also eliminated a cause of complexity in network design.

For the pilot, assessment of both bridge and router support for FDDI was required. Five universities participated, each focusing on particular aspects of products supporting FDDI. Edinburgh and Glasgow assessed routers; Birmingham and Manchester focused on bridges; and Queen Mary & Westfield (QMW)[39] concentrated on end systems and applications. By 1991 the pilot had finished, having demonstrated that FDDI worked and, with care, was practical as a backbone LAN; the main programme of equipping universities with FDDI backbones then took place over the next two years, so that by 1993 many universities had campus networks with 100Mbps backbones. The only significant qualifications were in regard to packet sizes and interconnecting differing varieties of 802 LAN, specifically, token ring and Ethernet. The packet size issue was simply a consequence of each LAN having a different maximum packet size: it was necessary to configure all workstations and routers not to exceed the smallest maximum packet in use on site. (Between sites linked by routers, the routers will take care of differences in maximum packet size among sites by fragmenting packets, which may be inefficient but will work.) The other problem of bridging between LANs of different types was simple, fundamental, trivial – and entertaining for onlookers. In discussing the order of components in names, we mentioned the 'big-endian' and 'little-endian' conventions (Section 3.3). Names are not the only place where ordering conventions are needed. When transmitting the bits of a byte or word in memory onto a communications link, they have to be sent bit by bit, one at a time. It is conventional to start at one end of the byte or word and transmit the bits in order: but which end to start? There is no universal convention. Of course, it really doesn't matter until one has to communicate with a device using the opposite convention. In the early 1980s, when token ring was being standardised in one IEEE committee and Ethernet in another, transparent bridging between dissimilar networks – and hence order of bit transmission – was not a consideration for either network in isolation. Token ring had its origins in IBM whose machines at the time were largely big-endian, whereas Ethernet had chosen the little-endian convention. This might merely have been tiresome for bridging, in that hosts would have had to agree about reordering bits in the data. Unfortunately, the problem also applied to the address fields in LAN

---

39 Known since approximately 2000 as Queen Mary, University of London (QMUL).

frames, and the differing conventions caused addresses to be misinterpreted with corresponding failure of the learning algorithm for forwarding. The problem was eventually fixed but it was one of the issues revealed by practical experience during the work of the FDDI Advisory Group and the pilot programme.

The FDDI programme achieved its purpose in many ways. It equipped campuses with appropriate capacity to match what would shortly begin to be delivered by SuperJANET. It helped to provide suitable infrastructure for the increasingly ubiquitous deployment of workstations and PCs; but it was also the last of the cabled shared-media LANs: switched networks were taking over, somewhat helped by the complexity of FDDI preventing cheap interfaces from ever being developed. The problem for all the original LANs was that they used shared media; when a shared cable runs out of capacity it has to be replaced by new technology, and not just where the extra capacity is needed but throughout. Using switched technology, adding support for more stations is a matter of adding another switch and connecting it to an existing switch. When a switch reaches capacity it is replaced with a larger switch, but can continue to provide service in a less demanding part of the infrastructure, perhaps simply serving fewer hosts. Designing switches and links to operate faster turns out ultimately to be more tractable than designing ever faster shared access schemes. Increments in capacity can be made easily by adding switches, possibly in conjunction with modest upgrades in selected link capacity. No such incremental upgrade path is available for shared media. Although the switched technology associated with B-ISDN made an appearance in the local area at this point, both it and FDDI were by 1994 in competition with the successor to shared Ethernet.

In describing shared Ethernet earlier, the point was made that in order to detect collisions reliably, making the efficiency viable, the size of the LAN, its transmission rate and the minimum packet size are all related. What if a way round this could be found? What if few or no collisions occurred? The key was to exploit bridging technology. When first developed, bridges were large, cumbersome and expensive. Some products in 1982/83 were based on three circuit boards: essentially a microcomputer with two LAN interface cards, enclosed in a substantial crate with power supply, all of which could cost upwards of £2000 at 1982 prices. In the intervening decade all this had been miniaturised, enabling a simple development: instead of interconnecting just two LANs, now 10, 20 or 50 or so could be connected. The algorithms for a bridge are in no way dependent on the number of LANs being interconnected. And now, suppose further that each LAN segment just connects the multi-port bridge, or switch, to a single station. Technically, the cable is still a shared media Ethernet segment but very few collisions will occur, making it very efficient. However, it is also possible to use two signal paths, one in each direction. Now, collisions can never occur, removing the relation between rate, packet size and length of cable; moreover, there is now no impediment to use of optical fibre.

So, in 1993 switched Ethernet was born. It was an instant success. Care was taken to achieve the maximum in compatibility with shared Ethernet and with every new generation of equipment, each of which provided a factor of 10 improvement in speed; and within a decade, full-duplex Ethernet would be available for use in wide-area packet transmission links. While this spelled the end of wired shared-media LANs it was not the end of shared media: wireless LANs appeared within a very few years, bringing back all the same issues – and a few extra – but that is a later story.

6

Protocol Debates and Emergence of the Internet

From the earliest days of the Network Unit, we have seen that the major plank in UK academic service network strategy was to base construction of the eventual network on the use of standards, preferably international since they might be supposed to promise the greatest leverage. Since, apart from X.25, there weren't any around at that time, common cause was made at national level with others having similar need to create interim national standards and at the same time ensure full participation in international standards creation. The result by the early 1980s was a national service network which, with the addition of email, provided all the services originally required of it in the mid-1970s.

Of course, requirements had moved on by then, but if this assessment seems a little two-edged it should not be forgotten that JANET was the envy of many of those in the wider academic and research community, including in the USA, who did not have access to the incipient Internet – which was the majority.[1] JANET was among the earlier network initiatives in Europe. A major strength was that during the mid-1980s, essentially the whole of UK academia was connected, enabling the exploitation of JANET as the newest addition to the general university infrastructure. Nevertheless, in the intervening decade or so between its initial inception in the early 1970s and the official beginning of operational service in 1984, a good deal had happened. Notably, three technological developments would contribute to a major shift in the future direction of networking in general and JANET in particular.

Firstly, there was the introduction of optical fibre-based digital wide-area communications. By 1980 or so, four countries in particular which had been active in the research and development of optical fibre – Canada, Japan, the UK and the USA – had reached the stage of early deployment in the trunk network of the telephone system. Much of the basic research had been pursued in international collaboration, with several UK universities participating.[2] This overlapped with the initial introduction of digital technology into the telephone system. Neither of these developments happened overnight: quite the contrary, they were both spread over many years; nevertheless, the effect of both began to be felt at this time, the most significant initially being the start of a major improvement in bit-error rates in wide-area digital transmission. Whereas in the early 1970s an

---

1   The author worked in the USA part-time from 1984 through 1987 during which this point was made repeatedly by representatives of a wide range of universities and research institutes.

2   Southampton University in particular, which also participated in the development of the erbium-doped non-regenerative optical amplifier, enabling amplification of arbitrary optical signals transmitted over a fibre, avoiding translation to and from the electronic domain.

end-to-end bit-error rate of around 1 in 10^4 was good but 1 in 10^3 was common (corresponding to an average of a character error every couple of lines or so when typing at a terminal on an asynchronous dial-up line), by the mid-1980s, using the new digital services, error rates of 1 in 10^7 were being achieved. In any field of engineering, an improvement by a factor of a thousand or more cannot be ignored. In the case of networks it altered the basic design parameters for network protocols, as we shall see below.

Then there was the advent of LANs and distributed systems. We have seen already how the quest for a LAN of sufficient coverage for a large building or campus gave rise to the development of an interconnection technology which operated at the upper layer of the LAN itself, so that the interconnection was transparent to users of the network. At the same time, so few errors occurred in LAN transmission (bit-error rates of less than 1 in 10^{11} were not uncommon) that it was not necessary for error correction or recovery to be incorporated into the network protocols: they could be left to the application. Indeed, there was every advantage in doing so since compulsorily incorporating unnecessary mechanism into all network operations merely served to slow everything down at a time when it was still hard to exploit to the full the speed offered by LANs. In addition, the advent of distributed systems enabled new approaches to computation and the provision of information systems infrastructure, which in turn required not only higher performance but also new communications paradigms which would begin to allow the construction of software systems that spanned a number of hardware systems. The first such new paradigm was the remote procedure call but the general topic, which essentially explores the relation between software engineering and networked communication, has many ramifications and remains an active research topic at the time of writing, embracing the whole of what is sometimes termed middleware.

And thirdly, there was the evolution in concept about how networks might be concatenated to form a coherent whole: a concept and architecture which acquired the name 'internet'. Both the previous two developments contributed to confirming the emerging architecture because the lack of transmission errors in every part of the network contributed towards a trend in architecture in which the network did little other than route packets. That this was also conducive to the next development in packet switching, the support of multiservice networking, was no accident. To understand what happened next in the development of international networking, and hence JANET, we shall start with a brief sketch of the evolution of the internetworking concept in ARPANET.

6.1 Internetworking

The initial ARPANET was composed of a collection of what were termed Interface Message Processors, IMPs, which connected to each other to form the network and to hosts to provide access to the network. There was also a derivative of the basic IMP called a Terminal IMP, or TIP, which enabled interactive terminals to connect to the network, often via dial-in using the telephone network. (It was a TIP that was installed at UCL in 1973 which provided dial-in access from the UK to ARPANET as well as interfacing the IBM 360/195 at RHEL and the Computer Aided Design Centre at Cambridge to ARPANET.) End-to-end reliable transport was provided by a network control program (NCP) residing in a host, which communicated with its IMP; the IMPs effectively provided what would now be termed the data link and network layers.

In 1973, just four years after the beginning of ARPANET, came the first publicised suggestions that it would be a good idea to separate the routing and forwarding of packets from providing reliable delivery. There were two issues: network interconnection and real-time packet voice. At the time there were several experimental network technologies in use in the ARPA programme,[3] each with different packet schemes, and interconnection was awkward because changes made in one network could have repercussions on the others. If the various network technologies were to be concatenated to form some sort of larger network (sometimes referred to as a 'catenet' at the time), the essential aspect was to be able to route packets across the whole. There were also a number of experiments in the transmission of packetised real-time speech, for which timely delivery was (and remains) more important than delivery free of all bit errors. If some applications happened to need reliable delivery, that could be added on top but, since real-time applications such as voice not only did not require this but could be damaged by it, it should not be part of the basic internetwork protocol. Following discussions in 1973, the first proposal to make this separation was published in a paper by Bob Kahn and Vint Cerf (1974), and it was from this initial suggestion that TCP (Transmission Control Protocol) and IP (Internet Protocol) as they exist in the current Internet would emerge: IP would provide the internetworking network layer, which hosts would support as the means of forwarding packets onto the Internet; and routers (originally called gateways) would use IP to route packets across the constituent set of networks (generally known as subnets) forming the Internet. It didn't happen immediately but, in 1977, the first independent initial implementations of TCP and IP at Stanford University, BBN and UCL demonstrated operation of the new model architecture by interconnection of ARPANET, SATNET and the San Francisco Bay Area Packet Radio network.[4] In 1978 work by these groups, Vint Cerf, Danny Cohen, Peter Kirstein and Jon Postel came to fruition in the publication of the first versions of the current TCP and IP, together with papers by Cerf and Kirstein (1978) and Cerf (1978) explaining the issues in the interconnection of networks and what would now be termed the Internet model for network interconnection. By then, LANs were also among the growing list of packet network technologies and the IP model of interconnection was also evidently applicable in this case too – especially as bridges were not quite yet in deployment.

Thereafter things began to move rapidly, and by 1981 it had been decided that ARPANET should change to adopt IP (using the transition plan published as RFC 801 by Jon Postel in November 1981). The timescale was swift: everybody should begin by January 1982[5] and the transition would be complete by January 1983, when NCP support would be removed. Later in 1983 a section of ARPANET was split off to become MILNET, which would be incorporated into the US Defence Data Network; the remainder would become the Internet. IMPs effectively

---

[3] ARPANET, based on 56Kbps land lines; packet radio; a secure network; and SATNET, a satellite network which linked the US and Norway and which, with the help of a link from Norway to UCL, also provided the original UK link to ARPANET and subsequently the Internet: see, for example, Kirstein (1998).

[4] Peter Kirstein, private communication; see also Handley (2003).

[5] Indeed, UCL was the first to make the transition to IP, driven by the imperative of worn out PDP9s (upon which the UK-US gateway was implemented) being replaced by PDP11s which did not support NCP (Peter Kirstein, 2010, private communication).

disappeared on 1 January 1983, having been replaced by routers; and the terminal access function of a TIP was replaced by a Terminal Access Controller, or TAC, the equivalent of an X.29 PAD in an X.25 network.

All this happened just in time for the rapid deployment of LANs and LAN-based workstations. The growth of the network began to take off – but this would not be under the auspices of DARPA, which had a remit to support experiments and pilots. By 1984, plans were under way which would lead to the National Science Foundation (NSF) funding a backbone which would become NSFnet, supporting academic and research use more generally in the US. At the time of the MILNET split in 1983, ARPANET numbered less than 50 hosts; by 1984 there were some 1000 hosts interconnected using IP and the growth and spread of the Internet had begun.

6.2 Birth of the Internet

The previous section has sketched the technical emergence of the Internet architecture but the network was still, in 1984, largely confined to computer science or research institutions, many with participation in ARPA projects. At this point Europe was focused primarily on awaiting international standards, which were taking an uncomfortably long time to emerge. We shall see in Chapter 7 that a variety of international networks were appearing in the 1980s because 'needs must': nobody could wait any longer. The USA was not exempt. In 1981, following an initial proposal in 1979, the National Science Foundation (NSF) provided funds to initiate CSNET, a network to provide some basic network support to computer science departments not a part of ARPANET. The network primarily provided email facilities and was also connected to ARPANET using TCP/IP. As well as using leased or dial-up lines, it also used IP over X.25 – as UCL had done since 1979 for the UK-US gateway service, following the introduction of IPSS; and as JANET would do in 1991 – where there happened to be no more suitable links. While the operation of TCP over X.25 – involving two layers of flow control, each using somewhat different mechanisms – did nothing for performance, it worked.

In 1984, the NSF funded four new supercomputer centres, in addition to the one already existing at NCAR (National Center for Atmospheric Research), and it needed to provide access to these on a national scale. The official line in many government-funded quarters at the time, like the UK, was to endorse the eventual arrival of international standards, as promulgated in the US FIPS alluded to in Chapter 3. However, the spread of manufacturer and third party support for TCP/IP had begun. The ARPANET transition had been hugely successful, there was now a wealth of expertise available, and the NSF took the decision in 1985 to put in place a backbone network, using TCP/IP protocols, which was deployed in 1986. Just as in the UK, the official mantra was that these protocols had interim status: when international protocols were mature enough to perform as well and had sufficient support then a transition would be made.

Of course, underlying this was a technical debate about which network architecture was 'better': connection-oriented virtual-circuit networks like X.25 (or even pure circuit switched networks) or datagram networks like IP (and the Xerox PARC PUP[6] architecture or the original NPL network architecture before it)

---

6   PARC Universal Packet.

in which each packet contained source and destination addresses and was routed across the network individually. This was the way LANs worked. The network was as simple as possible (though the routing in the Internet is anything but simple) and pushed the complications associated with flow control and error recovery to the edge, to the users of the network. As already described, the improved quality of transmission made this approach regarding error recovery more attractive in the 1980s than it would have been in the 1970s. Then, a full-length Ethernet packet of 1500 bytes would have been quite likely to have contained a bit error when sent over a wide area link: by the 1980s the opposite was true. Flow control was another matter, and would require attention from time to time as the Internet developed.

As noted above, the 1980s would see a plethora of networks spring up, largely to fill the vacuum caused by the lack of anything else – 'Because It's Time NETwork' (BITNET) summed up the feeling of many. However, later initiatives from the mid-1980s onwards had all backed TCP/IP; indeed, the take-up of TCP/IP was by 1986 so great that the Internet Activities Board (IAB; it became the Internet Architecture Board in 1992), recognising the disparate activities of improving service networks and pursuing network research, reorganised its activities under two groups: the Internet Engineering Task Force (IETF) and the Internet Research Task Force (IRTF). By the end of the 1980s EARN, BITNET, CSNET and others were all in the process of vanishing: they had served their purpose but now the Internet had simply rendered them superfluous.

In the international standards arena, ISO protocols had included connectionless transport and connection-oriented transport over a connectionless network from the early 1980s. Nevertheless, certainly with hindsight – and to many at the time – it was by now questionable if ISO protocols could ever supplant TCP/IP. ISO protocols were more complicated but, even more importantly, they simply didn't have the weight of practical experience and operational knowledge behind them – to say nothing of installed market position. The protocols and networks which resisted TCP/IP for longest were of course those already well established and, within their own limits, working well: among others, they included DECnet (SPAN), several IBM protocols including those in use by EARN/BITNET and the JANET protocols. By the late 1980s, several were already making plans for a transition to ISO standards, but in the event this turned out really only to provide a postponement to the adoption of TCP/IP. Although in many cases it took longer, by 1990 it was apparent to essentially all that there was no alternative to adopting IETF protocols and joining the Internet – and the important thing by then for the communities being supported by these networks was the latter aspect: becoming a part of the growing international Internet community. Moreover, by then, there was well-established support on all platforms for the IETF protocols, along with a range of commercially available routers: this was in effect a consolidation of a position that had begun to develop since 1983, once ARPANET had effected the transition to IP. What was particularly telling in the UK was that, by 1990, availability and support for IETF protocols was better than for either Coloured Books or ISO (Day, 1992).

6.3 Adoption of IP by JANET

There were two overriding requirements now being voiced: access to new applications and integration with the world-wide research community. JANET was now suffering the classic consequences of having been overtaken by events: it had considerable legacy investment in an infrastructure which was not only not leading, it was out of step internationally, particularly with both Europe and the USA. Actually, this is a little harsh: while it was certainly arguably true regarding the commodity packet-switching service, the need to open up JANET to new fundamental possibilities had already been recognised and was under way in the SuperJANET programme, the subject of much of Part C. Moreover, JANET now had in place a well-developed national, operational support infrastructure which had taken years to build: this would stand it in good stead as it entered a period of intense evolution.

Indeed, the JNT was about to enter one of the most active phases in its history. Not only in SuperJANET was the largest activity it had ever undertaken in progress. JANET Mk II and the High Performance LAN Initiative were in full swing; and against a background of high-level debate about how many Research Councils there should be (a single super one or a proliferation of even more), reorganisation of funding in higher education (creation in 1992 of new, somewhat devolved, higher education funding councils, the HEFCs), whether polytechnics should become universities (they did, with consequent abolition of the PCFC), and the future of the Computer Board (the birth of the JISC by way of the ISC[7]), on top of all this it had become clear that something more than just a team funded by the Computer Board but employed by a Research Council was needed to run JANET. Discussions were in progress about what form this should take (*see* Chapter 9).

Apart from an overriding need for UK network infrastructure to match that now in dominant use internationally at a time of increasing international collaboration,[8] the other pressing need was to gain untrammelled access to new network software and applications, which were by now almost exclusively written exploiting IETF protocols: operating with the handicap of different infrastructure was no longer acceptable. Actually, it hadn't been acceptable for a while: the application that achieved most prominence as an example was X-Windows (mentioned in Chapter 3). Originating from MIT in the early 1980s, it had by the late 1980s reached the stage where it was stable enough to be used quite widely. X-terminal access was being assumed by other applications and some university departments had already adopted it as part of the local infrastructure. While it was not designed to be used beyond the local area, it was nevertheless convenient occasionally to do so. The same could be said of Sun's Network File System (NFS), which had also been adopted into most university infrastructure by then. Shortly, there would be several more instances: at CERN, Tim Berners-Lee was starting development of Web technology.[9] Multicast extensions to IP had just been made by Steve Deering at Stanford University and multi-party desktop videoconferencing applications would soon follow, pioneered initially by Van Jacobson at LBNL, quickly joined by Peter Kirstein's group at UCL. Though these latter developments were not in time to have any influence on the decision to take the initial steps towards TCP/IP, they were in use soon afterwards and had a substantial *de facto* bearing on subsequent decisions on infrastructure in JANET.

---

7   *See* Glossary and Chapter 9.

8   Encouraged by the beginning of EU-funded collaborative research.

9   Stimulated at least in part by preparations for use of the Large Electron-Positron Collider (LEP) beginning in 1989 and the urgent need to exchange collaboration information and data world-wide: *see*, for example Close (2009), p.98.

Shoestring

In July 1990 the process of bringing JANET into line with existing international wide area and university internal infrastructures was begun by convening a group with the unlikely name of the Department of Defense Advisory Group[10] to advise on the feasibility of a JANET IP service using existing infrastructure (Day, 1992). This didn't take long: the DoDAG reported in the affirmative in November, by January 1991 the Computer Board had endorsed the plan to proceed, and the technology pilot – known by the endearing sobriquet of 'Shoestring', a reference to the paucity of its resources – began in February 1991, led by Bob Day, at that time on secondment to the JNT from the Informatics Division at RAL.

The initial IP backbone consisted of Cisco routers at Edinburgh (Edinburgh University Computing Services, EUCS), RAL and ULCC, the routers at EUCS and RAL being on loan to JANET from those institutions. Interconnection of the backbone and access from other sites was by 'tunnelling', that is, encapsulation of IP packets over X.25 virtual circuits – just as had been used by the UK-US gateway in 1979 and CSNET in 1984: a venerable example of the technique of creating an 'overlay' (still much in use in creating, dynamically or otherwise, new networks over old, whether as an aid to transition, the engineering of special properties, controlling scope, or configuring middleware). It really didn't take long to establish that this all worked as expected and was a practical proposition for the initial establishment of an IP service. By August 1991, routers were in place at all the NOC sites (except Belfast, which was added in time for the opening of the service). The additional sites participating in the initial trial included Brunel, Durham, Glasgow, Lancaster, London (i.e., Imperial, QMW and UCL), Newcastle, Olivetti Research, Warwick and X-Tel (Figure 6.1).

On the technical side, there was every incentive to keep things as simple as possible initially. So, although the NOC backbone routers participated in dynamic routing protocol exchanges, in order to retain connectivity if an underlying X.25 connection should fail, a site was connected via a single X.25 tunnel provided by its regular NOC and used static routing. Trials were made of various applications and protocols, partly to determine whether any presented problems, partly to get some idea of performance. In terms of pure network infrastructure, one of the more substantial elements was the establishment of the Domain Name Service (DNS) for JANET – of which more below.

JIPS

So, a year after the DoDAG reported that mounting a prototype IP service over the existing X.25 infrastructure was feasible and provided a plan for doing so, in November 1991 the pilot JANET IP Service (JIPS) went live. It is remarkable how quickly this happened, and at how little initial expense to the Information

---

10 ARPA had always been funded as part of the US Department of Defense (DoD); since about 1980 it had been referred to as the Defense Advanced Research Projects Agency, DARPA. Although the formal relationship of the Internet (as the former ARPANET) with its DoD sponsor did not end until 1991 (when responsibility was handed over to NSF), its protocols were now managed by the IETF, and involvement in this respect with the DoD had in practice largely ended in 1984, after the split from MILNET.

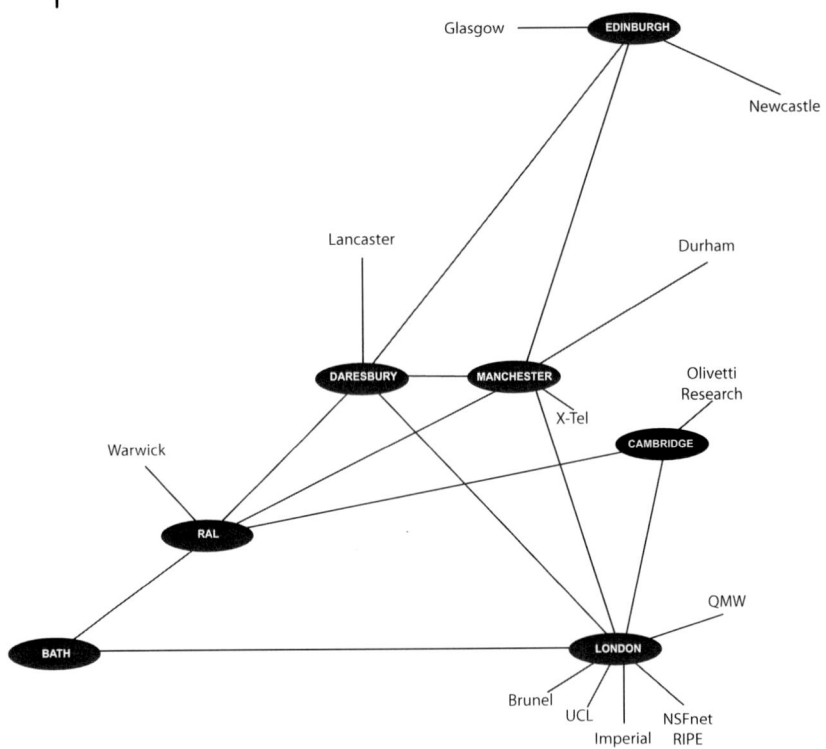

6. Protocol Debates and Emergence of the Internet

Figure 6.1. Shoestring pilot connectivity, August 1991.

Systems Committee (ISC), the successor to the Computer Board[11] – a testament to the voluntary effort at the Shoestring sites, who already had the expertise and most of the equipment, and couldn't wait to get it going!

For JIPS, Cisco AGS/3 routers were chosen and these were installed at all the NOC sites. As with the original X.25 network, a single operations centre was considered desirable and this role was undertaken by one of the original three Shoestring sites, this time ULCC – which had been operating the UK-US gateway service for some years[12] using Cisco equipment – under contract to the JNT: it became the JIPS Network Operations and Service Centre (NOSC). Subsequently, the role gradually expanded, especially as SuperJANET developed; later, with the demise of the X.25 network, the NOSC[13] became responsible for the whole of JANET network operation. ULCC was also the site where international connections were now made: to RIPE and NSFnet initially.

In approving JIPS the ISC had laid down some guidelines on its use in terms of protocols and applications, primarily to protect the existing X.25 service but also because it was not yet ready to formalise alteration of the future protocol strategy for JANET, which still at this point envisaged convergence with or transition to ISO protocols. In effect, use of JIPS was to proceed with caution: the

---

11 The Computer Board became the ISC in 1991, with slightly updated terms of reference; its major responsibility at this time was JANET.

12 *See* Chapter 7.

13 The NOSC became part of UKERNA in 2007: *see* Chapter 12.

general attitude was that use of any of the well-established protocols in the TCP/IP stack was acceptable, provided it didn't lead to trouble!

There was initially some discussion about NFS and SMTP. The former, which was proprietary to Sun but in the public domain, was known not to scale and to have security deficiencies in an open environment – though the risks were to users of the NFS application, not the network. The issue with SMTP was that adoption could lead to loss of email communication with other communities still using Grey Book or X.400 (also known as ISO 10021 by now). In practice, use of SMTP was introduced rapidly, with sites ensuring that conversion facilities were available to handle the other mail protocols. Use of NFS never became particularly widespread in the wide area for the reasons already noted.

Once JIPS was running, an advisory group with a longer-term remit was needed, and the IP Technical Advisory Group (IPTAG) was created for this purpose at the beginning of 1992, with members drawn from the former DoDAG and participants in Shoestring. Two issues for immediate attention were identified: one to do with naming, which we will come to below; the other related to ISO OSI protocols. At this point JANET evolution entered what in retrospect appears a slightly strange phase for nearly two years. Even as the IP service was demonstrating its popularity in meeting people's requirements, it was still assumed that this, like Coloured Books, was in effect an interim, if unexpected, step on the road to adoption of ISO standards. Work on the ISO transition strategy, which had begun in the first half of 1980s (though its roots went back to the 1970s) and had resulted in the White Book transition strategy document in 1987, did not slow – TCP/IP was just added to the list for transition. In November 1992, for example, interest was canvassed in testing an FTAM-NIFTP converter, which became a service for a short period over JANET.[14] In retrospect this was ironic: within the year, the decision would be taken effectively to abandon not only Coloured Books but ISO protocols. Of course, it could be argued that the size of the engulfing tide had not quite become apparent: Web technology was in the making but the World Wide Web phenomenon was not yet a world-wide tsunami.

During 1992, Web technology began to spread. The CERN server had appeared in the summer of 1991 and by December it had spread outside Europe when the server code was mounted at Stanford Linear Accelerator Center (SLAC) in California. It became freely available to anyone in April 1993 – coinciding, ironically, with a decision elsewhere that the Gopher service would no longer be free, which merely gave the Web an added boost! Work at LBNL by Van Jacobson and the Network Research Group had also resulted in an initial set of tools for videoconferencing and desktop computer-supported real-time collaboration. The tools all exploited the new multicast addition to IP[15] so that a group of people, each situated anywhere on the Internet, could talk to each other, see each other[16] and exchange ideas with the help of a distributed whiteboard.[17] To participate,

---

14 This was an outcome of a project called General Internetwork File Transfer (GIFT) which came out of the particle physics community, with contributions from Italy, CERN and the UK. It ultimately provided conversion amongst CERN file transfer, DECnet, Internet FTP, ISO FTAM and NIFTP.

15 Initially defined in RFC 1112 (Deering, 1989).

16 This was not new: however, previous systems such as Cornell University's CUSeeMe used application-layer 'reflector' relay nodes for multipoint distribution, a potentially less efficient method in terms of network traffic loading; in practise, used over the MBone overlay network, there was probably little difference.

17 The initial set of tools were vat (audio), vic (video), wb (whiteboard) and sd (session directory); the latter was a multicast live directory facility on a dedicated channel to assist in connecting to a

support for the new multicast IP architecture had to be available. It would be almost a decade before multicast became generally available in routers. Initially, an overlay network called the MBone (for Multicast Backbone) was constructed by implementing the new multicast additions in a set of separate machines called MBone routers (often Sun workstations), which communicated multicast packets to each other by encapsulation in ordinary IP packets. IP multicast on an Ethernet LAN just used the native multicast facilities of the LAN, which was typically how a host communicated with its local MBone router.

Development of these multicast videoconferencing tools during 1992 through 2000 was led in the UK and Europe by Peter Kirstein's group at UCL[18] – see, for example, Handley *et al* (1993) and Chuang *et al* (1993). There were a number of groups in the UK who wished to participate in these videoconferencing experiments and an experimental MBone was mounted over JIPS in 1992, co-ordinated out of ULCC. (These same tools would later become the basis of Access Grid, developed at Argonne National Laboratory.) Notable landmarks in the use of this technology worldwide were the audio- and video-casts which began in 1992 of the regular IETF meetings; in 1992 and 1993, JANET carried these over the MBone over JIPS – a double overlay. Even in 1993 these multicasts, which were in effect early exemplars of Internet TV – if of appeal only to a specialist audience – were seen by thousands around the world. In the JANET context it is also pertinent to note that at this point, within Europe, the two largest users of IP were Germany and the UK. On the other hand, IP connectivity with Europe was desperately inadequate, being two 64Kbps X.25 links via IXI. Capacity to the US was somewhat better, at 384Kbps, but still needing augmenting.

As a result of the popularity of IP, by the end of 1992 almost all the sites by then connected to JANET Mk II were also using JIPS. Additionally, following an initiative by the PCFC to upgrade many of the former polytechnics (which had become universities in 1992) from 9.6Kbps to 64Kbps, many of these had also begun to sign up for JIPS connections. This rapid move to IP resulted, of course, in substantial loading of the network. There is a natural overhead on transmission produced by encapsulating packets. There is a considerable additional processing and transit delay introduced by having to fragment large IP packets into smaller X.25 packets and by having to route or switch everything in both the overlay and the underlying network. And lastly, there are 'hot-spots' of traffic produced in the links in any overlay network implementation between the overlay router and its local underlying switch or router – and these effects are all magnified in a double overlay like the JIPS MBone. So, it was not surprising that by March 1993 several links in JANET saw more traffic generated by IP than by native X.25 applications, and the most heavily loaded routers in the network were experiencing overload (largely because although IP processing benefitted from some assistance in hardware, X.25 processing was all done in software). Apart from stimulating router upgrades, this also triggered the first use of links for 'native' IP, i.e., without X.25 encapsulation. It will be recalled that JANET Mk II was concerned with access link upgrades from 64Kbps to 2Mbps; to carry the additional traffic, it was, of course, necessary to upgrade some of the links in the X.25 backbone: adding a second 2Mbps link was the option that was available and affordable in the interim before SuperJANET began operation. When such double links reached the stage that the majority of the traffic was IP, one of them was reconfigured to

---

session.

18   Through a series of EU-funded projects: MICE (and MICE II), MERCI and MECCANO; within the UK, the enhanced versions of these tools became known as the MICE tools.

carry all the IP traffic directly between the relevant routers, by-passing the X.25 network altogether and at a stroke relieving the load on X.25 and producing as much as a tenfold improvement in the IP transit delay (reductions in round-trip times from 100 ms to 10 ms and 200 ms to 20 ms were recorded).

DNS

Brief reference was made at the end of Chapter 3 to the investigations and systems developments in the field of naming and addressing at the beginning of the 1980s. One of the most oft-quoted and seminal papers at the time was that by John Shoch (1978) distinguishing naming, addressing and routing. The UK Name Registration Scheme development mirrored much earlier work in ARPANET and was stimulated by essentially the same considerations. The starting point for wanting names tends to be that numeric addresses are basically unmemorable: alphabetic names for network services are much more memorable and friendly. So, lists are made and code is written to look up and translate from one to the other. The commonest early application requiring this facility was email. John Shoch pointed out that there is a semantic difference between the two concepts and that one should distinguish between the name of a service and where it is on a network, that is, its address. Moreover, there may be added operational convenience in this distinction because a service may change its network location but remain the same service. The name of a service might be a localised name but on the whole it tends to be more useful if everybody uses the same name for a given service. Both to maintain naming consistency and to ensure that the address is correct, as we've seen already (Chapter 5), it becomes convenient if there is a single authority and system organising this table. This was the original approach taken in the ARPANET, as it was later in JANET with the Name Registration Scheme. In ARPANET, hosts had a local copy of the central table in a file typically named 'host.txt'. While there are not too many services, they don't move around too often, and the network is all managed out of one office, this more or less works. As the network gets bigger, new services keep appearing, hosts break and services change address more often, and all the hosts keep reloading the table more often to remain up-to-date, the system becomes unmanageable because it doesn't scale properly. ARPANET encountered this in the late 1970s, JANET encountered the same thing with the Name Registration Scheme in the late 1980s.

Apart from experimental work at Xerox PARC on Grapevine (Birrell *et al.*, 1982) and the Clearinghouse (Oppen and Dalal, 1983), there was work at Sun which resulted in its Network Information System (NIS) and work in the ARPANET community which resulted in the DNS – a convenient acronym sometimes used to mean Domain Name System, the concept and worldwide implementation; Domain Name Service, the service provided by the system; or Domain Name Server, an instance of the distributed set of servers which provide the service. Although the ideas were developing in the late 1970s and early 1980s, the DNS did not form part of the initial set of protocols promulgated by the Network Information Center (1982) in support of Jon Postel's 1981 Transition Plan. Its design, specification and implementation were followed soon afterwards by Paul Mockapetris; by 1986, this implementation was in operation on all the Internet root servers. The ubiquitous Unix implementation was (and is) BIND (for Berkeley Internet Name Domain), originally written at the University of California at Berkeley as a student project in the mid-1980s.

The DNS supports a hierarchical naming system, using a set of distributed servers responsible at each level in the hierarchy, and with alternative servers available in case of failure either of a server or a part of the network rendering a server inaccessible. The system is a complex one in reliable distributed database design. By the latter part of the 1980s it was in general use all over the Internet. Several root servers exist in different parts of the world.

In joining the Internet, JANET needed to deploy its own set of servers in support of the UK part of the name space. Initially, during the Shoestring trial, UCL had provided the UK node linking UK names into the main system with its root in the USA. As JIPS came into service, a JANET DNS was mounted at ULCC under the auspices of the new NOSC. To begin with, following the original arrangement set up by UCL, the UK was attached to a root server in the USA: this changed subsequently to avoid DNS failure in the UK merely because the transatlantic link was down. As had been done with the Name Registration Scheme, each institution was responsible for looking after its own part of the namespace (its delegated namespace) and to do this it needed to mount its own server, configure it correctly, obtain approval to connect it into the UK system and negotiate with at least one other institution to provide it with a shadow server, for resilience against failure. All this was referred to as 'delegation' and, in the earlier stages of the JIPS, it was perhaps the most onerous part of implementing the service on a site.

IP rules – OK

In effect, during the period 1990 to 1993 the JNT observed the consequences of establishing JIPS as well as the progress of the Internet worldwide. By 1993 it was universally apparent that the Internet was here to stay, certainly in academia worldwide, and it was now beginning to penetrate other sectors. The first real signs of this came somewhat earlier when X.400 email found itself in competition with Internet email. As the Internet spread, take-up of X.400 stalled, perhaps encouraged by Internet email having a now long-established form of addressing which was relatively easy to remember, especially when compared with X.400's clumsy, long-winded, key-word based form. But the new application which would turn out to be pivotal was the World Wide Web. Its emergence in 1993 would be almost single-handedly responsible for maintaining the exponential growth of the Internet during the 1990s.

Observing the beginning of this, coupled with the rush to TCP/IP on JANET, away from X.25 and Coloured Books, it was hardly surprising that in October 1993 the JISC[19] confirmed that IP should become the preferred network protocol and sites should now be offered the option of native IP connections. Equally significantly, it also signalled the end of ISO protocols on JANET, although it would be almost a decade before the last ISO (and ITU-T[20]) protocols were finally abandoned. Specifically, at the same meeting in 1993, the JISC confirmed that ISO CONS (Connection-Oriented Network Service) would no longer be strategic; and the requirement for ISO CLNS (Connectionless Network Service) would be reviewed regularly. As if confirming the need for the move to TCP/IP, the January 1994 issue of *Network News* (41), which reported the JISC's decision, also carried

---

19   The ISC had been renamed following the formation of the Higher Education Funding Councils.

20   The CCITT became the ITU-T in 1992; *see* Glossary.

the announcement that Digital (formerly DEC) had in October 1993 terminated support for Coloured Books in its VMS operating system.

This was a major transition to manage and it would take some time; on the other hand, it was not like creating a network from scratch. There was a well-established distributed network operations support structure within JANET and virtually everybody now had experience of TCP/IP, which by then existed on all systems, frequently supported by the manufacturer but typically also available from several third parties. And the incentives were considerable: just a single family of protocols to manage; the (eventual) elimination of potentially all conversion services; the removal of old X.25 equipment; and, with the coming of SuperJANET, access to a substantial upgrade in transmission capacity, coupled with the removal of many 2Mbps links.

Of course, it wasn't quite as easy as that but the main technical outline of what to do was fairly simple to state: just turn the JIPS overlay arrangements 'upside-down': that is, run IP directly on all the transmission links, providing X.25 network service, where needed, as an overlay over IP. A way to do this had already been defined and implemented as a proprietary mechanism by Cisco, generally referred to as XOT (X.25 over TCP). Technically, this was a way to send X.25 layer 3 packets over TCP; following its adoption for use in JANET, the details were published as RFC 1613 (Forster *et al*, 1994).

The complicating factors, which would be responsible for the transition taking so long to complete before all signs of X.25 and ISO protocols had vanished from JANET, were twofold: 'native' X.25 services for which there were either no equivalents or for which the transition to TCP/IP equivalents would take time; and interconnection with other networks, especially international. In the latter case JANET naturally had little or no influence on timescales, though in practice it would not be long before most others had either made a similar transition or closed.

As it would turn out, the two major application services and protocols for which special action was required to enable transition from X.25 to TCP/IP were interactive terminal access and email. What these two services had in common in this transition context was interworking between 'old' and 'new'. And the first issue in both cases was naming. It will be recalled that JANET had inherited a big-endian service naming convention from the end of the 1970s. This had originally given rise to debate in 1980 in the context of email, when the UK had opted to adopt the ARPANET email format but retain the UK naming convention for consistency with other UK services. During the 1980s this had become increasingly tiresome to those habitually corresponding internationally, because a user had to know what convention was in use at a site or host and yet all the rest of the email looked the same. Once the decision had been taken early in 1993 that the IETF protocols were the strategic future, there was immediate consultation and assessment carried out by Sandy Shaw and George Howat of EUCS about changing the UK convention. The upshot was to permit use of little-endian naming in Grey Book mail and X.29. The latter was included because this was the other main area where users needed to enter service names which crossed the boundary between 'old' and 'new'. This decision was greeted with universal relief. In many cases, sites and users rapidly made the change and by the beginning of 1994 people were quoting email addresses in Internet order. This finally laid to rest what had been a bone of contention in UK networking for nigh on 15 years.

Apart from the naming issue there was the more substantive question of how best to enable, indeed, encourage transition while maintaining service. For interactive terminal services, the principal Internet terminal protocol was

TELNET, one of the earliest from ARPANET; on JANET, there was considerable use of X.29 terminal protocol and there were PADs throughout the community, as well as PAD software on various host systems. In September 1994, after a trial period, a central X.29-TELNET conversion service was provided, called the Terminal Access Conversion Service (TACS), which enabled terminal connections to be made in either direction.

Email was a little more complicated. There were by now three email protocols and accompanying systems in use in the UK community: the original Grey Book, which used the same email content and formatting as Internet email but transferred the email using NIFTP; X.400/ISO 10021 Message Handling Systems (MHS); and Internet email, which differed from Grey Book in using Simple Mail Transfer Protocol (SMTP) to transfer the email. And, of course, while Grey Book email used the big-endian addressing convention, Internet email was little-endian. Conversion between the two was technically straightforward; unfortunately, with two conventions in use and conversion occurring in a variety of places, both locally and in gateways, attempts were made to recognise which might be in use on the basis of top-level domain recognition, regardless of 'endianarity'. This was an heuristic doomed to fail unpredictably under such circumstances as email to or from the former Czechoslovakia (CS, before it split in 1993), which caused several computer science departments to change from CS to DCS, and legacy systems known as GB being mistaken for top-level UK by mailers equating the two: *see below 'Naming – the last word'*.[21] X.400 had initially appeared on the scene in 1984, developed in conjunction with ISO; it was substantially revised in 1988 and had been anticipated in standards circles to be the international standard towards which other systems would converge. Although take-up in the UK had been patchy, it enlarged the email community accessible to JANET users because there were by this time communities in the UK and internationally who had adopted it exclusively. Use of the converter introduced in May 1994, together with suitable entries in both the NRS and the DNS, enabled an institution to change to sole use of Internet email while maintaining email contact with correspondents in both the non-Internet (i.e., X.400 and Grey Book) email communities.

By November 1994, the requirement for X.25 was being referred to as a 'residual need' – this within a year of the official change of policy on the major supported protocol stack for JANET. However, since the new SuperJANET service supported IP only, and there remained some parts of the community, as well as some applications, based on X.25, support for X.25 over IP was retained as a transition service. Indeed, one of the important higher education services for which X.25 remained essential was UCAS, the Universities and Colleges Admissions Service, a service of obviously prime importance to the sector, institutions and prospective students alike.

At this point no new X.25 connections were being made and X.25 over IP was only guaranteed until March 1996 (around 18 months). Protocol policy continued to evolve in favour of Internet protocols: X.400 was no longer the single convergence target for email. The official stance was that while several stacks were acceptable pro tem for email (SMTP and X.400 over TCP/IP), Grey Book over X.25 and X.400 over either ISO CONS or X.25 were now obsolete.

In 1995 a different sort of development occurred: a regulatory development brought about not so much by continued growth in Internet protocol use as by evolution in what the Internet and now JANET were being used for – indeed how networking in general and JANET in particular were perceived. JANET was

---

21 A problem of this era experienced by the author!

now an integral part of life in post-school – further and higher – education in the UK. It was no longer confined to use by academic staff for research purposes but was in use by the whole student body and all staff; usage was now integral to teaching and administration throughout the sector. From its inception, and for much of the 1980s, its unique position in supporting higher education and research had meant that communications suppliers had been willing to provide services to the sector on advantageous financial terms but under fairly restrictive terms regarding use. By the mid-1990s, providers had become less sensitive to use – except in respect of, rather naturally, not countenancing reselling of facilities in competition with those who had provided them on special terms in the first place. General purpose use, including, for example, leisure and use by research spin-off companies or other bona fide associated commercial partner concerns was now formally regarded as acceptable, although the technical terms of connection for such organisations was not the same as for education and academic research institutions.[22] While this may have been part of a general shift in the climate of opinion brought about by increased access to networking more generally, it was also no doubt influenced by the so-called duopoly review of telecommunication provision which had taken place at the beginning of the 1990s and had almost completed the process of deregulation in the UK.[23] The general tenor of the altered stance on use of JANET was that the nature of use to which JANET could be put was now delegated to subscribing institutions except in so far as use should be legal and not detrimental to JANET or others' use of JANET; one of the more tricky areas was (and remains) use which may be considered offensive, since that may be a matter of opinion or context. Commercial use in association with specific research or educational activities was now generally acceptable. JANET was not alone in finding it appropriate to revise its Acceptable Use Policy[24] (AUP) in this way: NSF had had to do roughly the same thing in respect of the NSFnet backbone in 1991, for much the same reasons. It may be remarked in passing that the revision of JANET's AUP in 1995 was in the same year that the first of the significantly effective and successful search engines, AltaVista (developed by researchers at Digital), became available, indicative of just how much of a general and widespread resource the Internet was evolving into.

The next and most significant landmark in JANET's transition was in 1997. It began in February of that year with a further reduction in support for X.400; but the major event was the closure of the X.25 service from August 1997. This had a knock-on effect on a number of associated services. Since this also meant that use of Coloured Book protocols was no longer supported, all support for Grey Book mail now ended and the NIFTP-FTAM conversion service was no longer needed. There was some residual support for external X.25 communication, primarily with BT's PSS (now called Gold Network Service), for access to X.29 terminal services. The TACS, which had been provided for conversion between TELNET and X.29 in 1994, continued but with a much reduced role: with the internal JANET X.25

---

22 Typically, a non-education institution would be sponsored by an education institution which would become responsible for service provision (including the point of connection – the line being paid for by the non-education institution); the service would essentially consist of connectivity and would not include the complete portfolio of JANET services. In more recent times this has become more formalised and is known as a Sponsored Connection. It should be noted that a commercial educational content provider is typically connected to the backbone, in effect sponsored by JANET(UK); the commercial provider would still pay for the connection, though of course in this case that would be recovered through the cost of the service.

23 *See* Chapter 8.

24 *See* Chapter 13.

service closed, it was now only needed for outward calls from JANET using TELNET to access external X.29 services: external X.29 access to JANET was no longer supported. (This service was eventually closed at the end of 1999 because use had dropped to such a level that it was not worth the effort to ensure it was year-2000 compliant.) And with the closure of X.25 and Coloured Book services, there was no longer any need for the Name Registration Scheme since the central support of registration and maintenance of the database of such services had been its principal *raison d'être*. The Name Registration Scheme had also been used to support the initial introduction of ISO services; however, with the exception of X.400, there was now little use of these, so support was withdrawn and the NRS closed. It was suggested that use of X.400 could be supported by the X.500 directory service but, in practice, from now on use of these services dwindled. The X.500 service closed in August 1999 – another service which never needed to be checked for Year 2000 compliance! And, lastly, in 2002, support for X.400 ceased with the closure of the X.400–SMTP conversion service.

X.25 had formed the basis of JANET since its official birth in 1984, a period of about 13 years; however, it had actually served networking in the UK academic community for rather longer: nearer 20 years. Outside the JANET community, particularly in enterprise networks, where there might be a larger investment in specific business related applications and protocols, there was often an evolution to frame relay from X.25: a means of gaining increased performance through exploitation of the much improved transmission quality by using a network-layer protocol without error recovery but retaining the connection-oriented network property. The debate about the relative merits of the connectionless versus the connection-oriented network layer would continue, as we shall see in the next stage in JANET's development.

Naming – the last word

Having reversed the order of its names and joined the Internet, one might have supposed that JANET was finished with issues of naming: but, no, there was one last spasm to come. Over time, Britain has adopted a number of official names for itself, that since 1922 being the United Kingdom of Great Britain and Northern Ireland. ISO standard IS 3166 lists a set of two- and three-letter codes for the countries of the world. Britain's entry, GB, evidently reflected the name Great Britain, like the car stickers for foreign travel, which originate from the 1940s. The modern 3-letter code for currencies is another ISO standard, derivative from 3166, so uses GBP.

But people change and usage changes, and since sometime in the second half of the twentieth century, United Kingdom and UK have been the increasingly prevalent title and acronym used in most contexts, official or otherwise. By the time the need arose for a British top-level domain in the DNS, Peter Kirstein[25] adopted '.UK' in 1983 on the basis of common usage. In 1985, when Jon Postel was introducing order into the DNS, he pointed out the existence of the GB ISO code for Britain. Since some Government Departments wanted to use the ISO code, '.GB' was introduced alongside '.UK'; later, the issue was raised again as use of the Internet became more widespread: indeed, it was suggested at one point that the UK be given five years to change.

---

25 Peter Kirstein, April 2010, private communication.

However, the code has far wider implications: 'UK', like 'US', is in common use worldwide in many contexts beyond the Internet. There was considerable discussion in Government circles, in which the BSI participated. However, any benefits of such a change were far outweighed by its cost which, even in 1988, was estimated at £25m. As might be supposed, nobody was willing to pay such a sum for this almost nugatory change: thus was '.UK' established *de facto*. '.GB' continued to be maintained by UCL for another decade or so, in particular for the benefit of some MoD hosts. To prevent the possibility of further confusion, ISO has reserved 'UK' and '.GB' continues to be controlled from within the UK.[26]

An oft-repeated aphorism of computer science is that there are only two topics: on the one hand, naming and addressing, and, on the other, algorithms; and that of the two, the former is the harder, whether in software, middleware or networks. Certainly, the generalised topic of binding in software engineering continues to attract considerable research and development effort. It is to be hoped, however, that the ghost of naming that haunted JANET for the first decade of its life has now been laid to rest!

---

26 James Hutton, April 2010, private communication.

7

INTERNATIONAL DIMENSIONS

A curious aspect of international computer networking from a UK perspective has been the consistency with which the early UK pioneering efforts seemed to meet with a warmer reception abroad than at home. Donald Davies's NPL group had been early pioneers in the UK: EIN, led by Derek Barber, achieved operation linking five countries (including the UK), which was more promising than the fate of Donald Davies's earlier, rejected proposal to the UK Government for a national network. Peter Kirstein's ARPANET link provided UK academia and the MoD with access to the USA for a decade before JANET officially began. And even IPSS preceded PSS by three years.

From a JANET perspective, international connectivity was for almost a decade provided by a series of ad hoc initiatives, all different and nowhere near global. The reasons are not hard to find: funding, management and technical foundations are hard enough to agree nationally; internationally, the only approach was to concentrate on the most important bits first. In practice this meant the USA and Europe, and it is the provision of UK research and education access to these two regions which is briefly sketched here, primarily during the 1980s through to the mid-1990s. Before that, JANET did not exist; by the early 1990s, the Internet had arrived everywhere and the issues were becoming much more focused. In Europe, as in the USA, networks would now be based on explicit backbone architectures managed by consortia organisations: DANTE in Europe, Internet2 (following NSF) in the USA.

Technically, for this period there was something of a vacuum – or perhaps it might more accurately be characterised as a form of 'planning blight'. Within Europe,[1] whether in the context of government funding or the telecommunications industry, the only universally agreed position was that international standards should form the technical foundations. However, there was no consistent range of products available to support these. So, various groupings appeared, driven by a combination of necessity and mutual interest, out of which a variety of networks appeared. Without exception, all agreed that they would adopt international standards – when stable products became available. Actually, this mantra was adopted widely outside Europe as well, especially by those who had something that already worked; and that included the USA. The National Bureau of Standards, NBS,[2] advocated the adoption of international standards throughout this period through its Federal Information Processing Standards (FIPS),[3] and the

---

1 For those interested in more detail on the history of international networking to support research, particularly in Europe, the book edited by Howard Davies and Beatrice Bressan (2010) contains a wealth of detail; for those interested in a less detailed overview from a different perspective, particularly during the period 1984–2004, the article by Peter Kirstein (2004) is recommended.

2 NBS was the predecessor of NIST: *see* Glossary.

3 *See* Glossary.

US computer networking industry, whether providing manufacturer proprietary or IETF TCP/IP solutions, adopted the same position. Of course, working solutions were always more mature and frequently performed better in a given context than the early implementations of international standards and, ultimately, this deadlock was only broken by the universal, unregulated but generally popular adoption of TCP/IP.[4]

7.1 Transatlantic connections

The first ARPANET node established outside the USA was that created by Peter Kirstein's group at UCL,[5] which came into operation in the autumn of 1973. For about 15 years, until TAT8, the first transatlantic optical fibre cable, came into operation, UCL would provide US ARPANET and then Internet access for UK education and research.[6]

The node was a Terminal IMP (TIP) which offered dial-in terminal access to TELNET for UK MoD and university users. The UCL ARPANET host attached to the TIP was originally a PDP9. However, UCL had an RJE link to the IBM 360/195 at RHEL and the PDP9 was programmed to appear to the 360/195 as an RJE workstation using HASP multileaving.[7] RHEL had extended the HASP protocol so that traffic from line-mode interactive terminals attached to its RJE stations could be carried along with the standard control console, lineprinter, card reader and card punch streams. The effect was that interactive terminals at RHEL or at any RHEL RJE station could access ARPANET and authorised users could access the RHEL terminal system from anywhere on ARPANET. It was also possible to effect file transfers by logging onto the PDP9 and using ARPANET FTP; file transfer to and from the 360/195 was effected by automated batch jobs via the RJE interface. Subsequently mail facilities were provided, and these evolved into a mail gateway between JANET and the early Internet.

The link from UCL to the US was originally provided by a 9.6Kbps line from UCL to Oslo and a 50Kbps satellite link to Washington. Initially the link to Oslo was paid for by NPL and the British Post Office; subsequently, it was jointly funded by the SRC and MoD. The UK-US link stayed at 9.6Kbps until 1978/9. By 1978, experimental use of SATNET had formally ended and the UK-US gateway services moved to use of that channel which (using slotted Aloha[8]) provided about 20Kbps capacity in each direction. Shortly thereafter IPSS became available and a 9.6Kbps access link was installed, which the UCL gateway could also use.

---

4   For those interested in a little more detail on the international situation in academic networking in Europe during the period 1984–2004, the article by Peter Kirstein (2004) is recommended.

5   As mentioned in Chapter 1, to be strictly accurate, Peter Kirstein's group was at that time part of the University of London Institute of Computer Science; it would become the Department of Computer Science at UCL later.

6   I am grateful to Peter Kirstein for providing (and correcting) much of the information relating to the first 15 years of the UK-US link. Sources relating to the establishment of the NSFnet gateway at ULCC include *Network News* (26–29, 34).

7   IBM terminology of the time for application-layer multiplexing.

8   Pure Aloha (originally used in the packet radio system at the University of Hawaii in 1971) is random or stochastic access and is about 18% efficient at best in its use of a channel. Slotted Aloha divides time into slots equal to the (maximum) length of a packet and packet transmission may only begin at the beginning of a slot boundary. This halves the chance of maximum fixed length packets interfering (colliding) and so doubles the efficiency obtainable.

In 1981, SATNET was phased out and the possibility of a 48Kbps IPSS link to the US was explored, the upshot of which was that BT gave UCL a free 48Kbps IPSS channel in exchange for performance measurements. When this ended, in 1984, joint funding arrangements began: in the UK, initially by the Alvey joint DTI/ SERC IT research Programme[9] and later by the JNT and MoD; in the US, funds were contributed by NASA, DARPA and NSF. This regime continued until 1989; it covered a period of transition on both sides of the transatlantic link, during which, in the US, NSFnet was established as the Internet backbone to support research; and in the UK, the JANET-NSFnet gateway was established at ULCC.

Planning the establishment of a US national backbone was under way at NSF in 1984.[10] When NSFnet came into operation in 1986, it provided the opportunity for a gateway between JANET and the US network having a community and service remit similar to that of JANET. The plan was to colocate a JANET-NSFnet gateway with the JANET NOC for London and the South East at ULCC and to install a new, 64Kbps link between ULCC and the John von Neumann Supercomputer Center at Princeton, New Jersey.[11] UCL would retain its independent link for research purposes after the transition. As a result of a proposal made in 1987 by JNT, UCL, ULCC & USC ISI,[12] agreement was reached in 1988 that NSF would fund the US end of the new link and the UK end would be funded jointly by the Computer Board and SERC.

The new 64Kbps link was to be provided out of the new, optical fibre transatlantic cable, TAT8, and was scheduled to come into operation before the end of 1988. However, this was delayed and, courtesy of UCL and the MoD, use of the UCL link continued for a further six months. However, when further delays occurred, a temporary 56Kbps satellite circuit was brought into operation between ULCC and Princeton in April 1989. Eventually, the 64Kbps circuit came into operation towards the end of 1989.

Initially, the NSFnet gateway only handled mail from the UK to the US: mail in the reverse direction, together with TELNET and (ARPA/Internet) FTP remained with UCL. However, once established, the NSFnet gateway at ULCC took over the complete service functionality which had been provided for over 16 years by UCL. Then the following year saw the beginning of the transition of JANET to IP with the Shoestring project. As service traffic moved over to use of the NSFnet gateway, so funding associated with that service traffic at the US end was transferred to the new link, enabling a series of upgrades. By 1990, the link capacity had reached 256Kbps, 64Kbps reserved for and funded by NASA. The term 'fat-pipe' was now used to refer to this new multi-agency link.

As with growth elsewhere in network use, demands on the transatlantic link continued to grow and, by the beginning of 1991, a new 512Kbps link was in operation. Then, as JIPS began operation, JANET-NSFnet traffic began to consist partly of native JANET IP Internet traffic and partly of the historical application gateway services: mail, file transfer and TELNET terminal access from non-IP

---

9   1984–89.

10 Conversation with Dennis Jennings (Dublin, EARN) in 1984, then at NSF. The author was then at UCAR, Boulder, working on communications for the US atmospheric sciences community, also funded by NSF.

11 A part of the original stimulus for NSFnet was the NSF supercomputing programme launched in 1984, which created four new centers at Cornell, Illinois, Princeton and San Diego. In the event, the pre-existing center at the National Center for Atmospheric Research, NCAR, Boulder (separately funded by the Atmospheric Sciences Division of NSF) was also included.

12 University of Southern California Information Sciences Institute, where Jon Postel was based.

parts of JANET. As JIPS and the JANET Mk II infrastructure came into operation, pressure on this link continued to grow and it would now be only a short time before IP became the main protocol for JANET. By March 1993 the fat pipe had been upgraded to a 1.5Mbps T1 link, 1Mbps of which was used for the JANET-NSFnet connection, and already plans were under way for the latter to be upgraded to 2Mbps. This was achieved in the summer of 1994, with the JANET-NSFnet connection now having its own 2Mbps E1 circuit – and immediately negotiations moved on to planning the next upgrade! A year later, in the summer of 1995, a second E1 circuit was in operation.

It will be clear from the foregoing that JANET-US traffic was subject to the same general growth in demand as JANET itself in earlier years. Now, with the TEN-34 project (see below) beginning to plan a 34Mbps European backbone interconnect, it was time to step up for a similar leap in capacity in the JANET-US link. However, there was now another aspect requiring attention: ULCC had become a single point of failure for international connections. In 1994, in an initiative to move some of the international terminations away from ULCC, UKERNA had participated in discussions with BT, Demon, Pipex, UKnet and other ISPs which resulted in the creation of a non-profit organisation – LINX, the London Internet Exchange – and the setting up of an Internet interconnection point in Telehouse in London's Docklands.

Funding for the US link continued to be an issue. Partly, it was an expensive item; and partly there was always debate, particularly with it being international, about who should pay. During 1987, with the prospect of connecting to the newly created NSFnet backbone, negotiations had succeeded in achieving agreement for NSF to fund the US end and a combination of the Computer Board and parts of SERC[13] to fund the UK end. However, the continuing growth in demand would later cause the JISC to institute a charging regime for this item.[14]

7.2 European angle

The pioneering European packet network, EIN,[15] has been mentioned earlier. It had five primary centres located in Italy (Politecnico di Milano and the European Communities Joint Research Centre, Ispra), Switzerland (ETH,[16] Zürich), France (IRIA,[17] Paris) and the UK (NPL). Although EIN began experimental operation in 1975–76, following the adoption of X.25 by the CCITT, it was succeeded by Euronet in 1978; however, the latter never achieved widespread service deployment. Instead IPSS, which came into operation in the same year, backed by the Post Office in the UK, began offering an X.25 service to North America and Europe. Like PSS later (see Chapter 5), IPSS was a commercial service and use required a chargeable account, a situation which ultimately acted as a disincentive to all

---

13 The Astronomy, Engineering and Science Boards.

14 The view generally espoused for JANET has been that it is not effective use of resources to institute usage charging. The resources which have to be expended on billing by any commercial telecommunications operation are considerable; in the context of JANET, the preference has generally been to fund by 'top-slicing', for the service to be free at the point of use, and for all funds to be directed to the benefit of the service; see Chapter 14.

15 For further information, see Helms (1978).

16 Eidgenössische Technische Hochschule.

17 Institut de Recherche en Informatique et Automatique; see Glossary.

except those having an essential need and with funding to back usage. Moreover, even for those in a position to pay, authorisation and billing procedures consumed time and effort. Nevertheless IPSS, accessed via PSS once it came into operation in 1981, provided an important means of access to Europe during the early 1980s. (It also provided access to commercial organisations in North America.) It provided X.29 terminal access and, later, X.400 mail. Although use of IPSS lasted through much of the 1990s, once the ascendancy of IP became clear it was never really more than a supplementary service for organisations reachable no other way.

In 1981, while ARPANET was still not generally accessible to the academic community in the USA, IBM inaugurated BITNET – originally 'Because It's There Network', later this became 'Because It's Time Network', perhaps reflecting the sense of frustration experienced by those not enjoying access to the facilities of the incipient Internet. Based on IBM's proprietary RJE protocol in its peer-to-peer form,[18] it operated on the basis of staged file forwarding between adjacent (that is, directly connected) hosts using IBM operating systems.[19] Like Unix-based UUCP, BITNET spread on a community club basis, encouraged and supported by IBM which sponsored the cost of the links between systems. Part of the commitment to join included a willingness to forward files, including mail, on behalf of others. Subsequently the network was open to any system which supported the same protocols. Particularly for mail, gateways appeared between BITNET and other networks, including ARPANET and UUCP.

In 1983 a European arm of this network appeared, called EARN (for European Academic Research Network), also supported by IBM. Since the network performed third-party switching (in this case, at file rather than packet level), it was technically in breach of the monopolistic telecommunications licence regulations in most European countries. However CEPT,[20] a European PTT consultative organisation, suggested that it should be allowed on condition that it undertook to migrate to ISO protocols in the future, and European PTTs generally followed this advice, if on widely differing timescales. RAL joined EARN and became the UK gateway between JANET and EARN for mail. In line with its policy for PSS which made it so unattractive for academic consideration in the years immediately preceding the creation of JANET, BT imposed volume-related charges for EARN links, which meant that, even at its inception, its future was uncertain in view of IBM's fixed-sum sponsored funding. The network came into operation late in 1985 in the UK; at the time, IBM sponsorship was agreed until 1987. For the next few years, until the Internet became more generally available throughout Europe, from a UK perspective, EARN provided a useful service within Europe amongst some of the research organisations.

Late in 1984 a meeting was convened at RAL's Coseners House, Abingdon, of representatives of organisations concerned with national academic and research networking across (Western) Europe. This led to a European Networkshop in Luxembourg in May 1985. It was recognised that a Europe-wide association was needed to create a pan-European communications infrastructure for the research and education community; the name RARE, Réseaux Associés pour la Recherche Européenne, was chosen in August; and, with the help initially of funding from The Netherlands and then from the European Commission, RARE was

---

18 Network Job Entry (NJE); probably better known by the system name RSCS, Remote Spooling Communication System, the peer-to-peer derivative of the earlier HASP (*see* Glossary).

19 Subsequently, non-IBM operating systems which could emulate RSCS protocols also joined.

20 European Conference of Postal and Telecommunications Administrations.

established as a formal legal entity[21] a year later, following the second European Networkshop which was held in Copenhagen. The major initial activity for RARE was to undertake the specification of the COSINE[22] Project to develop the basis for a Europe-wide network infrastructure for the education and research community. The first three years of COSINE were devoted to requirements, definition, planning and negotiation among the various organisations involved. Implementation began in 1989, the first objective being the establishment of an X.25 network of European extent, upon which further facilities (X.400 mail services, X.500 directory services, ISO FTAM file transfer, information services, etc.) would be based. The plan was to provide 64Kbps X.25 access points for the connection of national networks to this International X.25 Interconnect (IXI). The one-year pilot phase began in 1990, provided by PTT Telecom of The Netherlands under the auspices of COSINE and a contract with the European Commission, which provided the majority of the funding. The JANET connection to IXI was via a 64Kbps link to ULCC, which came into operation early in 1990.

However, it will be recalled that it was in just this timeframe that JANET began the Shoestring IP project, the first step in the transition of JANET to IP. Albeit that the whole weight of European thinking – and funding – was still firmly committed to ISO OSI, and had been so for over a decade, within a year there was increasing demand for improved IP connectivity within Europe (and with the rest of the world). Early signs of this occurred with the formation of the Réseaux IP Européens (RIPE), effectively a club of European IP network operators, which provided a forum for sharing operational experience and undertaking technical coordination; it held its first meeting in 1989.

IXI, which by 1991 was completing its first year of pilot operation, now became the European Multi-Protocol Backbone (EMPB), with substantially upgraded infrastructure (2Mbps instead of 64Kbps) and offering both X.25 and IP services; parts of the transition were initially made, as in JANET, by use of IP tunnelling over X.25. By 1992, parts of the IP service were native IP.

In almost the same timeframe, late in 1991, a small group of representatives from European research and education networks came together to pool resources and concentrate on the specific task of providing a European IP backbone through coherent organisation and rationalisation of existing resources. After just over a year of sustained effort the European Backbone, EBONE, came into operation in September 1992 – providing considerable relief at the time, in terms of capacity and alternative or complementary interconnection routes. Its topology was a loop composed of 512 or 256Kbps links connecting Amsterdam, Stockholm, London (ULCC), Montpelier and CERN. There were also links to the US via CERN and Stockholm, the latter also providing EBONE access for NORDUnet,[23] the Nordic universities network. Although the JANET-EBONE interconnection was at the NOSC at ULCC, the JANET 'fatpipe' was not part of EBONE, partly because it was itself funded by a different consortium and partly because the capacity available on it for JANET-US traffic was already saturated at this point.

Having established IXI in 1990, COSINE recognised that a permanent organisation was needed to support and develop the network and, in 1991,

---

21 Dr Peter Linington was elected the first President of RARE, at about the time he moved from being Head of the JNT to become Professor of Computer Communications at the Computing Laboratory, University of Kent.

22 Cooperation for OSI Networking in Europe; *see* Glossary.

23 Since 1988 NORDUnet has served Denmark, Finland, Iceland, Norway and Sweden; its story may be found in Lehtisalo (2005).

initiated its creation. Two years later, in July 1993, DANTE[24] was launched as a not-for-profit company based in Cambridge, UK – not least because the UK Government granted attractive tax-exemption status. For the first year DANTE was owned by its sole shareholder, RARE; then, in 1994, it was transferred to the joint ownership of the European national research and education networks (NRENs), at that time numbering 11. From its inception, DANTE took over responsibility for EMPB, which was now a service provided by Unisource, a consortium of three PTTs. EMPB formed a part of EuropaNet, a name coined as the umbrella term for DANTE's original portfolio of European backbone services, which by 1993 also included EBONE.[25] Since then, continuing at the time of writing, DANTE has been responsible for all subsequent international European networks in the academic, research and education sectors.

And what of EARN? From the beginning, EARN faced two issues: the IBM sponsorship for the cost of the links was fixed in both amount and period of time; and, in order to gain approval to operate at all, it had had to give an undertaking to move to use of ISO OSI protocols. IBM funding for EARN had only been agreed until the end of 1987. Only specific parts of the community benefited from the service and, accordingly, only parts of SERC were willing to contribute towards its future support, together with some support from the Computer Board. The latter agreed support for 1988 and 1989, but wished to see costs reduced. Support would continue to be provided by RAL, but now the organisation and management of EARN in the UK would come under the JNT Network Executive and the Network Advisory Committee, just like JANET.

By 1989, EARN was looking to begin moving to OSI protocols. In doing so, it was looking to exploit X.25 to create network connections between hosts over which it would continue to operate NJE/RSCS. Like the NRENs it encountered the problems of using the public international services, which were organised as a 'mesh' of bilateral international agreements and imposed volume charging. The latter made such a solution unaffordable. So, IXI not yet having materialised, EARN was looking to create an initial private X.25 international interconnect to which the national components of EARN could connect. In the UK, the only use of EARN by this time was for mail to some hosts outside the UK. Within the UK, mail used JANET and the NJE/RSCS services were not required. As a result the interconnection with EARN was now via a (mail) gateway. In the event, the Computer Board (ISC) funded the UK contribution for another two years but, by 1992, there was dissatisfaction that the cost of UK participation was too high considering the small fraction of the community which benefited; the point of view now was that EARN should find a way to meet its own costs as a network and JANET would negotiate a peering arrangement. Interestingly, by this time BITNET in the US had already been discontinued: having merged with CSNET in 1989, by 1991 its services had been rendered redundant by the Internet and it ceased operation.

In the event, EARN continued as a network for a little while longer in Europe; then in 1994, the network ceased and it merged with RARE to form TERENA, the Trans-European Research and Education Networking Association.[26] TERENA's purpose was to 'promote and participate in the development of a high quality

---

24 The name is derived from the acronym: Delivery of Advanced Networking Technology to Europe. *See* Glossary.

25 EBONE was subsequently taken over, eventually ending up in the ownership of KPNQwest, and was ultimately closed down in the aftermath of the failure of that company in 2002.

26 *See* Glossary.

international information and telecommunications infrastructure for the benefit of research and education'. It became the European forum for international networking technical discussion, coordination, education and the development of new services.

Since 1994 the two organisations, TERENA and DANTE have pursued the development, support, coordination and operation of international European networking services for education and research in tandem. Both are funded by NREN subscriptions though specific activities, particularly the European international backbone, receive European Union funding. And by this time, like JANET five years earlier, it had become evident that the lack of adequate capacity in the European interconnect was now acting as a hindrance to progress: it was time to step up capacity by one or two orders of magnitude from the 2Mbps links still in use. So began TEN-34 – but that came after the initial achievements of SuperJANET and we shall take up the story in Chapter 10.

7.3 Other community networks

As we have already seen, international connectivity was pursued in a variety of ways, using a variety of networks and protocols, providing a variety of services, before the ubiquity of the Internet reduced the issue effectively to one of providing links of ever increasing capacity with other constituent parts of the Internet. We have given a brief outline of the main threads of development of the US and European connections up to about 1994, by which time the Internet was established as the universal user network of choice and international connections were achieved by backbone interconnects. However, there were one or two other networks having an international dimension which also contributed to developments in the UK.

UUCP, mentioned briefly in Chapter 3, formed the basis of Usenet and its European arm known as EUnet. Like BITNET/EARN, it was based on informal cooperation amongst hosts using the facilities originally available in a particular operating system, in this case Unix and UUCP. As already described in Chapter 5, use of Usenet for news dissemination in the UK predated JANET, subsequently constituting one of the earliest information dissemination systems deployed across JANET.

Other communities that had an influence on academic networking in the UK were typically associated with science programmes which were themselves large enough to be international and sufficiently well-funded to afford dedicated networks where these were deemed necessary for the scientific programme. There were two notable examples in the UK during the 1980s: space science and particle or high energy physics.

In the case of space physics, we have already seen that UK astronomers found a need for network facilities some years before JANET began and created STARLINK, which included its own network. A powerful influence in the creation of this facility was that, in order to gain the maximum from international collaboration, the same systems as in general use in the international community were chosen: DEC VAX using the DEC's VMS operating system. VMS included DECnet, DEC's proprietary network system. The latter included support for terminals, mail and file transfer and had good performance over the wide area and Ethernet – it will be recalled that DEC was one of the companies which participated in the DIX Ethernet standard, as well as being responsible for the spanning-tree algorithm for layer 2 networks incorporating loops for resilience. On the international scene at this time was SPAN, the Space Physics Analysis Network, which was essentially spawned out of NASA, the US National Aeronautics and Space Administration. This had links all over the world, including particularly multiple countries in Europe. DEC, like most others in the 1980s, adopted the standard mantra of committing to ISO OSI protocols when implementations were available and of suitable quality – indeed, DEC did actually make substantial progress in adopting OSI protocols (Phase V, 1987). DEC had also made available the Cardiff implementation of the Coloured Book protocols.[27] However, by the late 1980s TCP/IP was also available. As the STARLINK community moved to use of JANET and the international community moved to TCP/IP, so links dedicated to space physics could gradually be removed, with corresponding cost savings. However, even as late as the early 1990s, in the time of the 'fat-pipe', the link to NASA was of sufficient importance for this community that a part of the 'fat-pipe' capacity was funded by and dedicated to this community.

The particle physics community[28] has always been international: like space physics – space voyages, big telescopes and telescopes situated in space – particle physics accelerators are large and expensive and ceased to be generally affordable (except in the US) on other than an international basis in the mid-1970s. In Europe, high energy or particle physics has been focused on CERN since the mid-1950s; and since the late 1970s, DESY[29] in Hamburg also provided an additional focus. In the case of high energy physics, HEP, the arguments advanced for space physics apply, except that in addition vast amounts of data need to be moved around and, by contrast, the community was in the 1980s not in a position to opt for a single type of system for its work. At that time it needed the largest available machines to perform its analysis: initially IBM (or compatible, like Amdahl or Fujitsu), subsequently CDC, then Cray. But it also needed data-gathering systems and local systems to support individual groups; this caused the HEP community to use everything from micros to supercomputers and everything in between – including small IBM systems and VAXes. This in turn meant that it also used almost all the proprietary network systems, including DECnet and IBM RSCS, which is part of the reason why RAL and other HEP sites participated in EARN. Since CERN was the major hub of European HEP experimental activity, it is unsurprising that it was also the hub of the community's networking activity. Many of the major HEP centres in Europe, including RAL, had lines to CERN, because use could saturate whatever was affordable, which meant that it was also unsharable. CERN was correspondingly faced with a plethora of nationally influenced network initiatives. This combination caused the HEP community to expend considerable effort, including the creation of a senior, international management team to coordinate particle physics computing and networking. The UK HEP community funded leased lines to both CERN and DESY and effectively extended SERCnet – and, by inheritance, JANET – to these sites by placing X.25

---

27 Initially, having adopted the Cardiff Coloured Book implementation, DEC then bundled it with its system, which originated in the USA. The company then claimed that the Coloured Book software was covered by the US limitations on the export of strategic technologies and so should not be made available to all international students. The JNT negotiated the company out of this position. (Peter Linington, April 2010, private communication.)

28 I am indebted to James Hutton not only for sight of a draft (2008) of his HEPnet contribution to the history project now published as Davies and Bressan (2010) but also for further information in 2010.

29 *See* Glossary.

switches at both centres in the early 1980s.[30] This provided seamless links between university researchers and the data acquisition systems, as well as enabling the summary experimental results to be retrieved. At this time, during the 1980s, the network capacity available was only sufficient for small data volumes: the raw experimental data was transferred by van-loads of magnetic tapes for processing by national teams.

Its research support requirements caused the HEP community to expend considerable effort in basing its own networking on international standards, such as X.25 (and X.400), as well as contributing substantially to the initiatives towards establishing a European backbone for research. Initially it was hoped that the international public X.25 service would provide the required service but, in practice, as for JANET, there was a mismatch between PTT public service provision and research requirements, which could only be overcome by the research community taking matters into its own hands. The particle physics community took an active interest in RARE and later, as TCP/IP became the preferred technology in the early 1990s, in supporting RIPE and then EBONE, with CERN acting as one of the EBONE nodes with its own link to the US. Within the UK, HEP was always an active user of the JANET network (even at one period using DECnet over JANET because of the dominance of the DEC VAX in data acquisition). Internationally, with the roll-out of TEN-34 (*see* Chapter 10) there began to be sufficient European capacity for particle physics. Particle physics has remained a very active network user community, vigilant of service performance and maintaining its own network monitoring well into the current decade and, latterly, becoming a major participant in Grid monitoring.

---

30 Evidently by 1983 at the latest (conversation with James Hutton, May 2010).

Part C

SuperJANET Era

Introduction

What is or was SuperJANET? To some it was the name of a project conceived in 1989/90, born partly of the necessity to escape the apparently endless round of upgrades in capacity which never did more than marginally relieve the immediate need, leaving none to spare for the new, innovative and 'performance-hungry' applications then appearing. The name held promise that the renewed network would enable exploration of these new applications. Then, as the new infrastructure slowly started to appear, the name began to be used to refer to the new network – or perhaps parts of the new network. The latter usage also somewhat reflected the fact that those parts of the new infrastructure which were constructed of high-capacity new links were the parts where the new application trials, like video, were under way: would this be the 'real' SuperJANET? Initially, this part of the infrastructure only covered trial sites, no more than ten percent of HEIs – though the original SDH/PDH/ATM pilot sites were also an attempt to seed the network on a national scale. But then this pilot infrastructure began to be thought of as the backbone for the new network, encouraged by the simultaneous emergence of a new, explicitly hierarchical network and organisational architecture. Perhaps SuperJANET, as a name for a network, only referred to the backbone? And indeed this was incidentally encouraged by the various phases and upgrades of the JANET backbone being referred to as, for example, 'SuperJANET4' or 'SuperJANET5' (sometimes with 'project' attached). Moreover, funding for the regional access networks began to be handled separately and upgrade cycles did not occur in unison for all parts of the network.

All this was a diversion: the name had more to do with the vision of what JANET might be if only it could be released from the cycle of upgrades to cope with growth. As a personal view, as a member of the original SuperJANET project team who also later worked on SuperJANET5, the lasting contribution made by SuperJANET was its vision, which was finally realised in 2007 in a form recognisably similar to the concept of 1990 or so. Accordingly, the period covered here is this period of realisation: the 18 years from 1989 to 2007.

With the opening up of the transmission potential of optical fibre came a major shift in how the network was used, the sort of traffic it could carry and the community it served. It is these threads which we take up in this part. In 1989 it was still not entirely clear what protocols would carry the day – the choice being between the ISO family of Open Systems Interconnection protocols, then still being promulgated actively by governments worldwide, especially in the USA, Europe and Japan, and the *de facto* Internet protocols, defined by the IETF – and arguably even more open. However, since the result was largely the outcome of protocol debates which occupied much of the previous decade, this part of the story has been covered in Part B, though the formal adoption of IP by JANET was not until 1993. At the time the issues of capacity and protocol evolution were almost entirely independent; only later did the two intersect as engineering issues of implementation detail affected packet handling equipment specification.

The salient feature with which JANET was to cope was keeping up with developments on all fronts: size and diversity of community; range of application, exploiting multiple forms of communication; information access; and audiovisual communication. Some applications would demand new capabilities of the network. Image-based applications, moving or still, the sheer increase in number of users, and the length of time for which people and machines used the network, would all converge to produce an explosion in the capacity required of it. The combined effects would not only demand a new approach to capacity provision but also signal the need for a new organisation to manage the network.

We begin this part in Chapter 8 by looking at some of the key developments that had altered the technological environment in which JANET was now operating at the beginning of the 1990s. Some of these network developments have already been partly covered, including their exploitation by JANET, but more were shortly to emerge and would help to shape JANET's development. The World Wide Web also appeared on the scene at this time and ushered in the information revolution: networking comparatively suddenly began to influence everyday life, including how the JANET community grew. In Chapter 9, we look at how the network gradually and quite naturally evolved to embrace more and more of the education sector. In doing so, JANET also had to elaborate more explicit policies for managing its interconnections with the other sectors with which education and research naturally had working partnerships. The chapter focuses on how the organisation, management, operation, and funding of the network all altered substantially in order to deal with running what had become not just an option but an essential part of the national education and research infrastructure.

Then in Chapter 10 we add the detail of how SuperJANET developed over roughly its first decade from its initial concept in 1989/90. This coincided with a time when there was considerable technical development in low-level network technology, which caused a number of the developments confidently expected to herald the future to be abandoned. By the mid-1990s a more devolved model of how JANET would be organised was emerging and this had become an embedded part of the network architecture by the end of the 1990s – as the FE sector was connected and discussions got under way about incorporating the schools sector. Chapter 11 describes this new, devolved organisation, together with how schools have been incorporated and enabled to begin exploiting JANET services

In Chapter 12 we return to the further development of JANET, gradually exploiting technologies as each reached deployable maturity, finally achieving realisation of the original SuperJANET concept in 2007. International developments as they affected JANET since about 1997/8 are incorporated into Chapters 10 and 12, since from the mid-1990s onwards there are parallels in the way technology was exploited nationally and internationally. And in the scientific research sector, the UK now embarked upon the quarter billion pound national e-Science programme, arguably the largest scientific engineering programme it had ever undertaken and requiring substantial network support compared to what had been available a decade before: in the latter part of the story, this programme effectively helped the final realisation of SuperJANET by demonstrating first the need and then what could be achieved through its exploitation.

8

MULTISERVICE AND UBIQUITOUS NETWORKING

During the 1980s much of the development – certainly the funding – of JANET was devoted primarily to keeping up with demand for capacity. This had its origins partly in JANET's original construction over SERCnet, a network originally at capacity serving a fraction of the community; but there had also been a prolonged period of some five or more years before the formation of JANET during which pent-up demand for network services had steadily grown. The subsequent five years, from 1984 to 1989, saw rapid expansion of sites connected, coupled with capacity upgrades which tended to be saturated almost immediately. By the end of the 1980s the demand had reached the stage where the network needed more capacity than it could either afford or obtain from the regulated tariffed services then on offer. Moreover the perception was growing that capacity was becoming an inhibitor. Chapter 5 described the technological development of the shared LAN and its exploitation in JANET campus support, including the expanded performance upgrade of the High Performance LAN Initiative. But, it was recognised at the time that the interim – even stop-gap – JANET Mk II upgrade was just that: the time had arrived for a leap in capacity, not merely to satisfy the ever-growing demand but to enable new, 'hungry' applications.

However, before describing the SuperJANET programme which was the response to this situation, we need to look at some of the technology research and development background which had been taking place over the previous decade (some with roots of even greater longevity), which had already altered the environment in which JANET was now operating and was now emerging for exploitation within JANET. Video (and audio) applications required new capabilities of the network: it had now to develop from a single-service data network to a multi-service network combining support for audio-video[1] and data traffic. Later in this period of JANET's development, aspects such as ubiquity and nomadic computing began to influence the JANET infrastructure. These areas had already been in research for over a decade and, benefiting from the continued advance of supporting technology, had now arrived at the point where parts of what was on offer were now appropriate for service exploitation.

8.1 EVOLVING HORIZONS

By the beginning of the 1990s a major shift in the perception of networking could be observed. Instead of being regarded as something that only some – or even a

---

1   The latter is sometimes termed 'continuous media', alluding to its 'non-stop' properties.

lot of – people used, it began to achieve the status of a commodity service which all could be expected to have: perhaps not at home, certainly not yet 'on the move' or when visiting, but typically it was expected to be available at work. In JANET constituency terms at the time, this meant that students and academic and administrative staff at a university, polytechnic or research laboratory could now expect the network to form part of the working environment. As we have already seen, it would now be only a year or so before JANET moved to use of the Internet protocols, enabling easy, rapid adoption of new applications emerging in the international community. Many had already had experience of these applications, had observed the opportunities opening up, and were impatient to be in a position to exploit these at the earliest opportunity.

We have noted how, during the 1980s, provision of computing facilities changed from terminals accessing large mainframes to workstations and PCs in a campus network environment, accessing a combination of local resources for printing and storage to remote resources of a more specialised or substantial nature, whether for computation or data and information. The Alto workstation, developed in the 1970s at Xerox PARC, had been the touchstone for the development of personal computers and workstations generally, as well as the windows environment ubiquitously available from 1980 onwards. However, in some respects it was but a step on the way to realising a concept due to Alan Kay, conceived in 1968 – two years before Xerox PARC came into being – and first articulated in a Xerox PARC document (Kay, 1972; see also Kay, 1977) describing what was termed the 'Dynabook', generally considered to be the original concept from which the laptop and tablet PC developed. By the end of the 1980s[2] the laptop[3] was a reality – and computing was portable and on the move, though its environment had yet to be taught to keep up with it.

And, of course, just around the corner was the World Wide Web, which had its origin in a proposal made by Tim Berners-Lee in 1989 at CERN,[4] the European particle physics laboratory near Geneva. The original server was developed in the early 1990s and, as we have seen in Section 7.3 (under JIPS), by 1993 it was rapidly spreading worldwide – effectively sealing the fate of any protocol stack but TCP/IP and ensuring the continued exponential growth of the Internet. However, there were a number of other developments which occurred during the 1980s and early 1990s which set the scene for JANET's development in the 1990s and beyond.

Telecommunications

So far, in the 1980s, we have seen how JANET concentrated on constructing a packet network for data applications. In doing so, all it really required of the telecommunications industry were basic transmission links between its packet switching nodes, initially using modems but gradually making use of

---

2   Kay was right about the timescale in his 1977 Scientific American article (Kay, 1977): the abstract reads 'Rates of progress in microelectronics suggest that in about a decade many people will possess a notebook-size computer with the capacity of a large computer of today.'

3   Here, by 'laptop' is meant a portable general-purpose computer. There had been other highly portable devices a little earlier: for example, in the mid-1980s, there was a DTI trial of a solid-state word processor called a 'Liberator' made by Ferranti, in which Peter Linington as Head of the JNT participated. (Private communication, Peter Linington, April 2010.)

4   *See* Glossary for the origin of the acronym.

the new digital links as these became available. So what was happening in the telecommunications industry and, indeed beyond?

'Digital links' is one clue: telecommunications had already committed to becoming increasingly digital in the 1970s. During the 1980s it steadily deployed digital exchanges and digital transmission throughout the network, and digital services began to be available. The broadcast industry of radio and television, until then almost entirely analogue-based, was now considering exploiting the possibilities of digitisation. Indeed, there was talk of 'convergence': if 'everything was digital' (whatever that might turn out to mean!), could common technology be exploited? Digital processing technology was becoming powerful enough to cope with real-time image compression, making business videoconferencing a reality.[5] And while the possibility of distributing TV via the telephone network was still a long way off, the question of how to develop the telecommunications infrastructure so that it could carry a rich variety of services was beginning to be addressed. A curiosity at the beginning of the 1980s was that the importance of packet data was not fully appreciated in the telecommunications industry: early ISDN developments were all circuit-oriented. Indeed, whether broadband ISDN should be circuit- or packet-oriented was initially hotly debated.

On the other hand, by the end of the 1980s a great deal had changed. Digital compression technology had advanced considerably; optical fibre links had now permeated the telecommunications network sufficiently to support business applications at 2Mbps and beyond, though the cost was still very high. Moreover the importance of providing packet data services was now beginning to be appreciated and such services were now beginning to be offered. Broadband ISDN, the means by which the telecommunications industry sought to unify the transmission and switching of data and real-time (continuous media) services, had chosen to use packet-oriented technology and achieved initial standardisation. In other words, technological developments had reached the stage where integration of audio, video and data was looking possible, both within end-user terminals and in communication systems, though considerable engineering would still be needed to develop deployable services. What was missing was a convincing business case for a service with sufficient revenue to pay for the considerable investment still needed.

The situation was a classic chicken-and-egg one. The telecommunications industry would not fund the necessary development and deployment without such a case. Videoconferencing systems were available (often via satellite because of cost-distance factors) but they were almost exclusively the domain of commerce because of the costs. Although the necessary R&D to support domestic videophone services as an adjunct to the circuit-switched telephone service had been done by the early 1980s (including the development of prototype small flat-panel displays), such a service was seen as something of an expensive gimmick and never saw widespread deployment. Moreover, since packet data networking was not predicted to outstrip telephony and had hardly penetrated the domestic market, suggestions that integrated services might be made to ride on packet data services were barely credible. And video services themselves were still regarded as very much a niche market. Essentially, deploying a new service or a new infrastructure separately might pose an acceptable business risk but to do the two together, each effectively dependent on the other, was too big a step into the unknown.

---

5   Particularly using VSAT (see Glossary) satellite systems.

In the event it was the Web-enabled Internet, and its rapid take-up in both the business and domestic markets, which, during the 1990s, provided the demand for much greater capacity in the transmission infrastructure. This would not only contribute to altering the tariff structures but also triggered the appearance of a new industry of fibre infrastructure providers during the 1990s – not early enough for SuperJANET at its start but exploited fully by its end.

Deregulation continued: 'duopoly' review

Within the UK, another event occurred around this time that would also be influential in shaping the future of UK telecommunications. We have seen how, shortly before JANET began, the UK telephone service had first been privatised, then its monopoly position altered. This effectively created a 'duopoly' in 1983 consisting of BT and Mercury Telecommunications (latterly part of Cable & Wireless), which was to be reviewed after seven years. In 1990 the Government initiated this duopoly review (DTI, 1990), and in 1991 ended the duopoly. The result was to enable other providers in both local and national contexts, and subsequently JANET would take advantage of both.

The more far-reaching consequences of deregulation and the appearance of the fibre infrastructure provisioning industry was the beginning of the release of communications provision from the hitherto conventional model of traditional telecommunications. By the early part of the 21st century, ownership by enterprises of very low-level underpinning of enterprise communication, be it physical fibre or collections of wavelengths, would become familiar; and we shall see that, in effect, JANET attempted to do this very early, right at the beginning of the 1990s.

New services pull?

By the early 1990s, the effect of the deployment in the wide area of fibre and digital technology was making itself felt beyond the telecommunications industry. Fibre was by now commonly deployed in campuses; and in the wide area, although the cost on a tariffed basis was prohibitive, 10 to several 100Mbps was now technically feasible. In the public network the business model for communications services had not yet begun to evolve away from prices based on the telephone, so that the price of a transmission line was still comparable, on a capacity basis, with the price of an equivalent number of telephone lines, not reflecting the substantial reduction in the underlying cost of the transmission capacity and with negligible reduction for quantity. In the academic sector, the rapid growth in computing power now available on most desktops was now poised to enable a new generation of application possibilities. Some of these were already capable of demonstration in the local area and all that was needed to develop widespread deployment was a way to open up capacity in the wide area.

Several trends can be identified. The handling of data and large scientific calculations were both showing signs of becoming increasingly interactive and in some cases collaborative. Study and synthesis of molecules needed visualisation of the modelling phase and could with advantage be performed collaboratively in real-time with live discussion. Data visualisation usually required remote access to large amounts of data; performed interactively, it required high-capacity communication of data, whether application-level or ready rendered for display. Steering of scientific calculations, usually remotely, increasingly needed visualisation, with the concomitant requirement for high-performance communication.

Remote participation in scientific experiments, either by access to samples of the data measurements in progress or perhaps through remote access to instrumentation, offered another exciting possibility, particularly as it would enable closer participation of experts who had many other calls upon their time. In healthcare also there were inviting possibilities: remote viewing, consultation and assistance in diagnosis and operations showed promise of being realised. Real-time, visual collaboration was on the horizon in many contexts. And underpinning all this, in every discipline, was the possibility of ubiquitous videoconferencing technology as an aid to teaching and collaboration at all levels, independent of location. Individual special provision had occurred previously to share teaching across university-sector as well as other community sites or institutions, such as Livenet in the University of London in the late 1980s and the 2Mbps video network connecting the campuses of the University of Ulster at the beginning of the 1990s; however, these were relatively isolated examples at the time. To enable ubiquitous deployment throughout education needed a higher performance network of substantially higher capacity.

Now, with the promise of increased network capacity, all this became a real possibility for service. This not only offered the vision of potentially worldwide opportunities for real-time audiovisual collaboration, with the options of it being either in a studio-like environment or to the personal desktop, but the digital format meant it could be stored, edited, replayed and made generally accessible on computer workstations by virtue of being integrated into an ubiquitous computer communications environment.

Technology drivers

The foregoing has attempted to sketch some of the possibilities for network development which would, over nearly two decades, evolve from experimental demonstrations into facilities increasingly taken for granted. The degree to which technical innovation is driven by identified requirements or research curiosity is much debated. From roughly the end of the 1980s, several technical developments began to move into exploitation after a research phase going back, in some cases, well over a decade.

Mention has already been made of the deployment of optical fibre for communication. Originating from work in the 1960s and 1970s, it saw worldwide service deployment in telephone trunk networks from the beginning of the 1980s and the first transatlantic deployment in 1988.[6] These applications used expensive single-mode[7] fibre and laser diodes for the light source; in the same period, much

---

6   TAT-8: *see* Glossary.

7   The inner glass-fibre core is approximately 10μm in diameter (or less) and the glass has to be very pure: the combination is expensive. Multimode fibre typically has a core diameter of 50–62.5μm and need not be so pure because modal distortion prevents use over long distances anyway. *See* Glossary for explanation of single-mode and multimode fibre transmission.

cheaper systems using multimode fibre and LED[8] light sources were deployed for use in the local area and saw rapid take-up for use across campus and in linking campuses dispersed over an abbreviated urban area. However it would take the next developments, enabling the use of multiple wavelengths in a single fibre, before the much publicised great capacity of fibre would show signs of being realised.

The digitisation and encoding of audio, particularly speech, has its origins in the late 1960s, and experiments in transmitting packetised audio are almost as early as packet networking itself. Experiments over the ARPANET began in the early 1970s, particularly at MIT Lincoln Laboratories.[9] As we have already seen (Chapter 6), the appreciation that for audio purposes the occasional corrupt bit is not detectable to the ear, whereas the time taken to obtain redelivery of a packet with even a single error can destroy the sound, became one of the motivations for the separation of packet routing and reliable delivery into the familiar IP and TCP, respectively. The next question was to explore sending moving pictures. Here, simple sampling of a video signal immediately produces bit rates in the order of 10s to 100s of megabits per second, even for relatively small or low-resolution pictures, 100 to 1000 times more than simple encoding of sound. Much more powerful processors (whether special-purpose digital processors or general-purpose computers) were needed to handle the sheer rate and volume; likewise, much greater capacity was needed in the network – unless the signal could be compressed in some way. By the beginning of the 1980s, substantial study of compression for both still and moving images was in progress. Communication was one motivation and the ability to store a movie, for example, on a disk was another. The two requirements are not the same. For storage, compression does not need to be done in real time whereas it is an advantage if decompression for playing only requires relatively simple processing so that players can be made cheaply. For interactive, conversational use over a communication channel, both have to be done in real time. As we shall see, during the 1980s advances in both compression and the capacity of networks were such that by the start of the 1990s 'network video' was becoming a practical possibility.

Of course, once packets containing continuous media like audio and video began to appear on a network, mixed up with data packets for traditional applications not sensitive to timing delays, the next problem was how to control and schedule the different sorts of packet traffic for transmission. Networks carrying more than one type of traffic are generally referred to as either multiservice networks or integrated service networks, depending upon whether the emphasis is on the multiple services being offered or the fact that they are all carried on the one network. Although there had been experiments in the 1970s, the more widespread study of the problem began in the 1980s. The circuit-oriented telecommunications industry approached it from the perspective of how to generalise the telephone network so that it could properly carry data, while the packet-oriented computer data network studied how to introduce different treatment for packets belonging to different application streams so that timely delivery could be provided for those needing it. As it turned out, both parties would end up taking a packet-oriented approach and the debate then moved to how much knowledge the network should have about packet flows, if indeed any. This all gave rise to a variety of developments: Asynchronous Transfer Mode

---

8   Light-emitting diode.

9   See, for example, Gray (2007) and Weinstein and Forgie (1983).

(ATM) in the 1980s and 1990s; the integrated services architecture for the Internet in the first part of the 1990s; some long-since vanished experimental protocols for the Internet; and the study of quality of service (QoS) in packet networks.

One of the most significant developments of the 1990s was the development of switched Ethernet, which was mentioned briefly at the end of Chapter 5. This development would sweep away all other cabled network technologies on a campus during the 1990s – indeed, in the following decade it would begin to threaten traditional transmission in the wide area too. However, no sooner had shared media, with all its scaling problems, disappeared in one guise than it reappeared in the form of the wireless LAN (WLAN) – with all the old problems back plus a few more. By mid-1990, laptops were common and people began to carry them about wherever they went. Battery technology improved sufficiently during the 1990s such that they could be used for several hours without a power cable. Of course, by now everybody was used to being connected to the Internet. If really necessary, a cable might be found, an Ethernet socket plugged into and, (usually) with the aid of the local network administration, a network connection made even when visiting.

Had true ubiquitous networking arrived? Well, not quite: there were just a couple of problems. The first was security: it was bad enough people coming along with disks wanting to insert them into other people's machines, but at least you could get software to scan the disk for known malicious software (sometimes known as 'malware'). But putting a visiting machine on your network was another matter altogether: who knew what it might do, intentionally or otherwise? We shall return to this aspect in Chapter 13; for now, it is enough to note that the issue has nothing to do with wireless *per se*. It is simply the ability to go visiting with one's computer and connecting to a host network which is not the 'home' institution: wireless merely made this much easier and more convenient, encouraging its adoption, and the presence of a wireless link then also increased the risk of unauthorised connection. The other little problem was that typically there was a collection of services provided by the home institution, at least some of which the peripatetic user would like to be able to access while away 'roaming'. Providing this sort of facility in all its generality remains an area of research. Depending upon which aspects of the overall field one focuses upon and how restricted a solution is sufficient, the terms 'mobile', 'location independent', 'roaming', 'ubiquitous' and 'pervasive' are used: here, the term 'roaming' is adopted since it has recently been adopted both for the corresponding JANET service and more generally. While it still required a cable to attach to a 'foreign' host network, it never became particularly prevalent. However, as soon as wireless arrived, all that changed. What could almost be considered a bonus was that the intrinsic vulnerability of a wireless network finally caused the authentication, authorisation and security issues associated with foreign computers attaching to a network to be taken seriously!

All of these developments have had major influences on the way in which JANET has developed since 1990. In the next section, some of the key technologies are explored in a little more detail in order to provide an introduction to the technical evolution of JANET over the next 15 years or so, which we shall look at in subsequent chapters, after a look at the major administrative, funding and management reorganisations which happened in the early 1990s, not only as a result of the major expansion of the community but also partially in consequence of governmental devolution in the UK.

8.2 Technological developments

From the beginning of the 1990s onwards, the technological focus of JANET would begin to change. Hitherto, development had been concentrated primarily on packet networking and application layers. Of course, LANs had a crucial role in the development of campus networks but the LAN technology, like the transmission technology in the wide area, was, from the mid-1980s, largely standard. Although JANET would no longer develop its own technology as in the past, it would now begin to seek ways of providing a more general form of service than simply a packet-switched service. While that packet-mode service would remain of paramount importance, strategic considerations would begin to influence JANET more and more toward exploitation of new technologies at a much lower level than in the past. This section provides a brief outline of the technologies which would now influence JANET's course of development for two decades.

Fibre

During the mid-1980s, as fibre continued to be deployed, there was an increasing need to raise the transmission rates which could be supported by telecommunications networks, particularly in the trunk network. In a trunk link, individual channels or circuits are multiplexed together into a higher-rate circuit by taking one or a group of bits from each channel in turn and transmitting them on the higher-rate circuit. The cyclic pattern of groups of bits (slots) generated on the higher-rate link forms a series of frames and the process is referred to as time division multiplexing (TDM). Demultiplexing of an individual channel is achieved by extracting and concatenating the contents of the same slot in every time-frame. In order to provide the transmission capacity needed in trunk links, a hierarchy of transmission rates (known as the plesiochronous digital hierarchy, PDH, *see* Appendix G) had been developed, with each rate being defined to enable it to carry a multiplex of a number of circuits of the rate immediately lower in the hierarchy. Apart from the need to extend the hierarchy to higher transmission rates, PDH had several shortcomings. Because links were not quite synchronised with each other, extracting a single channel from a high-rate multiplex required the whole hierarchical multiplex to be demultiplexed, which was expensive; in addition, the multiplex structure provided insufficient facilities for operation and management; and the multiplex framing structure was a fixed one, with no flexibility in how lower rate sub-multiplexes were accommodated. To overcome all these issues, during 1985 to 1987, a new hierarchy (known internationally as the synchronous digital hierarchy, SDH, *see* Appendix G) was defined.

The information-carrying capacity of optical transmission is much greater than copper cable-based transmission, by virtue of the higher carrier frequency. The fastest rate defined in SDH was originally 2.5Gbps, about five times the top rate for PDH; currently, the highest rate for SDH is 160Gbps. Although this 100- to 1000-fold increase in two decades was substantial, it was only one way of exploiting the transmission capacity of fibre. Implicitly, the foregoing relates to transmission of a single (digital) signal along a fibre using a single carrier. In the radio part of the electromagnetic spectrum the transmission of many signals simultaneously, using different frequencies or wavelengths, has been familiar for more than a century. One can do the same thing at optical wavelengths: a process generally called wave division multiplexing (WDM). And, just as the encoding rate for a single channel has steadily increased with time, so has the number of WDM channels on a fibre, with over a hundred currently possible.

The other essential component of any network is switching. Switching in a TDM system can be thought of as redistributing slots among time frames: a form of switching known as time slot interchange. In SDH it is technically possible to switch a single channel or a particular sub-multiplex channel between any multiplex on any fibre; however, for many purposes substantially cheaper switches with more limited capabilities are sufficient. SDH switching equipment began to become available during the early 1990s, just as the original SuperJANET was being deployed. WDM switching is performed by converting the digital signal to the electronic domain, performing the switching and regenerating the signal, if necessary on a different frequency.

As with any transmission system, there is noise and signal attenuation to contend with and, after some distance, the signal needs amplification or complete regeneration, as just described for WDM switching. Optical wideband amplification became possible in the late 1980s,[10] reducing the need for signal regeneration; however, as in any amplifier, additional noise is introduced, as well as the existing noise being amplified along with the signal, and eventually regeneration is needed. Depending on the lengths of links in the network, the regeneration present in WDM switches may be sufficient to eliminate the need for at least some of the regenerators in links. However, it should be noted that regeneration requires knowledge of the digital signal transmission format, SDH or (non-SDH) Ethernet, for example, so that the format generally has to be chosen and fixed for a network. During the 1990s, improvements in fibre and transmission techniques resulted in deployable WDM systems at affordable costs.

Compression

The process of digitising a continuous media signal, such as audio or video, consists of measuring or sampling the analogue signal over time and then encoding the measurements in digital form, which implies that only a finite range of values is available for representing the samples. Generally, the more frequently the samples are taken and the greater their accuracy, the closer the reconstructed signal will be to the original and the higher its fidelity will appear to the viewer or listener – and the greater the volume of data that will be accumulated for recording, transmission, reception or playback. Not only does the greater volume imply a higher rate of transmission but also a higher rate of processing during encoding or decoding of the signal.

For audio there is a single parameter to sample: the amplitude of the waveform. For a moving picture, each sample in time is a whole picture, which is itself a two-dimensional continuous array of colours. As in TV, any colour at any point in the picture may be constructed from three parameters, such as the amplitudes of red, green and blue. More usually, three linear combinations of these are used: a measure of overall amplitude (called luminance) and two other colour combinations (chrominance), because less resolution is needed to record the two chrominance combinations acceptably than either the luminance

---

10 In particular, the development of the erbium-doped fibre amplifier (EDFA) by groups at the University of Southampton and AT&T Bell Laboratories.

or the red, green and blue amplitudes. And, since the picture is continuous in two space dimensions, it must be divided up into a finite number of picture elements (pixels), at each of which, in each picture (frame), the three colour parameters are recorded to arrive at the complete digitisation of the moving picture.

For telephony, speech is sampled every 125µs, that is, at a frequency of 8kHz;[11] using 8 bits to record each sample results in a signal bit rate of 64Kbps. If a much better quality of sound is required, as for example for music, the samples may be recorded more accurately and more often: stereo CD quality is typically obtained by using 14 bits for each sample, taken at a frequency of about 45kHz, which results in a bit rate of about 1.3Mbps. On the other hand, if transmission (or recording) capacity is at a premium, it is desirable to compress the signal, losing fidelity, to obtain a lower bit rate: a common example of the latter being the compression used for mobile phones.

The same general remarks apply to still and moving pictures, with the qualification that, in both cases, the uncompressed representations are typically too voluminous to be practical for most still or motion picture applications: thus compression is needed, for example, for DVD recordings, digital broadcast TV and videoconferencing applications. (For a brief outline of how pictures are compressed, see the panel 'Basic principles of picture compression'.) In the context of video compression, the publication in 1990 of H.261 by the CCITT marked a turning point. There had been other previous compression schemes, in particular

---

11   Which is the length of the multiplexing time frame in SDH, so that, in principle, one sample from every channel is present in the time frame of a multiple at any level of the hierarchy.

Basic principles of picture compression

Compression is typically based on a combination of there being both redundancy present in the signal and a choice of ways to represent the signal, some of which are more advantageous when choosing how to quantify the image information. For speech, the most obvious example of redundancy is in the periods of silence which occur: any samples taken in such periods will all have the same value. Likewise, a picture of a (uniformly) blue sky would have the same values of all three colour parameters associated with every pixel, and a digital video of a stationary scene would have the same values for the same pixel in every frame. These examples contain some hints about how to go about compressing a picture. For the blue sky still image, it might be desirable if successive pixels in the picture which all have the same colour values were associated with a single instance of those values, together with a count. Similarly, if successive frames of a movie differ little, recording only the differences could be beneficial; furthermore, if there has been little change for several frames, the likelihood is that there will be little change for a few subsequent frames – indeed, by observing which parts of the picture seem to be changing, and in what way, predictions may be made about pixel sample values in succeeding frames. All of these observations have been exploited in various ways for video compression. However, there is one other observation which contributes to all the major forms of video compression in current use: in natural scenes, the variation in sample values for pixels in the same neighbourhood tends to be relatively modest. Under these circumstances it is productive to represent the values across the two dimensional picture in terms of a family of spatial waveforms of increasing (spatial) frequency. Mathematically,

H.120, published in 1984, but H.261 was the first to see widespread and rapid take-up in videoconferencing service products. It incorporated all the techniques alluded to in the panel (and more), including use of the DCT. However, it was intended for use over synchronous circuits, such as ISDN, and required accurate synchronisation, which also meant that its performance over a packet network with, for example, substantial variation in delay could be (very) badly affected. Nevertheless, in the early 1990s, software H.261 codecs (coder-decoders) became available for use on general-purpose workstations and were incorporated into the desktop video tool, Vic, part of the LBL/UCL/MICE videoconferencing suite of tools introduced in Chapter 6.

For real-time, videoconferencing-style applications it is important that neither compression nor decompression should take too long. By contrast, for compression of a still picture, which may generally be compressed once, stored and viewed many times, substantial processing may be invested in the compression provided that decompression requires relatively little processing, enabling rapid viewing using either cheap devices or software not requiring high-performance computing. In 1986 the Joint Photographic Experts Group (JPEG) began work on compression of still images and published the original standard in 1992, which became an ISO standard in 1994. In more recent times JPEG has become a joint group of ITU-T and ISO. A number of improvements and derivative standards have appeared including, as a result of further research, additional lossless compression of existing compressed images. Although intended for still images, by projecting a series of JPEG images a simple form of moving image encoding can be achieved

this just amounts to representing the image in a different way; however, the foregoing observation about modest variation between nearby spatial parts of the picture translates to the major components of the picture coming from the lower frequencies in the transformed representation: so much so that it is possible, for many purposes, to ignore the contributions from the higher frequencies above a cut-off, the value of the cut-off depending on the nature and purpose of the picture and the quality required. This turns out to be a productive approach: a typical family of functions with increasing spatial frequency chosen for this purpose is based on a set of sine[1] waves, the specific form being known as the 'discrete cosine transform' (DCT), which is used in ISO JPEG and MPEG, as well as in the ITU H-family.

It will be observed that a picture with a lot of detail, music with a large dynamic range, or rapidly changing movie scenes are all likely to be less susceptible of compression. Not only is this generally true but it also implies that compressed forms of audio and video will produce bit streams with considerable variations in bit rate. It is by varying digitisation and compression parameters in such a way as to trade off quality against bit rate that it is possible to constrain a compressed continuous media signal to have a constant rate. Such video encoding is often referred to as a 'constant bit rate' or CBR encoding, or just CBR video. The other feature of all compressed formats is that they are more susceptible to error than uncompressed formats: the very elimination of redundancy which contributes to the reduction in bits in the signal representation also means that each bit is more crucial.

---

1   Technically, cosine – but the shape is the same.

which is of quite good quality at the expense of substantial bit rates. This is the simplest form of Motion JPEG (MJPEG), which was used over parts of JANET for teaching, particularly Scotland,[12] at the time of SuperJANET III.

If the purpose of video compression is for either storage or broadcast TV-like applications then the same comments apply as for JPEG with regard to the balance of processing for compression and decompression. The Motion Picture Experts Group, MPEG, was formed in 1988 under the auspices of ISO to address this area. In the next few years, building on knowledge acquired in both H.261 and JPEG work, a large family of standards was developed which included the DVD[13] formats used in domestic recordings and the MPEG-2 format used in digital TV broadcasting.

H.261 was succeeded by H.263 in 1996,[14] defined by the Video Coding Experts Group (VCEG) of the ITU-T. The latter incorporated substantial enhancements, many as a result of what had been learned in the H.261, MPEG-1 and MPEG-2 work. In particular, H.263 included much improved error recovery in the presence of data loss, as a result of which it had much better performance over packet networks and was soon adopted for use over the Internet in both streaming (TV-style) and videoconferencing applications.

Continuous media transport in a packet network

Commonly, a multiservice network combines carrying computer data traffic with continuous media audio and video traffic.[15] For the user or subscriber, the idea of having a single network to which any device may be attached to access whatever services are required has considerable attraction. From a network operator perspective, to have a single bearer service over which all subscriber services operate is a desirable goal because a major cost in providing telecommunications services is the operation, management and billing, including the capital investment in software to support these. A single unified scheme for delivering all services would have considerable benefit. However, if the underlying basic delivery mechanism is unified, a way has to be found to provide a range of quality of delivery appropriate to the various services: a topic often referred to as quality of service or QoS.

Although initial efforts in this direction in the telecommunications industry explored generalised forms of circuit switching and TDM, in the mid-1980s it was decided that a form of packet transport called ATM would be adopted as the basis of the future broadband ISDN service. Since the packet-switched computer data network industry was also seeking to offer continuous media transport service, both industries were attacking the same problem. In a packet network there is always the possibility in a switch or router of the (more or less) simultaneous arrival on several input links of packets all heading for the same output link: a situation known as congestion and somewhat akin to a number of people all

---

12 Begun in 1996, during the latter part of SuperJANET III, using Cellstack equipment supplied by K-Net.

13 Digital Versatile Disc.

14 Ratification by the ITU was in 1996: the specification was produced in 1995: there have been several versions since.

15 Such a network may also include time-critical data traffic, giving rise to an even greater range of services.

trying simultaneously to exit a room through the same door. If time is of the essence for some packets, scheduling packets for transmission becomes a more complicated matter than just some variation of the traditional 'first come, first served' (FCFS) scheme, in which packets are transmitted in the order in which they arrive.

A packet belonging to a flow sensitive to time ultimately has some sort of time for delivery associated with it. Eventually it must be given priority over others or fail to be delivered in time; however, if the majority of packets are of this sort, then in the presence of congestion few or none may have priority over others – they have all become urgent – and the behaviour may degenerate to 'best effort'. The hint here is that in the statistical world of scheduling in a packet network, not too much of the traffic must be of this time sensitive variety: that is, only a proportion of this type of traffic should be admitted and this should be applied at all points along the path where there is potential for congestion. This in itself implies enough state in the network to impose admission control: that is, to know when to refuse further time-sensitive traffic.

A continuing debate remains regarding the closely related issues of whether resource reservation is necessary in such a network and how much information or state the network should have regarding the traffic flows it is handling. The description above implies that reservation is necessary in order to avoid over-committing the network so that it cannot cope with the time-sensitive traffic when congestion occurs. It has been suggested that if the capacity of the network (its links and routers) is made sufficiently large compared to the anticipated volume of time-sensitive traffic (sometimes referred to as 'over-provisioning') then, in practice, congestion will be sufficiently rare that it can be ignored – and the great advantage of this is that the introduction of considerable complexity into the network will be avoided. This is sometimes known as the 'infinite bandwidth' argument. The counterargument to this is that there will always be links which are not sufficiently provisioned and that, in any case, unanticipated congestion may still occur, malicious or otherwise. Although the foregoing refers to time-sensitive traffic, there is also the situation where a subscriber pays for (or otherwise books) a proportion of the network capacity (the simplest example being a share of a link) and similar issues arise to those described above in ensuring the subscriber receives the purchased allocation. This analysis was articulated in the classic RFC by Braden, Clark and Shenker (1994) in considering an integrated services architecture for the Internet – generally referred to as the IntServ model.

Early work in the late 1970s had experimented with a stream protocol[16] which was updated as ST-II[17] in 1990. Architecturally, it would have been an additional protocol alongside IP; it was connection-oriented and would have introduced both state and resource reservation into the network. Subsequently, following work begun in connection with implementing IntServ, a resource reservation signalling protocol, RSVP (Zhang *et al*, 1993; Braden *et al*, 1997), was proposed for the Internet. In the IntServ model, although resources would be reserved and some state would be introduced into the routers in the Internet, IP remained the network protocol and the reservations would attempt to re-establish, like the

---

16   For example, UCL had a videoconferencing terminal in use on its transatlantic link from 1982 which used ST-II.

17   Sometimes referred to as IPv5 by some sources, from the fact that both the packet format version field and the protocol ID field have the value 5. It is misleading: ST-II is a different protocol and not a version of IP.

network routing, should part of the network break. IntServ and RSVP[18] did not see widespread deployment in service because there were concerns associated with handling all the individual application flows (termed micro-flows) and it was realised that, as a consequence of the Internet having become the world network, the core routers were unlikely to be able to cope with the amount of state involved.

In consequence of both the complexity and doubtful scaling properties of the IntServ model using RSVP, an alternative, more coarse-grained model was developed, referred to as the DiffServ[19] model (Blake *et al*, 1998). In this, packets are classified into a small number of classes, each with an associated behaviour to be implemented by the routers on the path taken by any packet in the aggregate. To enable this, the router by which a packet enters a network (the ingress router) classifies and marks the packet with its class. Subsequent routers observe the packet and ensure the appropriate behaviour. Of course, all those packets belonging to a specific class form a flow, called an aggregate flow. Aggregate flows present the same issues regarding behaviour under congestion conditions as do micro-flows: however, if there is state with respect to resources available at each ingress router, policing of each packet is possible to ensure that it does not exceed the resources allocated for its aggregate. In other words, technically, resource allocation is still necessary and can be either static under administrative control or dynamic, in which case end-to-end signalling is required as in IntServ – though only the ingress routers for each DiffServ network traversed by an aggregate may need knowledge of the state: 'over-provisioning' may be adequate internally for known traffic patterns.

Owing to the statistical nature of packet multiplexing on any link, variations occur in the delay experienced by packets traversing the network. This can be exacerbated by the variable packet size if a small packet is held up by a large packet. For an application requiring regular delivery of data, this delay variation (also known as jitter) must be removed. This is possible provided the packet does not arrive too late; commonly, packets are placed in a buffer and 'played out' to the application as they are needed – which introduces extra delay. Packets may also be time-stamped at origin as an aid to timely delivery and playout.

As will already be clear, attempting to recover corrupted packets of continuous media traffic by requesting re-delivery is essentially useless in a real-time application (though it can be satisfactory for a streaming application where a large, fixed delay may be quite acceptable, even irrelevant). What is needed is a basic datagram transport, which is what the User Datagram Protocol (UDP) provides, by adding to IP a port number to enable sub-addressing or process stream demultiplexing within a host and a checksum to enable errored packets to be detected in case this is important. For the purpose of real-time, continuous media traffic the IETF also defined a Real Time Protocol, RTP, which operates over UDP and includes sequence numbering and time-stamping.

Beginning in 1996, video over ATM was established in Scotland for teaching; subsequently, IP-based video was established in Wales as the foundation of the Welsh video network. And, as we shall see, during 2001 to 2005, JANET undertook a substantial trial and assessment of QoS technology with a view to enabling IP-

---

18 Though RSVP has seen some deployment as a signalling protocol in other contexts, such as MPLS: *see* Chapter 10 and Glossary.

19 A contraction of 'Differentiated Services', referring to a (small) range of service categories differentiated from each other by the differing treatment accorded each by the network.

based video to be adopted more widely for education, particularly in the schools sector.

ATM

Asynchronous transfer mode, or ATM, is the name given to the technology standardised by the CCITT to implement broadband ISDN in the late 1980s. The overall objective was grossly similar to that for the integrated services Internet, the achievement of a multiservice network. Although, it was a packet-based technology, like the Internet, by contrast the packets were small (all the same size, 53 bytes, including header) and the network was connection oriented (in the latter respect, like X.25). Conversational speech is particularly sensitive to overall delay. The purpose of having the small packets, called cells, was to reduce the delay incurred in waiting for the cell to fill up with speech samples (packetisation delay). Having them of uniform small size helped to reduce the jitter during multiplexing and switching, because the problem of small packets being held up by large packets could not occur.

Although ATM was a potentially promising multiplexing technique, because of the small cell size, packets required segmentation and reassembly (SAR) for transport; the processing for this, particularly the simultaneous reassembly of several packets in parallel, was a disadvantage. SAR formed one component of what was termed the ATM adaptation layer (AAL), the purpose of which was to enhance or adapt the basic ATM cell transport for packet data or continuous media transport. Two of the AALs were intended for real-time continuous media, for constant and variable bit rates. Two more of the original AALs were intended for packet-mode transport, one for reliable stream and the other for datagram transfer. Subsequently a further, simpler AAL, AAL5, was defined for datagram transport, which became the one almost universally adopted for the purpose, including ATM signalling and IP transport. Broadband ISDN signalling, like that for ISDN, took place on a separate (virtual) channel and used the same signalling protocol[20] as would subsequently be adopted by H.323. For a very short period in the early 1990s, LAN emulation using ATM saw some deployment, effectively creating a switched LAN. Ultimately the need for very high rates of cell switching would contribute to the demise of ATM. However, ATM was exploited during the first half of SuperJANET's evolution.

Session control

It will be appreciated that for any multimedia session, like a videoconference, there may be a number of media streams contributing to the session. In the commonest case of a 'talking picture', some encodings intermix the transmission of audio and video within a single transmission channel; even where this is so, in any videoconference, for example, there will still be several of these composite audio-video streams. In any session composed of more than one stream, some control – generally known as 'session control' – is needed to organise the streams. In the same year, that H.263 was first published, 1996, the ITU-T also published the first version of H.323, an umbrella standard aimed at multimedia communication

---

20   CCITT / ITU-T standard Q.931.

over packet networks. Standards within the H.323 umbrella include IETF standards, notably RTP. Session control signalling is based on that defined for ATM[21] and includes specifications relating to use of the other protocols within the umbrellas. Applications of H.323 include videoconferencing, voice over IP (VoIP) and packet-mode telephony (including distributed PABX functionality). There has been widespread adoption of H.323 standards, particularly since around 2000 when a revised version of H.323 (and a number of its constituent standards) was approved (ITU, 2000). An alternative approach to session control is taken by the IETF Session Initiation Protocol,[22] SIP, which is capable of operation over several transport services and provides a general framework for unicast or multicast multimedia session initiation, modification and termination. Although many systems in current deployment are based on use of H.323, at the time of writing there is increasing deployment of SIP-based systems.

SMDS

As has already been noted, during the 1980s the telecommunications industry realised that data networking had not only become important but that there was demand for LAN interconnection over the wide area. Moreover, because LANs had enabled distributed systems applications and sharing of substantial volumes of data within an enterprise or collaboration, there was now demand for wide-area data services at speeds substantially higher than any then on offer, from 1–2Mbps to at least an order of magnitude faster. At the time, the MAN was being standardised by IEEE in its 802.6 committee and ATM was being standardised by CCITT for broadband ISDN. The 802.6 MAN was targeted at LAN interconnection (and, more general multiservice networking) in the urban area, and the IEEE 802.6 committee had been at some pains to track broadband ISDN and harmonise some aspects of its work with ATM, most notably the size and several other aspects of the DQDB slot structure with the ATM cell format. In 1989–90, Bellcore published the definition of its Switched Multi-megabit Data Service, SMDS, in a series of 'technical advisory' documents.[23] The SMDS specification was for a switched packet service access. It was based on minimal use of the QA format of DQDB used on a point-to-point basis, that is, with just two stations, one being the subscriber and the other the SMDS service provider. Among other things, this directly exploited that DQDB had been defined to operate over standard telecommunications transmission formats, allowing the possibility of several of these to be exploited. The definition also offered potential for several developments in the future. The first was a simple enhancement to enable more than one station within the subscriber premises to share access to SMDS using the same link, which could be done by enabling a small amount more of the DQDB MAC. Two of the other possibilities were of strategic value to an operator. Being a subset of the DQDB mechanism meant that the service would have a natural development path to incorporate 802.6 MAN service, should there be demand for this. Furthermore, because the DQDB slot format was compatible with the ATM cell, SMDS could be delivered straightforwardly over broadband ISDN, effectively exploiting ATM switches as SMDS switches. As with ATM, the

---

21   *Ibid.*

22   IETF RFC 3261; it was accepted for use in cellphone systems in 2000.

23   See, for example, Bellcore (1989) and Bellcore (1990).

small slot size of DQDB meant that SAR was necessary for packet transport; here too, SMDS hoped to gain potential leverage by adopting the same mechanism as DQDB, which was already essentially identical with one of those defined for ATM.[24]

Within the UK, SMDS was offered by BT over 34Mbps links.[25] However, a feature of SMDS was that by using a credit-based flow control system, a subscriber could be offered a range of packet access speeds (termed classes): 4, 10, 16 or 25Mbps.[26] In order that subscribers using lower-rate access classes could receive from higher rate subscribers without causing unnecessary congestion at the subscriber access point, all classes of subscriber could receive at the full rate for the line (which would have otherwise been under-utilised but with no revenue potential for the operator). We shall encounter this service as part of the initial phase of SuperJANET.

Switched Ethernet

Switched, fast Ethernet was briefly mentioned at the end of Chapter 5 as having arrived in 1993. As usual, it was the culmination of several developments. Traditional shared Ethernet had been based on the use of coaxial cable together with repeaters to connect segments together. Towards the end of the 1980s the multi-port repeater was developed, which was the key to introducing alternative forms of cable: fibre and twisted pair, like telephone cable. Being able to use telephone cable was immediately popular. It also heralded the adoption of the structured wiring paradigm mentioned in Chapter 5. This form of Ethernet was standardised in 1990. The next step was the switch, the multi-port bridge previously mentioned. Exploiting the processing speeds now achievable, together with several new transmission encoding formats, enabled the first 100Mbps, fast Ethernet products to ship in 1993. Several twisted-pair formats, as well as a fibre version (which, together with one of the copper cable versions, exploited the transmission formats developed for FDDI) all rapidly became available; 100Mbps switched Ethernet was standardised in 1995. People could use twisted-pair cable previously installed for 10Mbps Ethernet. Even more important, this was evidently a scalable solution and switches rapidly came down in price; interfaces which were, even initially, much cheaper than FDDI rapidly became cheap and could operate at either 10 or 100Mbps. ATM had seen some deployment in the local area but, although it was a switched solution, it was expensive by comparison and it did not operate any faster.[27]

In effect, neither FDDI nor ATM made any further headway generally in the local area network market. This was further reinforced by the development of 'gigabit' (1Gbps) Ethernet, standardised in 1998 and in production by 1999. Furthermore, as soon as the multiport repeater had made its appearance with twisted-pair and fibre media, it had been realised that there was essentially no impediment to removing the half-duplex restriction from Ethernet. Since there

---

24 AAL4.

25 A lower-rate service over 2Mbps links was also offered later.

26 25Mbps was the highest possible over a 34Mbps line because of the combined overheads imposed by the slot and segmentation formats.

27 Initially ATM in the local area used FDDI transmission format, i.e., 100Mbps effective rate. Local versions of the 155 SDH rate became available later but saw limited use in the local area.

could now be no collisions on a segment used simply to connect a station to a switch, this also removed the traditional Ethernet distance limitation imposed by the need to detect collisions reliably. The way was now open for Ethernet operation in the wide area – and it was not long in coming. The initial wide-area versions of gigabit Ethernet were available in 2000 and this was quickly followed by 10Gbps Ethernet, the initial version of which was standardised in 2002.

The rapid ascendancy gained by Ethernet can be gauged by its universal adoption in end-user personal computers: by the mid-1990s, Ethernet was standard on laptops and by 2000, laptops were appearing with interfaces which could operate at up to a gigabit per second.

Gigabit routers

A feature of network evolution over the years has been the regular appearance of forms of switching which achieve higher performance than rivals by 'stripped-down' functionality. In the early 1980s bridges could out-perform routers – which for a period led to wide-area bridged networks until routers began to catch up (and the chaos induced by two levels of routing, each ignorant of the other, was recognised and could be removed). Although not described here because it saw no exploitation in JANET, in the late 1980s frame relay, which was like X.25 without either error recovery or flow control, offered a temporary upgrade path for X.25 networks. The relative simplicity of cell switching, once a connection was set up, enabled ATM switches to compete successfully with routers for a brief period in the early 1990s. However, routers were able to exploit gigabit Ethernet the moment it was available – and the combined effect of having to switch cells around an order of magnitude faster than routing IP packets, coupled with the SAR overhead for hosts, effectively put an end to ATM for use in high-performance data networks.[28]

The experience gained in switch design from both ATM and Ethernet would now be combined in the multilayer switch which could act as a combined router and Ethernet switch.

Wireless

The use of wireless data communication is almost as old as that using cables: the original packet radio network being the Aloha network developed at the University of Hawaii, which came into operation in 1971 (Abramson, 1973). However, the wireless LAN has its origins in work in the early 1980s on use of spread-spectrum techniques. This has its origin in a desire to use radio bands which do not require the user to obtain a licence – which means there are other users of the bands adding to the general noise. The general idea is to arrange for the signal to be 'smeared' over a substantial fraction of the available band in such a way that, even if a part of the signal gets drowned in noise, enough will be received for the data to be reconstructed. By the mid-1980s, work was under way to exploit this for a wireless alternative to the shared-cable LANs just then enjoying such success.

---

28 However, its use in subscriber DSL networks continued; it was also to be found internally in parts of some telephony and special purpose networks.

As with the mobile phone, the frequency bands available are severely limited compared with the population of users, so a cellular approach to spatial re-use of frequencies is needed. By the beginning of the 1990s, wireless LAN standardisation was under way in a new committee of IEEE, 802.11. Initial products appeared in the 2.4GHz band in the mid-1990s, followed shortly by those in the 5GHz band, all with typical outdoor range of something less than 100 metres or so – less indoors. By the end of the 1990s WLAN products were in general deployment, built into laptops by about 2003 and in common use domestically by around 2005. 'WiFi'[29] was here to stay.

In deployment, wireless LANs may be used in a variety of configurations, the commonest in effect being to provide the last few metres of access for a device. In other than a domestic situation, because of the short range a number of WLAN access points (APs) will be deployed to cover the volume of space required, each AP being connected to the building or campus network. Because there are only 12 or 13 frequency bands available, each of quite limited capacity, providing adequate coverage and capacity over a campus or at a busy conference can present problems. Since around 2000, another range of radio technologies commonly referred to as WiMAX[30] has been in development and initial deployment, and is also the subject of standardisation by IEEE in its 802.16 committee. At the time of writing it is too early to say to what extent this may help with the capacity and coverage issues.

The enormous convenience of the 'cordless laptop' has ensured almost universal adoption for personal use, in spite of a considerable list of pitfalls compared to cabled LANs. The list of such pitfalls ranges from natural consequences of the technology to the malicious and is far too long to include here – but a few well-known examples are offered for illustration. Perhaps the simplest hazard is the radio noise produced by the microwave oven, to be found tucked away in various corners of both offices and homes. The power used by such devices is substantial compared with WiFi and enough escapes the shielding to interfere with WLAN operation in the same kitchen area: because the radiation is comprehensively 'smeared' over quite a wide band of frequencies, it completely defeats the spread spectrum technology and renders WiFi inoperative.

It is to be expected of any broadcast-radio-based technology that anyone can join in. With general experience of the relatively short effective range of standard equipment, what is perhaps less obvious is the substantial range which can be achieved quite simply by the inquisitive – not to say malicious – user. The small, inefficient antenna used in, say, a laptop implies that the power used needs to be high enough to compensate. Use of a very simple but much more efficient antenna, easily constructed from such items as foil-lined cylindrical snack tins (especially if of industrial catering size!) can achieve ranges at least a couple of orders of magnitude greater than typical use. So, the hacker[31] can sit comfortably on a hill in the countryside around 7 miles off 'tuning in' to a remote WLAN.

Within a building the effects of parts of the building on signal propagation can sometimes be disastrous. Commonly, signals may take many paths because of reflections; since the paths will be of different lengths, the receiver will perceive a distorted signal which, even if still capable of being decoded, is likely to cause the equipment to reduce its coding rate substantially, perhaps from 11Mbps to

---

29 *See* Glossary.

30 *See* Glossary.

31 *See* Glossary.

1Mbps when using the original 802.11b technology.[32] Unexpected, even puzzling effects have also been experienced. One well-known example was unpredictable on-off communication between a computer and an access point in the same room, eventually traced to whether a door was open or closed, the door in question being of a previous generation of safety glass having an embedded steel wire mesh of about 1cm spacing. At 2.5GHz, the radio wavelength is around 10cm, making the door an excellent mirror!

One effect of WiFi has been to stimulate a more serious approach to security. As already mentioned, the fundamental issues are not peculiar to wireless, but are exaggerated by it – the more so since the original security measures built into WLAN were flawed in principle. The basic problem was relatively well known quite early (at least by 2001) and stemmed in part from fixed bit-sequences inevitably being present in the signal, generated by known pieces of the frame format: these could be exploited in a variety of ways to break the encryption in use.[33] Better approaches to this specific problem have since been devised and are gradually seeing deployment. The more general issue we shall return to in the context of external roaming[34] in Chapter 14.

8.3 Effect of community growth

Over the next 20 years JANET was to see an expansion approaching a hundred-fold in its end-user population. As we shall see in the next chapter, this had already begun in the early 1990s and very soon had substantial effects in terms of reorganisation of JANET's administration and funding. However, it also had repercussions on the technical development of JANET. Of course, in this same period the Internet itself was undergoing rapid expansion and it had long since ceased to be a single, more or less homogeneous network. Technically it was based on the same protocols throughout, but there were now many different interests controlling various parts of the whole and the motivations behind those interests now covered an increasingly wide range.

To illustrate the nature of the changes one need hardly look further than when MILNET and ARPANET split in 1984, the latter to become the Internet. This was a natural effect brought about by a divergence of mission between military use and general purpose use by the research community, whether based in universities, government laboratories or commercial institutions. The fundamental effect of there being different interests and stakeholders was that different parts of the Internet would come under different management and this had the effect of dividing it into management domains, known as autonomous system (AS) domains. An AS would generally consist of a collection of subnets of the Internet. At the boundary between AS domains traffic might be subject to authorisation by the managements of the respective domains; the AS management might also impose admission control on traffic to ensure that its network did not carry more than its quota of a particular type of traffic or for a particular customer.

---

32 As of 2009 there are a number of technologies in use, several typically supported by a given access point or laptop. Currently the fastest, 802.11g, supports a transmission rate up to 54Mbps but, of course, this has to be shared among those using the channel.

33 For a review and further comprehensive method of attack, see for example, Bittau *et al.* (2006).

34 A device which disconnects from the Internet and then reconnects via a different network (AS) of the Internet run by a different operator is said to be roaming externally; *see* Glossary.

Within JANET, from 1994 onwards, within the HE sector the network would evolve (as we shall see in Chapter 11) a three-tier AS hierarchical structure composed of the JANET backbone, the regional access networks[35] and individual institutional campus networks. Broadly, this model would either incorporate or be followed by other educational sectors. The evolution of AS domains within JANET generally was related in part to the appearance of devolved regional and campus management. The latter would enable the management and organisation of the regional access network to develop in a way appropriate to the regional constituency. However, the extent to which elaborate AS boundaries were needed was the subject of considerable debate; it would add expense at some stages of JANET's development and sometimes complicate the introduction of new network technology on a JANET-wide basis.

The management autonomy also left unconstrained the regional choice of technical provision of the JANET services then in force. On occasion this non-uniformity in choice of technologies might have unforeseen impact when further network development took place which resulted in new JANET services being deployed. An example of this kind was provided by LeNSE, the Learning Network South East (managed out of Southampton), which at the beginning of SuperJANET4 bought in wide-area connectivity in the form of a virtual network provided using MPLS[36] technology. In effect, LeNSE relinquished control of the IP-layer provision for its network and this subsequently affected JANET-wide deployment of multicast. We shall see in subsequent chapters that there were several technologies for which deployment across multiple AS domains required additional components, not all of which had necessarily been developed: amongst these were multicast, DiffServ and MPLS.

A rather different example is provided by the need to make provision for roaming, alluded to earlier. To enable this, several components are necessary, but one aspect is that, as individuals move from one AS domain to another, the authorisation and authentication process for a visitor becomes split between the home AS and the AS being visited: we shall return to this topic in Chapter 13.

In the next chapter we shall see how the expansion during the 1980s of the JANET community to include the whole of higher education demonstrated just how essential network services had become, not just to higher education but potentially to the whole sector. With further expansion to potentially all of the education sector, it now became clear that some reorganisation was needed – but, as ever, this had ramifications and was influenced by all the other reorganisations in the sector.

---

35 *See* Appendix B.

36 Multiprotocol Label Switching; *see* Chapter 10 and Glossary.

9

ORGANISATIONAL EVOLUTION

In the early days of JANET its organisation was provided by a small core team, the JNT and Network Executive, which, with the aid of a secretary and finance officer, essentially managed everything. Through its success, JANET outgrew the capabilities of so compact and simple an arrangement. The team had been accommodated[1] in the premises of a larger organisation (RAL) whose brief, though related through research, was not rooted in overall support of education and was funded in a different way. By 1990 – even before the Web became ubiquitous – it had become clear that provision of the JANET service now required a dedicated, distinct organisation in order to cope with a rapidly expanding remit and perhaps potentially diverse sources of funding. It was already apparent that the education part of the brief was about to grow beyond the higher or tertiary education represented by polytechnics and universities; but the most significant aspect was that JANET had become – like buildings, laboratories and teaching space – part of the accepted and expected infrastructure upon which education and research had come to depend. The expansion into the whole of the education sector of the provision of the services so far provided only at higher level would be accompanied by growth and broadening of every aspect of JANET's organisation: funding, management, technical development of the network and the creation of new services. Choosing and creating a suitable form of organisation would evidently take some time: in the event, four years from 1990 until 1994.

Of course, the organisation responsible for JANET was not alone in changing during this time. The government departments responsible for funding UK research and education; the committees and councils with responsibilities for specific aspects and sectors of research and education; and some of the institutions which hosted operation and management of the network all experienced reorganisation and change in the same period, and an organisation positioned to benefit from such changes was needed. In the event, perhaps only two aspects remained constant: research funding in higher education remained separated from the main funding for the sector;[2] and the JNT and its successor organisation continued to be situated at roughly the same place in south Oxfordshire – albeit, since 2007, in a new building and with no formal relationship with its original host organisation, RAL.

The expansion of community and remit about to be experienced by JANET as a whole would also be experienced at regional level, provoking organisational

---

1   Originally a few offices; but by the mid-1980s, as JANET came into being and developed, the JNT occupied the whole of R31, a new wing of the Atlas building.

2   This separation is generally referred to as 'binary funding': the Research Councils fund academic research while overall university and further education funding is generally the responsibility of the various funding councils – the separation remains in force as of 2009. (However, Research Councils and funding councils have themselves over the years been funded from the same or different government departments.)

changes here too. In this case the key development would be devolved regional organisation, perceived as an integrated means of coping with growth; enabling participation in provision of regional services compatible with JANET but not within its immediate remit; being eligible to benefit from regional sources of funding; and exploiting regional communications opportunities available as a consequence of the further telecommunications deregulation begun in 1990.

9.1 Time for change

Since its formation in 1984, JANET had been 'run' by the JNT with operation managed by its Network Executive section. For the first ten years the simplicity of this arrangement had been an advantage. However, by the early 1990s, things were changing. JANET had been conceived in the 1970s as a network for universities. Research being an intrinsic part of the higher education sector remit almost automatically implied that any such network needed to encompass Research Council laboratories. This had, of course, been recognised in principle at the outset and very practically when SERCnet was made over to become the skeleton national segment of the infant JANET.

It has already been noted that expansion to encompass other sectors within education was always a possibility. Towards the end of the 1980s, it was generally anticipated: no longer 'if' but 'when'. Here the primary effect on organisation is scale. The connection of the polytechnics had by the early 1990s brought the number of HE institutions connected to JANET, in round terms, to about 200: connection of the FE sector would raise this to at least 700. And if schools were to join, the number of connected sites would be numbered in the tens of thousands rather than hundreds. Of course, scale was not the only aspect of such a development. HE institutions were typically well-provided in the way of in-house technical services: an enlarged community not so provided would require considerably expanded support services, including education and training.

However, the boundaries of the 'JANET community' were even at the outset not so easily defined as at first it might appear. We have seen in the previous chapter how the HE sector's interests cross over into other areas, particular examples being industry (collaborative research and support of spin-off companies) and medicine (teaching and research).[3] Apart from the evident funding differences, both exhibit security requirements beyond those typically required in a purely academic context[4]: in the case of industry, the issue is primarily company confidentiality; for healthcare, patient confidentiality. Now, in the early 1990s, with JANET part of the increasingly ubiquitous Internet community which was increasingly subject to a variety of malicious attacks, attaching either of these communities to JANET would properly require increased elaboration of management and access policies and the means to implement these on a community-wide basis..

A significant development in sight at this time, which contributed to the need to review the organisation and management of JANET, was the potential for increasing demands for access to an ever more diverse range of content. Initially,

---

3   Notably, however, JANET has never had any ties with the UK military, unlike the US ARPANET, the precursor of the Internet.

4   From the earliest days of multi-access systems, access had been controlled by identifier-password arrangements. As remote access became possible, such arrangements had, of course, been extended to include network and dial-in access both to networks and individual systems, including, for example, the application relay services of the US-UK gateway at UCL.

this was just to data and information repositories; later, with the advent of video and other real-time media, the breadth of information took on altogether new dimensions, both in volume and nature. While some of these sources reside within the education sector, it was already clear that as the performance and capacity of the network increased and the community broadened, so there would be demands for access to information and media beyond the boundaries of the community, and this would all need to be managed.

Yet another aspect was that the networking programme was now large: SuperJANET, which was in progress at the beginning of the 1990s (*see* Chapter 10), was a £20m project, far bigger than anything previously undertaken by the JNT. While its technical ramifications, particularly in respect of providing real-time continuous media transport, were in themselves considerable, the potential for new services based on its new capabilities would have substantial long-term effect on the diversity and range of services which could be provided in future.

It is this dimension of community, and the corresponding set of stakeholders, that has always influenced how JANET is organised; the prospect of a substantial change not merely in scale but also in the nature of the traffic content and community composition signalled the need for re-evaluation of the way in which JANET was organised and operated. For nearly a decade since it officially began, there was no question but that JANET was a part of the higher education scene and that its funding and management should be consonant with that. Thus the Computer Board initially funded around 90%, with the rest being contributed by SERC, with some assistance from NERC. In essence the Computer Board funding included JANET transmission lines and equipment; a small team to manage and operate JANET; and some development.

In practice, although links and equipment were funded by 'JANET', they were installed on university or Research Council premises and operated on behalf of JANET by HEI staff. Universities were recompensed for this (directly or through secondment) by 'JANET'[5] and the JNT Network Executive provided the operational organisation. Use by other organisations rapidly came to be acceptable in principle provided it was in association with an HE institution and demonstrably for education or research purposes. Costs associated with such use were approached on a marginal basis: typically, the costs of specific equipment and lines would need to be found by the non-HE institution but there was no usage charge made by JANET to contribute to connection[6] or operation.

In short the organisational framework that existed in 1990 was largely improvised, having been created over ten years previously to support the JNT as a relatively small collaborative networking venture by the Computer Board, SERC and the other Research Councils. The JNT was located at and employed by RAL (which legally meant SERC) but the major funding came from the Computer Board. Now, at the beginning of the 1990s, it was evident that the community and the services were about to change.

In addition to these strategic considerations, there were some rather more specific operational inconveniences which had become apparent. The JNT position was anomalous: it was sited at RAL, itself a SERC establishment, with operating procedures inherited from SERC. Almost all members of the JNT were employees of RAL, which actually meant SERC. Because of this the JNT was subject to the after-effects of SERC policy and management, which was

---

5   SERC contributed staff effort as part of its contribution to JANET..

6   This changed later, as the more formalised 'sponsored' connection category was introduced in the mid-1990s – *see* Chapters 6 and 13.

not always compatible with that of either the Computer Board or the JNT – for instance, if SERC had a ceiling on hiring new employees, the JNT could not recruit because its complement counted as part of RAL's. There were other increasingly inconvenient strictures, deriving from SERC operating procedures, on how the JNT could manage its budget, in spite of the budget being from the Computer Board, not SERC. Furthermore the Network Advisory Committee, RAL, SERC and the Computer Board all had a say in how the JNT was run. In short, through no particular fault of its own, the JNT had developed a byzantine management and reporting superstructure, which was now demonstrably inappropriate, wasteful of effort and increasingly ineffective. Partly as a consequence of this environment in which the JNT had operated during its first years of existence, and partly because of the increasingly diverse calls upon it as JANET itself grew and developed, it became increasingly clear that the organisational arrangements were acting as inhibitors to progress. The Computer Board felt that there was a danger that the JNT in its current form would not be adequate for the future management of the programme: not only were the demands on the network and its management likely to increase, but it was already clear that the next stage in the development of JANET would involve not only the beginnings of a substantial growth in community but also a significant paradigm shift in the nature of the services provided.

9.2 Community stakeholders

The review, discussion and eventual reorganisation of how JANET was run would occupy the period from 1990 to 1994. The miscellany of organisations by now involved in various ways with UK networking has already been glimpsed in earlier chapters and in the previous section. Now, along with the reorganisation of how JANET was run, all of these organisations were about to undergo their own reorganisations. Some were rearrangements reflecting changes in research areas; others could ultimately be traced to the devolutionary developments taking place in UK government.

Since 1919, overall funding for universities had been the responsibility of the University Grants Committee (UGC) which directly funded each one. This lasted until 1988 when the UGC was replaced by the Universities Funding Council (UFC) and, simultaneously, the Polytechnics and Colleges Funding Council (PCFC) was created to provide more coordinated funding for these groups. As we have already seen (Chapter 1), the Computer Board was created in 1966 under the UGC to handle funding of early, large, expensive, shared central mainframes. By extension, it naturally became responsible for funding JANET, the shared higher education network infrastructure, using the same 'top-sliced'[7] funding model as had been successfully demonstrated for shared computer provision. By the beginning of the 1990s the computer scene had changed almost out of recognition, with university provision becoming primarily based on workstations and PCs, with supporting networks of servers. Expensive high-end specialist computing facilities had become part of research facility provision and as such was now the preserve of the Research Councils. In consequence the Computer Board was replaced by the Information Systems Committee (ISC) of the UFC in April 1991, with a revised remit which emphasised more general information

---

7   *See* Glossary and Chapter 1.

systems infrastructure provision, which at the time primarily meant JANET in both the wide area and on campus.

In 1992 higher education underwent a further reorganisation. Polytechnics became universities. The UFC and the PCFC were abolished and in line with more general moves toward political devolution within the UK, university funding became the responsibility of three new funding councils: the Higher Education Funding Council for England (HEFCE), the Higher Education Funding Council for Wales (HEFCW), and the Scottish Higher Education Funding Council (SHEFC), all of which came into operation in April 1993. Responsibility for higher (and further) education in Northern Ireland is incorporated within the remit of the Department for Employment and Learning, Northern Ireland[8] (DELNI), which has responsibility for HE and FE. The ISC continued its remit for the sector but, to reflect its responsibility to the three HEFCs and DELNI, it was renamed the Joint Information Systems Committee (JISC); it would also take on an increasingly wider digital infrastructure role, including the beginnings of digital curation by the turn of the millennium.

Meanwhile, there was also reorganisation among the Research Councils. There had been some minor rearrangements in 1981 (SERC and AFRC[9]) and 1984 (ESRC[10]), but in the year following the HEFCs coming into operation, a more substantial reorganisation was begun. As in the past in the Research Councils, areas of research had changed, grown or consolidated; there was also the perennial issue of how best to encourage interdisciplinary research. There had been discussion of forming a single, 'super' Research Council but, not unexpectedly, this had not met with unanimous acclaim and instead, in 1994, another rearrangement and rationalisation took place. Two new Research Councils were formed: the Biotechnology and Biological Sciences Research Council (BBSRC), which took over the AFRC remit, together with the biological parts of the SERC remit; and the Particle Physics and Astronomy Research Council (PPARC), which took responsibility for these parts of the SERC remit. The remainder of SERC became the Engineering and Physical Sciences Research Council (EPSRC).

In parallel with the Research Council rearrangements the government had also instituted a review of its various research and development establishments. While this ultimately affected Ministry of Defence (MoD) establishments like the Royal Aircraft Establishment (RAE, Farnborough) and the Royal Signals and Radar Establishment (RSRE,[11] Malvern) rather more, RAL and Daresbury were caught up in the process. RAL and Daresbury Laboratory were merged – and spent a year as DRAL in an interim period in 1994 – to become the Central Laboratory of the Research Councils (CLRC) and the responsibility of a new Research Council, the Council for the Central Laboratory of the Research Councils (CCLRC) in 1995.

Since then, to date only two further Research Council reorganisation events have occurred: in 2005, after a number of years in gestation, the Arts and Humanities Research Council (AHRC) came into being; and, in 2007, CCLRC merged with PPARC to form the Science and Technology Facilities Council (STFC).[12] And in

---

8   Originally the Department for Further and Higher Education, Training and Development, Northern Ireland.

9   Agricultural and Food Research Council: *see* Glossary.

10 Economic and Social Research Council: *see* Glossary.

11 *See* Glossary.

12 With responsiblity for shared research facilities used by the UK research community, as well as research funding in astronomy, nuclear and particle physics – albeit that 'research' is omitted from its title.

Scotland, all funding for further and higher education was brought under a single council by merging the Scottish Further Education Funding Council (SFEFC) with SHEFC to form the Scottish Funding Council (SFC)[13] in 2005.

This brief tour through the plethora of administrative arcana of (mostly) the early 1990s has been included to provide a sketch of the background against which not only the reorganisation of the operation and management of JANET but also the introduction of campus FDDI (Chapter 5), the transition of JANET to IP (Chapter 6) and, as we shall see in Chapter 10, the fund raising and initial development of SuperJANET all took place.

9.3 The creation of a new organisation

The upshot of the discussion of those aspects of JANET organisation and evolution outlined in Section 9.1 was a general recognition of the need for an organisation and management structure which was sufficiently independent of any funding authority that it would remain free to act in the best interests of the community's networking requirements. The Computer Board agreed, in principle, to the formation of an autonomous association that would replace the JNT and take responsibility for the networking programme of the UK academic community. The organisation, initially referred to as the Networking Association, needed to be practical, legally constituted, and politically acceptable to the community of the universities, the Computer Board and the Research Councils. To this end, in 1990 the Computer Board set up the Networking Association Steering Group (NASG), with David Hartley as chair, to plan the formation of the Association. In doing so, NASG commissioned KPMG International to examine the forms of legal organisation which might be adopted. It also commissioned Brian Spratt[14] and Peter Linington of the University of Kent to advise on the objectives of the Association and the services it should provide.

Legal form of organisation

Four organisational forms were suggested and KPMG International was commissioned to assess these in meeting the Association's objectives, taking account of the need to provide clear management and accountability, and acceptability to funders and the user community. The four forms suggested were as follows.

- *Executive agency* An executive agency under the UFC. Although such an organisation at arm's length from the UFC would have provided a measure of independence, such an agency would have had limited capacity for growth. However, the arrangement was not possible because the UFC was itself such an agency and, as such, debarred from owning another.

- *Company* A company under the Companies Act, owned by the UFC and SERC. As it turned out, this would have been too restrictive, particularly

---

13   *See* Glossary.
14   Professor E. Brian Spratt, University of Kent at Canterbury.

in respect of supporting use of the network by a wider community. Indeed, at the time it might have prevented use by polytechnics (which were not funded by the UFC but by the PCFC) and might also have prevented use in furtherance of the business of other Research Councils.

- *Limited liability company* An independent, limited liability company, grant aided by the UFC, but owned or guaranteed by shareholders. The issue with this model as it stood was that future control could not be guaranteed because ownership could change through share trading.

- *Other* An organisation set up by an existing organisation in the academic community, for example a university. This was rejected because, apart from the matter of which organisation to choose, it would have introduced a third party – thereby reintroducing operational complications like those it was currently sought to remove – and also a set of additional third party interests.

In January 1991 KPMG (1991) recommended a variant of the 'limited liability company' model: a not-for-profit company, limited by guarantee, with no shareholders, with charitable status, and owned by its members. Such a structure has a number of potential advantages, including:

- no expectation of financial return, leaving it free to pursue its core objectives without the need to refer to shareholders

- charitable status, making it exempt from corporation tax, and potentially eligible for exemption from both capital gains tax and value-added tax – if its objectives are exclusively charitable

- many options for the composition of the directorate and membership of the company, allowing a flexible structure

- ability to incorporate into the Articles of Association specific objectives, such as dedication to research or education

- capability for ownership in equal parts by its members (which may be corporate or individual).

Assembling the Association

In January 1991 NASG submitted its report (Computer Board, 1991) recommending that the Association should be set up. The report included the KPMG report (KPMG, 1991) defining the recommended legal structure and the report from the University of Kent (Linington and Spratt, 1991) forming the basis of the SLA for the Association. The initial period of funding was for five years (reviewable after three). There remained much still to be done before the Association could be created and begin operation. Agreements were needed with the Computer Board and Research Councils regarding funding, operational arrangements and the overall nature of the services to be delivered by the incipient organisation. The question of charitable status needed more study. The composition of the company membership and its directorate needed definition; Articles of Association preparation; Members of the Association to be found; the company incorporated; and the directorate appointed. And, finally government approval was needed – meaning, ultimately, approval by the Treasury.

A year later, in February 1992, the ISC – as the Computer Board had now become – approved the creation of The JNT Association. KPMG wrote a business case for it for the Treasury in December of that year. In the event, charitable status was dropped: the UFC could not argue that the JNT Association would be of a philanthropic nature within Charity Commission rules without unduly constraining the operation of the Association.

The final steps needed were incorporation of the company, appointment of the company board and chief executive, and new employment arrangements for the SERC staff operating the JNT, all of which took a further year. The JISC – as the ISC had by now become – is not a legal entity, unlike the higher education funding councils: it cannot award grants, enter into contracts or be a party to incorporating companies. Where such powers are needed, often HEFCE will act on behalf of all three. On this occasion, all three higher education funding councils participated, and the new company was formed by HEFCE, HEFCW, SHEFC and the Office of Science and Technology[15] – the latter on behalf of the Research Councils. It was incorporated as a not-for-profit company limited by guarantee on 10 December 1993.

What's in a name?

The name chosen for the organisation was 'The JNT Association', which neatly captured its historical link with the team that had built and developed the network and its community over the first ten years. The title 'United Kingdom Education and Research Networking Association' – UKERNA for short – had first appeared in NASG reports early in 1991. It was considered important to maintain a separation between the 'brand' name or trademark under which the JNT Association would operate the network, the names associated with the network itself (JANET, SuperJANET), and the name of the organisational vehicle: this would maintain flexibility in respect of any associated rights in case of any further rearrangements in the future – for example, if another organisation were to take over the operation of the network in the future, the trademark name could be reassigned. Accordingly UKERNA was adopted as the trading name for The JNT Association, reflecting both the activity and the sector served by the Association. The scope of the name was intentionally left broad, allowing for the possibility that in time JANET would expand beyond higher education to encompass service to the whole of the UK education sector. To avoid any possibility of the sale of brand names by the Association, its articles prevent it from owning such names. The terms 'JANET', 'SuperJANET', and 'UKERNA' were trademarks of the higher education funding councils, with the JNT Association as their registered user.[16]

---

15   When the Department of Education and Science – which formerly funded both the UGC and the Research Councils – had been reorganised to become the Department of Education, responsibility for Research Council funding had been transferred to the Office of Science and Technology (OST), first under the Cabinet Office, then as part of the DTI.

16   Ben Jeapes: conversations with Malcolm Read, Secretary to JISC (since 1993), and David Hartley, who is credited with Bob Cooper for the name UKERNA.

UKERNA takes over

The JNT Association, trading as UKERNA, having come into being in December 1993, finally began operation on 1 April 1994, formally taking over from the JNT. Former JNT staff, who were on the SERC payroll, transferred to the new company. Contracts formerly with SERC on behalf of the JNT were replaced by contracts with UKERNA, including, for example, that with the University of London for ULCC providing the NOSC services set up in 1991 as part of the introduction of JIPS (Chapter 6). The position of Chairman of the Board of the JNT Association was taken by Ernest Morris, a former President of the British Computer Society; and David Hartley, who had chaired the Computer Board and FCC Joint Working Party on Networks back in 1980/81 during the creation of JANET, the subsequent JANET NMC, and the NASG, now became UKERNA's first Chief Executive.

9.4 THE JNT ASSOCIATION

A flavour of the objectives of UKERNA as formally stated in the Memorandum of Association drawn up in the early 1990s may be gained from the following main clause:

> 'The Association's objects are to take responsibility for the networking programme of the education and research community in the United Kingdom; and to research, develop and provide advanced electronic communication facilities for such use in that community and in industry; thereby facilitating the extension of many classes of trade through its own and the community's links with industry.'[17]

Among the sub-clauses are included:

> 'to carry out or support research, investigations and experimental work of every description in relation to networking and communication systems;'

and:

> 'to apply for, purchase or otherwise acquire any intellectual property rights, licences or know-how which may seem capable of being used for any of the purposes of the Association, or the acquisition of which may seem calculated directly or indirectly to benefit the Association; and to use, exercise, develop, licence or otherwise turn to account the property rights or information acquired.'[18]

However, to understand how UKERNA operated in pursuit of its objectives, we need to look at the organisation of its governance and finance.

---

17 Memorandum of Association, paragraph 3.
18 Universities Funding Council (1991), p.6.

Governance and operation of UKERNA

As with any company, UKERNA had a board (the Board of Management of The JNT Association) to which it was responsible, through its Chief Executive, for achieving the objectives set out in its Memorandum of Association. The Board, in turn, was responsible to the Members of the Association, who had a role similar to shareholders in a company which trades for profit. The members were individual or corporate, had subscribed to the objectives of the Association and, in the event of its being dissolved, had guaranteed to contribute £1 to the costs of that dissolution. The initial Board was constituted of the Chairman and one member appointed by HEFCE, one member appointed by the SHEFC, one member appointed by HEFCW, one member appointed by the Research Councils, two members elected by the Members of the JNT Association and two executives of UKERNA.

However, UKERNA was a not-for-profit organisation and the Members differed substantially from company shareholders in not having invested money in the company and hence not being entitled to any return. The powerful element of control deriving from the financial influence inherent in the relationship between investor and company was effectively replaced in the case of UKERNA with a consumer-supplier relationship with the JISC as consumer. The terms of JISC funding UKERNA were set out in a Financial Memorandum which effectively defined a grant to the JNT Association from HEFCE, the latter acting also for SHEFC, HEFCW and DELNI. Details of the services to be provided were defined in a service level agreement (SLA) forming part of the Memorandum of Understanding.[19] This was the first time the services expected of the organisation responsible for JANET had been formally articulated and was in keeping with the nature of restructuring arrangements being undertaken in the public sector at the time.[20]

In determining how funds should be allocated to provide the various services set out in the SLA, JISC took advice from community representation, as the Computer Board and JNT had done in the past. The Network Advisory Committee, constituted at the time of the creation of JANET, was now reconstituted as the JISC Advisory Committee on Networking (ACN)[21] with Alistair Chalmers as chair. To assist the JISC and the ACN with technical aspects of the JANET service to be provided by UKERNA under its SLA, a Technical Advisory Unit (TAU)[22] was set up at the University of Kent in late 1993 (Induruwa *et al*, 1999). Indeed, the initial task of the TAU was to develop the JANET SLA between JISC and UKERNA in time for it to start on 1 April 1994; following this, its continuing remit included providing independent performance monitoring and strategic technical advice on the development of the JANET service.

---

19   Technically, the original Memorandum was based on the Non-Departmental Public Body model as used, for example, for Research Councils. Subsequently this was updated to a Funding Agreement which is more like a contract and removed previous unnecessary constraints. Associated with this is one or more Service Descriptions. However, the term 'SLA' is used throughout this chapter in relation to the UKERNA services provided to the JISC.

20   For example, at this time RAL and Daresbury Laboratory became CLRC, the responsibility of CCLRC which provided core funding, but the major part of CLRC funding was now provided through SLAs with the Research Councils, principally PPARC and EPSRC at the time.

21   Later generally known as the JISC Committee on Networking or JCN.

22   Since subsumed as part of the JISC Monitoring Unit at the University of Kent.

It will be seen that there was community representation in respect of JANET and UKERNA through a combination of the JISC, the UKERNA Board and the JCN; formally, it was through these bodies that the Research Councils and other sectors of the education community were represented. There was, of course, representation on the various higher education funding councils and Research Councils, but these had much broader remits of which JANET, naturally, formed only a part. In general, the JISC was the forum where JANET policy was formulated, expressed through the items within the JANET SLA which was (and is) reviewed by the JCN and updated annually.

Two areas of particular concern to the community in general and the JISC in particular, acting as a focus for policy, are who can use JANET and for what purposes. The first is referred to as the JANET Connection Policy and the second as the Acceptable Use Policy (AUP). Both have effectively existed since JANET began but have evolved and of necessity become more formalised as the community has broadened and the range of use expanded. The Connection Policy has been touched upon in general terms at the beginning of this chapter, and the overall implicit stance taken has never really altered, namely that the network is for the benefit of those in education, whether teacher or taught, academic research, and those associated by collaboration or otherwise with these sectors. As suggested at the beginning of the chapter, there are occasions when the details of this policy have had to be clarified or even expanded a little. However, there have also been several occasions when the question has arisen as to whether JANET might provide service for other sectors entirely, for example, in the public service. Might it be used to enhance or supplant the NHS network? Could JANET provide or be a part of government network infrastructure? More recently there have been suggestions that government could save money by procuring goods and services on a government-wide basis (generally referred to as 'aggregation'). Should perhaps JANET, the NHS, local government, national government, even the emergency services, etc. all procure a common communications infrastructure? While it may be true that all involve 'communication' and 'network', the differences in such aspects as mission criticality, security, privacy and open-ended service all suggest that any such aggregation needs to proceed with considerable caution. However there can be opportunities here, as Wales demonstrated in 2007 with the evolution of its Lifelong Learning Network to its Public Sector Broadband Aggregation Network (*see* Chapter 11). Nevertheless, the central guiding principle that JANET is for the benefit of education and research has served it well in providing the essence of the Connection Policy for its first 25 years.

Acceptable use of JANET originally stemmed from prior policies which institutions naturally had in place regarding use of premises and equipment, including computers: fundamentally, much as for the Connection Policy, for purposes in connection with education and research. However (as already alluded to in Section 6.3) the ever-increasing ubiquity of information processing and communication systems, of which JANET is a part, has meant that the AUP has become a JANET-specific component of the overall regulatory regime which has grown up since the Internet became part of worldwide infrastructure. As such, consideration of the AUP is deferred until Chapter 13, on Regulation and Security.

Since UKERNA was publicly funded it needs to be publicly accountable. The public bodies funding JANET were already subject to such rules, essentially deriving from the Treasury. The JANET Memorandum and SLA effectively formed part of the mechanism by which public accountability was extended to UKERNA. When the business case for the creation of UKERNA was initially approved by the Treasury, it was for a period of three years and any continuation was subject to 'market testing'.

Accountability and value for money constitute the more obvious aspects of responsible use of public funds. However, in a situation where such funds are being used for specialist provision of private services within a public sector, it is also important to avoid undermining the open market. JISC supervision of the UKERNA SLA would also encompass this type of aspect, including careful scrutiny of both the Connection Policy (which JISC effectively owned as a component funding instrument for UKERNA) and the Acceptable Use Policy (the instrument by which UKERNA would regulate use of JANET).

Market testing

At this time in government-funded public service, it had become the adopted wisdom that organisations should be analysed in terms of their functions and services in such a way that costs could be apportioned to each and market-oriented testing applied in pursuit of achieving politically acceptable so-called 'value for money'. In practice, for UKERNA, this meant conducting a full-scale open tender for the services specified in the JANET SLA, and UKERNA bidding its apportioned costs in competition with commercial suppliers.

For the purpose of the market test, the JANET services being provided by UKERNA were divided into three parts: operational and development services; related support and liaison services; and CERT[23] services – CERT, short for Computer Emergency Response Team, provided support in combating malicious network and computer activity.

The test process began in 1995.[24] That for the JANET operational and development services, which constituted by far the major part of the JANET services, began in November. JISC issued an Invitation to Tender in January 1996 and UKERNA submitted its bid in April. After a detailed negotiation phase, a best and final offer was submitted in August which, following some clarifications, was approved by the JISC Tender Board in the third quarter of 1996. The testing of the JANET related services, an altogether smaller operation, was conducted over the following few months, in time for the SLA to be renewed for a further three years from 1 April 1997. The CERT service was a very new one, having begun only three months before UKERNA began operation in April 1994, and it was not in the event included in the market test; it was simply incorporated into the renewed SLA. Formal expressions of interest were submitted by BT[25] (supplier of the JANET transmission infrastructure at the time), Racal and RAL;[26] none led to a formal bid being submitted. The market testing process was lengthy, time consuming and costly, even when, as in this case, there were no other bids in

---

23 *See* Glossary and Chapter 13.

24 *UKERNA News* 1, October 1996.

25 Which might hypothetically have led to a situation with an element of irony in view of BT's stance in waiving its monopoly position a decade earlier in respect of JANET, provided the latter did not compete with BT (Chapter 4).

26 Private communication, David Hartley, April 2010.

Organisational structure chart

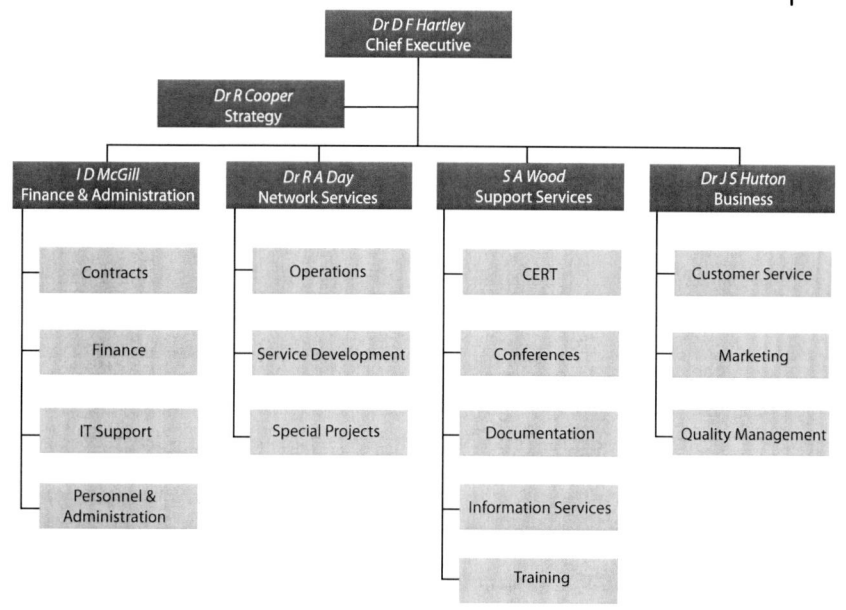

- **Dr D F Hartley** — Chief Executive
  - **Dr R Cooper** — Strategy
  - **I D McGill** — Finance & Administration
    - Contracts
    - Finance
    - IT Support
    - Personnel & Administration
  - **Dr R A Day** — Network Services
    - Operations
    - Service Development
    - Special Projects
  - **S A Wood** — Support Services
    - CERT
    - Conferences
    - Documentation
    - Information Services
    - Training
  - **Dr J S Hutton** — Business
    - Customer Service
    - Marketing
    - Quality Management

Figure 9.1. Organisational structure of UKERNA in 1996 (after market testing).

effect – though it seemingly also led to some constructive restructuring within UKERNA. Figure 9.1 shows the new structure which was adopted at the time.[27]

With this last step, the creation and structuring of the organisation that has run JANET since that time was finally confirmed as complete – with the singular omission that responsibility for operation of a vital part of JANET, the backbone (and subsequently some of the regional networks), remained subcontracted outside UKERNA to the NOSC at ULCC. It would be another decade before this changed, as we shall see at the end of SuperJANET4 in Chapter 12.

---

27 See 'Market Test Update' by David Hartley and 'Company Reorganization' by Shirley Wood in *UKERNA News* No.1, October 1996.

10

SuperJANET:
Birth of a Concept
and ATM Trials

From the moment packet technology proved successful in connecting computers, its continuing development has taken place against a background of relentless growth. JANET was no more immune to this than the Internet or ARPANET before it – indeed, a decade of pent-up demand ensured that JANET would experience this in concentrated form. We have seen how, throughout the 1980s, JANET was continually in pursuit of capacity. By the end of the decade, achievements in the local area had enabled glimpses of what might be possible if the whole community could be interconnected in the wide area at commensurate performance. Indeed, in the UK a taste of this had been achieved in the early 1980s when an experimental distributed system of the time had been extended over a footprint of European dimensions.[1] Until such wide area performance could be achieved, continual upgrades in capacity had been swallowed up by community growth: the goal now was to change the performance of the network by a sufficient amount as to escape the constraints of merely accommodating growth in numbers and instead enable the use in the wide area of all the applications already demonstrated using the higher performance available in the local area.

There was widespread realisation that achieving this level of performance – released from keeping up with numbers – would usher in a new age of visual communication and computation: in effect, a step up from handling data bits and bytes ever more quickly to dealing in 'information'. People began to talk of realising the 'information society' and the communication infrastructure began to be referred to as the 'information highway'.

Naturally, the UK was not the only place to explore what might be achieved were much higher network performance available. For example, in Germany, the BERKOM project was planning a fibre-based infrastructure delivering 140Mbps (the highest rate then available using PDH, outside the trunk network) to explore the applications which would be enabled by broadband ISDN. And in the USA, the Gigabit Testbed Initiative,[2] a major collaborative programme funded

---

[1] An extended version of the Cambridge distributed system, based on satellite-linked Cambridge Rings at eight sites (including a demonstration site at Geneva); remote filestore access, remote booting, remote device control (of a robot arm), packet speech and slow-scan TV had all been demonstrated by project UNIVERSE (Burren & Cooper, 1989).

[2] The Gigabit Testbed Initiative ran from 1990–95. There were five testbeds: Aurora, Blanca, Casa, Nectar and Vistanet. It was funded by NSF and DARPA, with contributions from industry (CNRI, 1996).

by government and industry, was beginning with participants from university, industry and government organisations. The Gigabit testbeds stemmed from the observation that wide-area communication performance had fallen behind computer performance and the programme was aimed at both communications and applications research.

The UK took a different approach. What would subsequently be called the SuperJANET project had its roots firmly in service networking. In March 1989, following a policy review meeting, Bob Cooper, then Head of the JNT, was invited by the Computer Board to propose a programme to address the major issue of network capacity for JANET, both as then being experienced and as anticipated in the future. We have seen already (Chapter 5) that the general form of the proposed three-part programme was approved in principle in April 1989. The interim wide-area JANET upgrade, JANET Mk II, took place from mid-1989 until the beginning of 1993, by which time SuperJANET was imminent. The matching initiative to bring campus access up to 100Mbps, the High Performance LAN Initiative, was formally approved in April 1989 and, following the FDDI pilot programme, installation ran from 1991 until the beginning of 1994 – when 100Mbps, 'fast' Ethernet appeared. SuperJANET would take over in the wide area from JANET Mk II and link sites at rates commensurate with those of the newly installed campus networks of the FDDI programme.

As achieving a factor of a hundred or more improvement in wide-area capacity would evidently take time to investigate and cost a considerable amount, at this stage approval was given to undertake sufficient investigation and planning to enable a case to be made for funding. As already mentioned, the amount anticipated would be so large as not only to exceed the Computer Board's budget but also to require the DES to bid for additional funds in the Government Public Expenditure Survey (PES) which occurred each year in the autumn. This chapter tells the story of how what started as a simply-stated urgent necessity for more capacity not only transformed the architecture of JANET but ultimately resulted in the basis of a national facility serving every part of the research, education and learning sectors of Britain, laying the foundation for the provision of ubiquitous user services which were as yet still emerging from the formative stages in research laboratories.

10.1 SuperJANET: breaking the log-jam 1989–1994

At the beginning of the SuperJANET project it was the necessity to escape the restrictions of tariffed services available in 1989, at that time still based on telephony-equivalent tariffs,[3] that stimulated the exploration of other options than those offered at the time by the duopoly of BT and Mercury for a private transmission infrastructure. Until then 2Mbps had been the highest rate available; shortly thereafter, 8Mbps became available but was expensive. JANET did install some 8Mbps links later, initially as a further stop-gap before SuperJANET became available and later as part of the early development of the current regional access networks, but this was incidental rather than strategic. The possibility of using microwave transmission was briefly investigated: 140Mbps links were available and 500Mbps links looked possible. However, quite apart from whether such

---

3   See Chapter 8.

links might suffer under adverse weather conditions, microwave transmission did not hold out the promise of transmission rates that could compare with the potential of optical fibre, where SONET/SDH rates of 2.5Gbps had already been standardised for telecommunications use. Investigation of a fibre infrastructure was almost immediately the main focus. Although Wave Division Multiplexing was discussed, it had not then reached the stage where JANET could have considered it for deployment.

Although extricating JANET from the threat of strangulation through lack of affordable capacity was the initial motivation, it coincided with the maturing of many of the developments described in Chapter 8. Principal amongst these was the promise of commodity real-time video applications. Although the initial H.261 video compression codecs for use at 2Mbps just then coming onto the market were heavy, 19-inch rack-mountable crates (comparable with GPO 2400/4800bps modems of a decade and a half earlier), they nevertheless signalled the prospect of a commodity service more ubiquitous than hitherto. The thrust of broadband ISDN was also toward support of video and image-based communication generally. At the same time IP multicast, video and audio streaming, and desktop videoconferencing tools were signalling the arrival of packet video for both real-time collaboration and, ultimately, IPTV. And then there was the Web, which heralded the elevation of computer networking to the medium of choice for ubiquitous information access. All this surfaced in the space of two or three years and the start of it coincided with JANET's search for a new way to obtain network transmission capacity. From the technological perspective, all the components were in place to support the development of real-time network collaboration and teaching in every aspect of education and research, coupled with ubiquitous access to data and information resources to support everything from experimental data acquisition to teaching and learning material. In fact there was only one thing missing: network-based, general purpose information search and retrieval. Aids to managing and finding information on the Internet had been explored for some years and had resulted in systems such as WAIS[4], Archie[5] and Gopher[6] during the late 1980s and early 1990s, immediately preceding early Web-oriented systems in the early 1990s. However, really effective search engines were still a little way off: DEC's Altavista would not appear until 1995, followed by the Stanford University spin-off, Google, in 1998.

Within a year, the prospect of enabling these sorts of services for research and education transformed what risked being perceived as an intrinsically boring and seemingly endless whinge about lack of capacity – or sufficient funds to do anything effective about it – into a gleaming promise of a brave new world of real-time audio-video communication for all and an information superhighway infrastructure to support the whole of research and education. And it might even lead the world: the age-old lure of overtaking a rival, chiefly the US in this context, was real at that moment because the Gigabit testbeds were for research: 'Super' JANET would be for service. Of course, it wasn't going to be cheap – no one ever thought it was – but now there was an alluring promise of great things: a

---

4   WAIS (Wide Area Information Servers) was a client-server system originally created in 1988 to index, search and retrieve text-based databases.

5   Archie was created at McGill University, Montreal, in 1990; it periodically built a searchable index of the contents of open FTP sites.

6   Gopher was a protocol and a set of application systems created in 1991 for organising, distributing, searching and retrieving text files.

better prospect for releasing major funds than merely maintaining the status quo at considerable cost.

Seeking funds

So how much was all this going to cost? Or perhaps the more relevant question was, how much funding should be requested? This wasn't like buying a commodity item out of a catalogue: indeed, one hardly knew what it might really be possible to purchase. It was all very well aiming for a 'fibre-based network' but what did that really mean? Even were 50 to 100 institutions to be provided with a private fibre-to-the-door network dedicated to JANET – and only licensed utilities are allowed to run ducts, pipes, rails, canals, cables or anything else around the country – a considerable amount of additional switching and routing equipment, together with operation and management infrastructure, would be needed to turn this into anything useful. The vision would need to be tempered by the art of the possible. Making a case is much like any other buying and selling negotiation in one respect: once a figure has been mentioned, it may be reduced but it can never be increased! The process of arriving at this figure for SuperJANET would be an iterative one of investigating technologies, having preliminary conversations with potential suppliers and exploring the possibility of there being other enterprises with which JANET might form a partnership in seeking suitable infrastructure and its means of provision. And there was one event which assisted in the later stages of this: the duopoly review was imminent. It would not take long and the removal of the duopoly prompted the formation of an immediate queue of applicants for telecommunication licences at the door of the DTI.

As already described, the decision to target a fibre-based infrastructure was taken very early and was straightforward;[7] moreover, not only was it never challenged, it received strong, indeed, enthusiastic support in all quarters.[8] Single-mode fibre had been in use in the telecommunications industry for almost a decade; and, in the cheaper multi-mode version, it had been in use in campuses and large buildings nearly as long. Following Computer Board approval to proceed with investigation and planning to the stage of making the case for funding, the first year or so was taken up with two activities which proceeded in parallel: technically-oriented exploratory discussions with suppliers together with informing and lobbying the higher education establishment about the proposed project and its benefits. The latter took on more than usual importance because it was clear from an early stage that whatever the eventual figure turned out to be, it would be well beyond the Computer Board's annual budget; moreover, the UFC would not be able to absorb all the cost and so it would have the knock-on effect of causing the DES to request additional funds as part of the Science Vote for this special case. The process for this required that the case be submitted as part

---

7   Mention of a fibre infrastructure first occurs in a briefing note by Bob Cooper to the Technical Options Group of the Joint Policy Committee on National Facilities for Advanced Research Computing, dated 15 December 1989.

8   Amongst those subsequently lobbied were The Rt Hon Jeremy Bray, MP, whose area of interest included provision of national infrastructure to encourage take-up and use of Information Technology, and Dr Nigel W. Horne, of KPMG Peat Marwick McLintock, Chairman of the DTI/SERC Information Technology Advisory Board (ITAB): the latter was the senior committee responsible for research funding of information technology at a time when this was being undertaken jointly by DTI and SERC, in an effort to encourage knowledge transfer.

of the Public Expenditure Survey in the autumn. The effort needed to prepare such a case, brief the community and gain support, coupled with the timing of the duopoly review, would have the effect of postponing submission of the case until the 1991 PES round. The significance of the duopoly review was not to be underestimated. It was carried out in 1990, and although the formal outcome was only announced in 1991, it was clear from an early stage that it would result in other providers than BT or Mercury being available. The timing was such as to provide an ideal window of opportunity for the project: even if no others were eventually in a position to undertake the construction of SuperJANET, the mere possibility of competition held promise of obtaining an affordable deal even from the existing incumbents, particularly since such a high-profile national enterprise was unique in respect of its being for education.

The major components of the case would be based primarily on the nature of future services which could be provided to the education and research community; and the cost would depend on the transmission and networking technologies chosen. Since many of the contemplated technologies and projected future services were only just emerging, the JNT hired technical assistance from the network research part of its community to help both in assembling elements of the case and in discussions with potential suppliers. The team that helped Bob Cooper with this pre-funding stage consisted of John Burren, Steve Wilbur and the author.[9] To clothe the case in some semblance of reality, exemplars were identified which were a mixture of those already in emergent form but which could also evidently benefit from a substantial injection of performance, and a wish list of somewhat more speculative areas. However, a notable feature of the case was the inclusion of the notion that, more than likely, among the more significant benefits would be applications which had not yet been dreamed of but the conception of which would become possible in a substantially higher-performance environment. It will be recalled that an inability to be too specific about the benefits of the original JANET had caused difficulty and consequently delay in funding in the 1970s. By contrast, by the time it came to making the case for SuperJANET at the beginning of the 1990s, there was a widespread appreciation of the benefits of a well-founded infrastructure in which innovation could thrive.

The general nature of what was becoming technically possible at this time has been explored in Chapter 8. The exemplar application areas upon which the case was built could be broadly categorised as follows.[10]

- **Remote or distributed teaching** using high-quality videoconferencing in association with a still-image camera for viewing teaching material, artefacts, etc. It was envisaged that the type of facility trialled several years earlier by the prototype Livenet service in the University of London and also the facility installed, for example, at the University of Ulster (based on 2Mbps primary rate ISDN and the relatively new H.261 codecs) in the beginning of the 1990s should become routinely available over JANET.

---

9 Steve Wilbur was a member of the Department of Computer Science at UCL, later Head of Department and Dean of the Faculty; John Burren and the author were both from the Advanced Communications Unit at RAL, the former being its Head.

10 The topics of this list are taken from one of the slides in the author's archive collection of those used at the time to publicise the SuperJANET project and engender support for the case in the community, as well as briefing funding bodies and prospective partners or suppliers. Many of the slides in the collection originated with Bob Cooper, as well as being adapted by those helping in these briefings.

- **Group communication:** a term used here at the time to encompass multimedia support for both real-time 'synchronous' communication and 'off-line' or 'asynchronous' communication in support of group collaboration for research projects, administration, dispersed tutorial groups, learning and training. Though not articulated in the same terms and only elaborated rather vaguely, the concept nevertheless presaged a number of elements of the more recent virtual organisation. Particular examples at the time included real-time collaborative molecular modelling using a shared workspace and remote electron microscopy, both with audio- (possibly video-) conferencing support.

- **Information services:** with an emphasis on sufficient network performance to enable browsing of collections which included everything from high-quality images of art, medical imaging (including, for example, access to what were then still relatively new MRI[11] and PET[12] brain image scans), meteorological image data, engineering information, architecture and archaeological artefacts, to name just a tiny few. The network performance to enable this was one to two orders of magnitude beyond that then available. Achieving an ubiquitous information infrastructure was also foreseen as both a prerequisite and an essential stimulus to achieve a long-awaited electronic publishing industry.

- **Remote access:** historically, 'more of the same' in the sense of the original *raison d'être* for JANET. By now, everybody was familiar with using resources outwith the local computer – which was by now personal – and these were comprised of much more than merely remote login, file transfer and remote job entry. Remote execution, remote procedure invocation, transactional processing and remote file (or disk) access were well established: a goal now was to extend to the wide area as much as possible of what had been achieved in distributed processing in the local area – bearing in mind the delay imposed by signal propagation. A further possibility which would be explored over one of the US gigabit testbeds[13] was that of 'network distributed memory'. At the application level, there was the vision of abolishing distance as a barrier to access to specialised instruments or experimental facilities: particle and nuclear accelerators, together with astronomical telescopes were among the more prominent speculative examples.

- **Remote consultation:** expertise has always been in short supply and therefore seldom available locally: can distance be abolished as a barrier to access? The medical field has always been the premier example, whether doctor to patient or consultant to surgeon in an operating theatre – particularly in an emergency context.

- **Supercomputer support:** fast processors need fast networks to support them – the point of departure for the US gigabit testbed initiative. This was essentially a recognition that powerful processors require access to

---

11 Magnetic resonance imaging.

12 Positron emission tomography.

13 Blanca.

substantial data collections and will generate a commensurate volume of results, and both aspects require high performance networks.

- **Visualisation:** consequential upon the increased volume of data, some of it time dependent, is the need to display it meaningfully for human appreciation and assessment; particularly important in this context is exploitation of the human talent for visual analysis. Enabling real-time visualisation of large, time-varying datasets independently of the location of the user requires a combination of high-performance processing and communication, whether the dataset is part of a large specialist resource, the output of high-performance modelling calculations, large-scale scientific experimental facilities, or the (remote) control and steering of the latter two activities.

SuperJANET was seen very much as a national pilot with potential impact both within and beyond academia: the existence of such a network would serve to stimulate the development of applications exploiting the freedom from the all-too-familiar shackles of inadequate transmission capacity. Moreover, it was considered that the existence of such an infrastructure would enhance the transfer of knowledge to encourage and enable commercial products having their origins in either research prototypes or advanced services developed for education and research (Cooper, 1990). It is also interesting to observe that this view regarding JANET's role is in part reflected formally in the subsequent Memorandum of The JNT Association (Section 9.4).

Having given a brief description of the nature of the vision which became the essence of the case for SuperJANET, it is now time to return to the question of how much money to request. It was from the discussions which led to identification of this figure that some of the enduring architectural concepts of SuperJANET emerged. Although the original SuperJANET would only realise a fraction of the vision, it would nevertheless provide a general, average increase in capacity of around five-fold over JANET Mk II, as well as initiating development of the eventual SuperJANET, which would turn out to be a composite of the original concepts and developments which occurred as time passed – and a great deal more time was to pass than anyone guessed in 1990.

And it should be remembered that, apart from the other technical developments taking place within JANET – including JANET Mk II, FDDI deployment, and the adoption of IP – throughout almost the whole of the original SuperJANET project, from 1990 until 1994, the events related in Chapter 9 associated with the reorganisation of the whole of the administration and funding of higher education, including JANET itself, were taking place, so that it was something of an undertaking even to gain sufficient attention for an activity such as SuperJANET. Perhaps it was a help that the project was big and bold, and was going to cost a lot!

Fibre or what?

1990 was an interesting year in which to be exploring the provision of a national transmission infrastructure. In expectation of new public telecommunications network operator licences being granted by the DTI, a number of international carriers, particularly from the US, were also exploring the potential for new business in the UK. A particular feature of this was the courting of various of the existing UK utilities by these potential new carriers, brought about by the need for 'way leaves'. The latter refers to the legal right of access to third-party property for the purpose of providing and maintaining the transmission infrastructure (in the most general sense) of the utility or public service in question. Utilising the way leave of an existing service by forming a partnership was perceived to be an advantageous route toward this part of the licence requirements. At various times there were suggestions that canals, the gas and water distribution networks, the railway network and the national electricity grid might all be suitable candidates. US companies were not enamoured of the railway because there had already been bad experiences in North America of vandalisation of cables by thieves in search of copper – though many of the cables were now fibre, this was typically not discovered until afterwards! Canal tow-paths offer similar opportunities for malicious damage, apart from the fact that there are substantial gaps in the network when considered as a basis for national distribution. At one time there looked to be serious possibilities of such a partnership with either water or gas, though not based on the rumour circulating at one time of actually running the fibres within the pipes: a suggestion oblivious of the fact that both networks have stop-cocks, valves or other means of shutting off the flow!

The National Grid, as it was then, seemed a more promising candidate. It needs its own communications network to monitor and control the grid. At the time, it was contemplating creating its own fibre infrastructure for this purpose, based on utilising its existing high-voltage distribution network to carry the fibres (*see* panel). In order to offset some of the costs of doing this, it was seeking a small number of enterprises with which to form some sort of partnership; SuperJANET and the television industry (for inter-studio links) might perhaps be plausible potential candidates.

The outcome of the technical exploration of transmission infrastructure options was, first and foremost, that it would be possible to obtain a fibre-based infrastructure and that there seemed the potential for perhaps half a dozen companies to bid to provide it. The other important outcome was that an estimated £20m would be enough to provide a significant amount of the required infrastructure, and this was the figure that went into the bid for support. So far as the JNT was concerned, this would be all 'new' money, in the sense that it would be in addition to its existing budget for operating the existing JANET; it would also be additional to the JANET Mk II and High Performance LAN funding. The knock-on effect was that the DES requested £9m[14] in its submission to the 1991 PES round, to top up its support of SuperJANET. Following the support of the Secretary of State for Education and Science, the Department successfully obtained the SuperJANET funds and the UFC approved the £20m funding for SuperJANET in December 1991. Now came the next stage of refining into a plan what had effectively been the outcome of preliminary enquiries into what might be possible: this plan would then form the basis of a tender action.

Procuring the basis of a network

Technically, the outcome of the preliminary discussions had suggested SDH/SONET products were sufficiently far advanced that SuperJANET should adopt this technology as part of its technical strategy for exploiting optical fibre

---

14   According to the author's recollection.

Exploiting the electricity grid pylon network to carry fibre

There were two ways in which the national high-voltage electricity pylon network might be utilised to carry a fibre infrastructure: as usual in such cases, each had its advantages and disadvantages.

The neutral cable in the pylon network is hollow: in effect, it is a metal pipe, albeit of not particularly large diameter. A possibility was to use it to carry some fibres, which would be well protected by the metal sheath. The disadvantages are twofold: there is only room for a few fibres – 24 was one estimate at the time; and installation is expensive because the fibres cannot be pulled (or blown) into the conductor, as would be done in ducts: a new metal cable with the fibres in place would have to be made and the existing cable replaced with it, all over the country, not as part of any long-term cable-maintenance replacement strategy – very expensive and with no further expansion in the number of fibres possible.

The other option has an almost Heath-Robinson feel about it: just wind the fibre cable in a helix around an existing cable – simple! One concern at the time was about the long-term durability of the fibre cable outer sheath, a particular aspect of which related to something which gained the sobriquet 'the beak problem'. The beak of a bird perched on a cable has some of the properties of a lighting conductor (essentially because the beak is pointed, so within an electric field will produce a region of local concentration of the field). In the region of high-voltage distribution lines there is an electric field. The discharge induced by the beak, while not harmful to the bird, was thought possibly sufficient to accelerate the deterioration of the sheath. However, the great advantage of this system is that a device exists which can travel along a cable *in situ*, rotating as it goes, which can install such a cable without the need to dismantle or replace anything. This is considerably cheaper than neutral conductor replacement, and has been used in the US and the UK.

transmission. SDH had several attractions, the most obvious being that it was already an international standard for optical fibre transmission up to 2.5Gbps, and products at 155 and 622Mbps were just emerging. Strategically, it had another potential attraction for SuperJANET. During the previous year, it had been suggested in a number of quarters, including DTI, SERC and a number of communications and networking research groups, that SuperJANET should form the basis of a national infrastructure for network research as well as service. While this suggestion was attractive it also contained a major potential pitfall: if experimental research use of the network was not to disrupt service to the whole community, a way had to be found to insulate one from the other. In the absence of either separate fibres or separate wavelengths in a WDM transmission system, the next best option was separate channels of a TDM system. SDH, as a substantially more sophisticated TDM multiplexing technology than PDH, had the potential to support a number of independently configurable transmission channels of varying capacity. It was suggested that this could be made the basis of a partitioned infrastructure, in which the major part of the capacity would be for the service network, whether X.25 or IP, while other channels could be used for new technology trials or packet network research. This potential ability to reconfigure the underlying transmission capacity seemed to offer the attractive possibility that SuperJANET could not only support the service network but might also provide testbeds for developing, trialling and helping to make the transition to new packet-level infrastructure, as well as supporting research use and perhaps also the provision of special circuit-oriented transmission facilities. By now the concept of such a network, inherently possible with the technology, was sufficiently encouraging that, in anticipation of success in obtaining funding, in August 1991 the SuperJANET Project Team was formed, initially composed of those already assisting with the pre-funding stage (listed above), with the addition of Chris Cheney, seconded from Cambridge University Computer Service, and James Hutton of the JNT. The remit of the project team was quite straightforward: plan the implementation of SuperJANET, starting with developing a specification of the transmission architecture which could be used as the basis for procurement of the network.

In addition to SDH, in anticipation of JANET becoming a multiservice network, the strategic decision was taken to move towards the CCITT broadband ISDN standards, in particular to adopt ATM switching and multiplexing. This was generally in line with JANET's long-term strategy of adopting international standards wherever possible and, in August 1991, at the same time as the creation of the project team, an ATM Technical Advisory Group, the ATAG, was set up; initially chaired by the author, later by Les Clyne of the JNT.

The 1980s had seen considerable debate over what form the bearer transmission service for broadband ISDN should take, but by 1988 it had become clear that the cell-based Asynchronous Transfer Mode, ATM, was the favoured technique. ATM was a connection-oriented packet-mode technology and the generic debate between connection-oriented and connectionless modes would continue, the proponents now being ATM and IP. ATM gained considerable momentum during the first half of the 1990s: it was seen by both the computer and telecommunications industries as a way towards a unified communications infrastructure. At the time this also received encouragement from its being a switched technique and therefore apparently more scalable to higher speed and capacity, unlike shared media networks which were running out of steam. JANET was not alone in adopting ATM as part of its strategy at this time, though it was among the leading adopters for service. The ATM Forum was constituted (in the US) in November 1991 as an international industry-led consortium to agree specifications relating to ATM implementation; it was notable for drawing participants from both the computer and telecommunications industries, as well as the academic and commercial research community. In particular it was responsible for defining user and public network interfaces for ATM, as well as use of the 100Mbps transmission format[15] used by FDDI, which became quite widely used in the local area. It would also not be long before details of how to operate IP over ATM were defined by working groups of the IETF.

So, during 1991 the concept of SuperJANET was gradually given form in terms of those technologies just emerging. The ideal would be an SDH transmission network, providing configurable SDH channels, over which JANET could (re)construct the service network(s) using X.25 switches and IP routers as necessary; it would also be able to embark on the construction in parallel of an ATM network, in anticipation of broadband ISDN becoming the universal technology of choice for a multiservice network. For SDH provision there seemed two possibilities: as a service direct from a supplier or, just possibly, as fibre provided by a supplier and with SDH equipment perhaps owned and operated in consortium – at this time JANET was rightly wary of attempting to operate its own SDH network,

---

15   Known as TAXI: *see* Glossary.

having no knowledge or skill to draw upon, potentially faced with equipment hardly out of beta-test, but nevertheless requiring it to support the production network service.

The SuperJANET team also gave some initial consideration to network topology at this point. Over the years of JANET development, this question of topology for national network coverage has been revisited at regular intervals. The general approach has been to cover the major (concentrations of) universities with substantial capacity and to reach the remainder with the aid of spurs. To gain some resilience, from the earliest days the central part of the network has been a simple mesh. By the early 1990s, with overall functioning of the community now increasingly dependent upon JANET, network resilience began to gain greater importance. One indicator that this was not peculiar to JANET but quite general for telecommunications was that SDH had adopted the counter-rotating ring[16] model of distribution because of its resilience against a single failure; indeed, one sample topology considered at the time of early SuperJANET discussions was three rings, covering northern, central and southern regions of the UK, with those not covered by the rings reached by spurs, including North Wales, South Wales and Northern Ireland.

Having gained a head start by being optimistic that approval would be forthcoming, the project team was able to issue a Request for Proposals (RFP) in March 1992. By this time, in order to concentrate full time on SuperJANET, Bob Cooper had relinquished his post as Head of the JNT (though he retained the title of Director of Networking, to which he had been appointed after Mike Wells's tenure ended in 1988). The ISC[17] and the UFC had by this stage approved the setting up of UKERNA, though it would be another couple of years before it was constituted and operational. Willie Black, already a member of the JNT, took over as Head, a position he would retain until UKERNA became operational. During this interim phase, under a consultancy contract, the UFC appointed Geoff Manning, a former Director of RAL,[18] chairman designate of the incipient UKERNA.[19] Technically the JNT was still part of SERC's RAL; however, in anticipation of UKERNA coming into operation, the contract for SuperJANET would be with the UFC and Geoff Manning, as ISC/UFC representative, led the formal procurement.

The formal Invitation to Tender (ITT) was issued in August 1992 to all the companies which had responded to the RFP. While most were not expected to respond, it was still hoped that perhaps as many as five or six might do so. However, it became clear that the process of obtaining a telecommunications licence had reduced the number so that, within a few days of the closing date, there were only two in the running. Then, the day before the close, the National Grid withdrew: it appeared that its business case for the venture was not strong enough at that time. This left BT. Fortunately, although what was offered was not exactly what had been envisaged, it was sufficient to provide a roughly five-fold upgrade in access speed over JANET Mk II, as well as enabling a start to be made

---

16 As had FDDI and QPSX/802.6 MAN before SDH, and 802.17 Resilient Packet Ring subsequently – although the mechanism of recovery is different in each case.

17 Information Systems Committee, which took over from the Computer Board in 1991.

18 1979–1986.

19 *Network News* 36 (March 1992). However, by 1994 the situation had altered and, in the event (see Chapter 9), David Hartley became the first CEO of UKERNA (and Ernest Morris Chairman of the Board).

on the SDH/ATM network for the future. The decision to award the contract to BT was approved by the UFC in November 1992. Following the original approval for £20m SuperJANET funding back in December 1991, the team had reserved £2m for contingencies and it was a four-year contract worth £18m which was signed by the UFC and BT on 27 January 1993.

So, after almost four years contemplation, discussion, fund raising, studying emerging technology and negotiation, what had JANET finally got for its money? The answer was in two parts: early access to BT's forthcoming new data service, SMDS,[20] for about 50 sites; and a start on an SDH-based network for around 16 sites – BT providing the transmission facilities, over which JANET could, in principle, implement whatever it required: X.25, IP, ATM, IP over ATM, etc. If this sounds imprecise, that was correct! All sides had been playing their cards relatively closely to their respective chests, partly because, from the provider perspective, much of the technology was sufficiently new that its real costs were not yet clear; and partly because, on the JANET side, until it was known what could be provided and on what scale for the amount of money available, it was not practical to specify in any detail which sites should be connected and in which order. The latter had a further effect on the providers: as already indicated above, some sites are relatively easy – inexpensive – to reach, others not: until it was known which sites were to be connected, costs could not be properly allocated. As it happened, although initially BT had treated the early stages as any other routine request, in the latter stages it became evident that the company had become interested in a more 'partner-oriented' approach, treating education and research as a unique special case. However, it also did not wish to create a precedent in respect of appearing to create a bespoke private network for JANET. So, what was offered was a mixture of service (SMDS) and transmission provision which, although it was new and happened not to be tariffed at the time, was not out of line with the sort of transmission facilities BT was accustomed to providing, and could not be construed as a private network. In signing the agreement it was understood that there would be further negotiation on both timescales and site detail: where and how many. One other thing, which was clear at the time of the tender, was that SDH was only just being installed for beta testing of various suppliers' products: SDH would not be available to begin with but PDH circuits would be made available initially. In effect, JANET was getting very early pilot access to both SMDS and SDH – and would effectively be helping BT to gain initial experience of operating both, though this was not articulated. Although formally a contract, there were some elements of partnership in the original SuperJANET deal with BT.

Technology trials and pilot networks

Constructing SuperJANET was quite a different proposition from the original JANET. For around a decade before JANET came into existence, there had been implementation efforts all over the country, all gradually converging on a single set of protocols, and the final step had been to connect them all together at the X.25 network layer. Now, there was a multiplicity of protocols and technologies at or below the network layer, all in a state of transition. In service were X.25 and IP, with a steady transfer to the latter. In support of these, in addition to Ethernet and

---

20 *See* **Chapter 8.**

the recent FDDI, there were now added SMDS (something of a wildcard), ATM and PDH, with SDH coming. A complicating factor in this was that the evolving position of ATM was not yet clear. In the local area, it might supplant Ethernet and, eventually, FDDI as a more scalable, switched technology; and in that case, might it not also become the end-to-end, universal bearer service?

For SuperJANET, 1993 was a major year for pilot trials of technology. Tactics, however, were not entirely straightforward. On the service network side, it was important to begin transition to the use of SMDS so that benefits could be obtained by as many as possible early in the life of the new, higher performance network. But, technically, there were several issues[21] needing resolution before a service could be offered. Then there were the initial technology trials to be mounted on the PDH pilot network: ATM, video over ATM and, at some stage, IP over ATM. And for PDH and SMDS, the initial set of sites needed to be selected. Of course, SuperJANET had been predicated on being an enabler of new applications, so it was also very important, after more than three years of planning and negotiation, to initiate pilot use of those new applications as soon as possible. While the lion's share of effort was to come from the sites participating in building SuperJANET, it was time to alter the composition of the SuperJANET team for the implementation stage. Chris Cheney returned to Cambridge and, with the strategic planning stage completed, John Burren and Steve Wilbur had also completed their contributions for now; Bob Day, who had moved to the JNT full-time in December 1992, joined the team to continue organising the newly enlarged IP programme; Les Clyne took charge of the ATM trials and piloting use of videoconferencing (over ATM) as a prospective JANET service; and John Dyer began assembling a pilot applications programme.

Choice of initial sites

From the perspective of using SMDS to enhance service performance, the choice of sites could be made largely on the basis of existing volume of use, coupled with ensuring that sites serving other sites were connected; some initial compromises were also made regarding timing of connecting some locations to fit in with BT's roll-out schedule. For the PDH/SDH network there were a number of aspects to take into account: availability of effort and expertise to participate in the pilots; potential for participation in research, developing new services and demonstrating new applications; and physical location.

In respect of location it was considered important that, even with the few sites which were going to be available on the PDH/SDH infrastructure, some sort of skeletal network of national extent should be established. This was not some expensive toy network in one corner of the country: it was intended to be the beginning of a next generation national network. In fact there was also a strong desire that as early as possible the network should reach all the constituent countries of the UK – an aspect with increasing political significance as moves toward devolution continued: the higher education funding councils had been formed in 1992 and they came into operation during 1993 (and the ISC was

---

21 These stemmed in part from JANET being an early adopter in Europe, as well as from SMDS being a non-broadcast multi-access (NBMA) network (in contrast to both Ethernet and FDDI) used as layer 2 support for IP.

renamed the JISC[22]). In the event, of course, it would prove too difficult to reach Northern Ireland initially and this would happen early in 1994.

The final choice regarding sites to be connected during the gradual development and construction of SuperJANET was something which the JNT – and subsequently UKERNA – advised the JISC on through the ACN (shortly to become the JCN). The ACN also commissioned a report from the University of Kent (Linington *et al.*, 1992), which provided a methodology for the SMDS site selection (which later helped to provide additional funds to complete the SMDS coverage). The first six sites selected for PDH connection were Cambridge, Edinburgh, Imperial College, Manchester, RAL and UCL (*see* Figure 10.1); the next six scheduled were Birmingham, Cardiff, Glasgow, Leeds, Newcastle and Nottingham. By March 1993 the first six plus Glasgow and Nottingham had four 34Mbps links to the BT manual cross-connect in Birmingham. For SMDS, in excess of 30 sites were scheduled to be available by the end of November (and these were generally delivered ahead of schedule).

IP service development

Right from the start of the JIPS pilot in November 1991, use of IP over JANET had exploded; by March 1993, most sites were using IP alongside the X.25 service and IP usage was overtaking X.25. Take-up and growth of IP had now reached the stage that it was already apparent that maintenance of X.25 was only for legacy reasons, and this would be confirmed by the formal decision of JISC in October that IP should now be the principal network protocol for JANET. Under these circumstances, the decision to get the IP service onto the higher performance infrastructure first – and worry about X.25 later – was an obvious one, on grounds of both strategy and providing urgently needed extra capacity.

The initial deployment of SuperJANET occurred during the first half of 1993, in the latter stages of this deployment of JIPS using X.25. Each of the sites with PDH infrastructure were also scheduled for SMDS connection but these were to a later schedule and more development work would be needed to bring them into operation. In order to provide early access for high-performance demonstrations, all of which were IP-based, it was decided to construct an initial six-site IP pilot network using part of the 34Mbps infrastructure. Two links at each of the first six sites (above) were configured at Birmingham to form a six-site ring. ULCC was connected as a spur to UCL using an independent 34Mbps link, enabling the NOSC to participate and gain experience of operation using the new transmission infrastructure, as well as providing international access from the pilot (*see* Figure 10.1). Cisco AGS+ routers were used at each site, connected to the local FDDI backbone. The pilot infrastructure was in operation by early summer, in time for the first formal, public demonstration in June 1993 of SuperJANET's capabilities. The IP pilot worked so well that in August it was incorporated into JIPS as a partial backbone, giving welcome additional capacity to the now hard-pressed IP service.

In parallel with the high-performance IP pilot there was also the task of bringing into operation IP over SMDS, which would begin to take over from the existing JIPS infrastructure, providing upgrades and releasing superfluous links. The initial stages consisted of router procurement and SMDS site acceptance tests.

---

22 *See* **Chapter 9.**

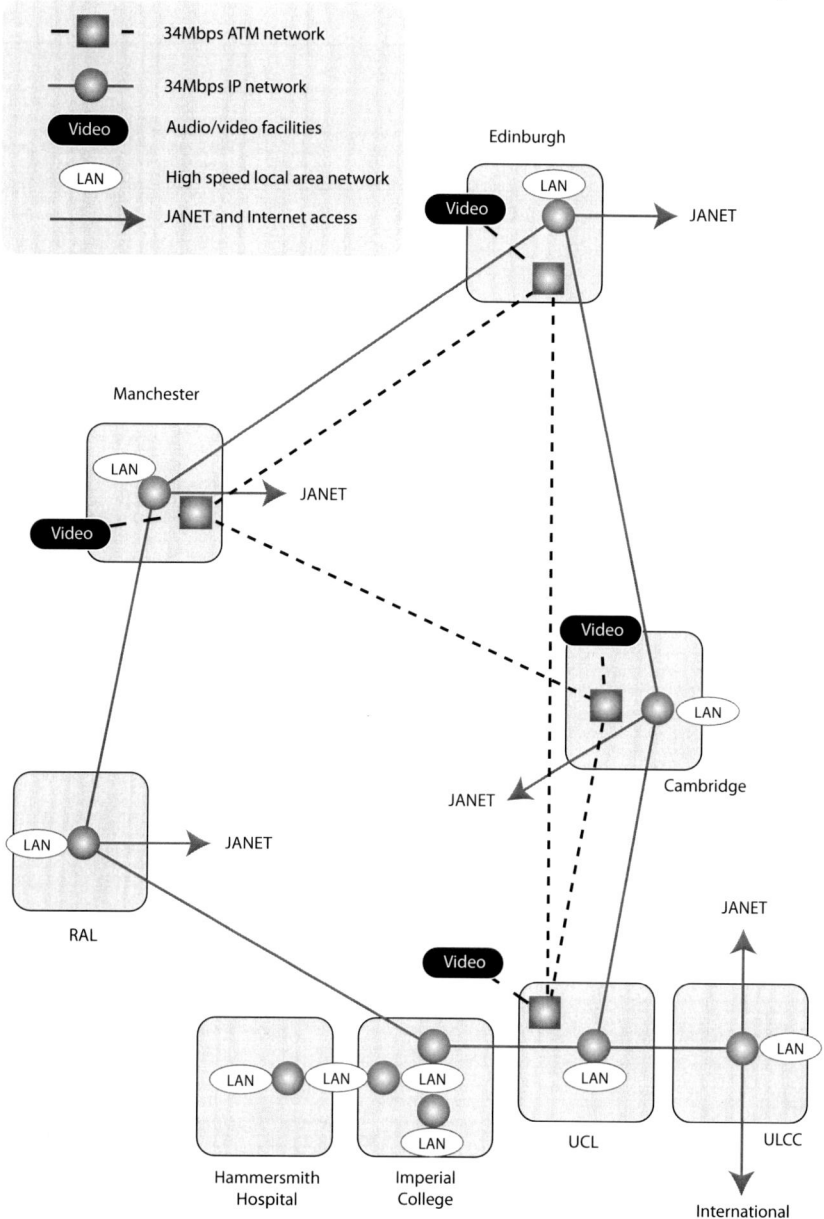

Figure 10.1. SuperJANET pilot network.

The latter were performed using IP, mostly with borrowed equipment. The essence of the acceptance tests was to check operation of the SMDS, its configuration and termination equipment[23] at each site, and it consisted of verifying connectivity;

---

23 Like 802.6, SMDS could operate over a variety of telecommunications transmission standards. For SMDS, having been defined and developed in the US, the original transmission format used had been US 45-Mbps T3; however, use of 34Mbps E3 had also been defined, initially by the European Telecommunications Standards Institute, ETSI. Mirroring the US SMDS Interest Group formed

measuring throughput using each of the access classes (4, 10, 16 and 25Mbps); and estimating point-to-point transit time between SMDS sites, specified in the service to be within 20ms. Although SMDS is not a broadcast technology, it supports multicast and this was used to support the address resolution protocol (ARP) used to translate IP addresses to layer 2 addresses. This enabled basic IP transmission between sites and was used for acceptance.

In the router procurement, Cisco and 3Com were competitive but Cisco could already support the high-speed serial interface (HSSI) needed to interface to the SMDS termination. In the latter half of 1993, as delivery and acceptance of the initial 30 sites neared completion, planning and development of JIPS to incorporate SMDS took place. The major technical accomplishment was the development of a routing architecture to exploit the SMDS multi-access architecture while maintaining the JANET connection policy (i.e., only allowing authorised traffic to use the JANET SMDS), and incorporating use of the PDH and existing JIPS backbones where necessary, including in case of failure. The routing architecture for SMDS needed a separation between the routing of data traffic and of routing protocol messages, and made innovative use of an 'exterior'[24] routing protocol called BGP (Border Gateway Protocol). By the end of 1993, all of the initial set of 30 sites were in pilot operation. Incorporation into the JIPS service then took place during the first half of 1994, at the same time as coverage expanded to over 50 sites, including those in Northern Ireland, with the PDH link being established to Queen's University Belfast in January 1994 and SMDS serving the University of Ulster and the remaining HEIs.

There were several notable features of this development, all connected with JANET being a very early adopter of this technology. At the time it was the largest SMDS IP network in Europe – possibly in the world. The routing issues which had to be solved for JIPS were new.[25] It was also new technology to BT, which meant that resolving configuration issues took longer than JANET was accustomed to because this was the first time it had not owned and operated the switching equipment: it was not possible to inspect, alter or correct such things directly – requests had to be passed across to BT operations by fax. There was also no traffic monitoring available for SMDS at the time,[26] something which would have been useful in a variety of contexts in developing and operating the SMDS segment of JIPS.

ATM pilot

Two of the 34Mbps links at each of the first six PDH sites had been used to construct the initial IP-over-PDH network. For the ATM pilot, the same topology

---

in February 1991, there was also a European SMDS Interest Group, ESIG, of which JANET was a member, formed in November 1991 to coordinate interface and service specifications for European SMDS.

24 Exterior routing protocol: one used to route between independently administered networks; BGP is an example.

25 In December 1993 the IETF Working Group on 'Routing over Large Clouds' (ROLC) was convened to address such routing in the generalised context of SMDS, ATM, X.25 and Frame Relay. This work was only just beginning at the time when JANET began using SMDS.

26 Subsequently, under a research contract with BT unconnected with JANET, Loughborough University undertook monitoring of SMDS for BT (Phillips et al, 2006), beginning in 1995 and continuing (latterly informally) until 2005 (Iain Phillips, private communication).

was planned for the initial network, exploiting the remaining two links at each site. The previous year, 1992, the first part of a tender process for ATM switches had been undertaken, in the form of a request for information, resulting in a shortlist of eight companies. For initial testing, switches were requested on loan; at the time, only Netcomm[27] could provide 34Mbps interfaces and it lent the project four ATM switches, which were sited at Cambridge, Edinburgh, Manchester and UCL (*see* Figure 10.1). At this time the major video compression and encoding standard of interest in the telecommunications industry was H.261,[28] and it was also chosen at this time for use over JANET. The standard was intended for use over ISDN, using multiple 64Kbps channels, or a non-ISDN 2Mbps link, the latter being that chosen for JANET. Being intended for operation over a synchronous bearer circuit, such a codec requires the bearer to conform to the fairly strict timing requirements of such a circuit (which include low jitter). While ATM had been designed having in mind the need for such circuit emulation, testing switch performance in this respect, especially in the presence of bursty packet-data traffic would be important.

Initial testing of 2Mbps H.261 video using ATM circuit emulation proved very encouraging, with loop-back video being established at UCL on the same day as hand-over of the PDH link took place. By early summer, 'ATM video' had reached the stage where a successful three-site demonstration could be given at the first public SuperJANET event on 8 June 1993. On this occasion the participating sites were BT in London, Edinburgh and UCL, the latter two using JANET ATM circuit emulation. One extra component is needed to implement multi-site videoconferencing using circuit-oriented video: a video switch, referred to as a multipoint control unit or MCU, and GPT lent one of these to JANET for the pilot phase of the video network. The trial was sufficiently successful that an initial procurement of ATM switches was made in the summer and 4-site videoconferencing demonstrated in October over the pilot ATM network. Following this, funds were released to expand both the ATM network (using Netcomm[29] switches) and the pilot JANET video service (using GPT codecs and MCUs) to 12 sites. Edinburgh University was contracted to run both the ATM network and the pilot video service – the origin, ultimately, of the JANET Videoconferencing Service, JVCS, being operated by Edinburgh. At this time some of the switches were also upgraded to have local (TAXI-based) ATM interfaces, so that initial testing of combining data and video traffic over ATM could begin.

Early application demonstrations

Already, during 1993, some of the new applications envisaged as exemplars for part of the case to fund SuperJANET began to appear in early demonstration form. Many of these new applications had been posited on the use of real-time interaction, most commonly exploiting human interaction, coupled with interactive still or moving image access. As had been anticipated, the research,

---

27 The same company as had supplied the X.25 switches for the 2Mbps JANET Mk II.

28 *See* Chapter 8.

29 During 1993, Netcomm (originally a spin-off company from Imperial College) first formed a technology partnership with General Datacomm, Inc., and then in December became a part of General Datacomm. The Netcomm ATM switch – known as the DV2 – used in SuperJANET became known as the General Datacomm APEX ATM switch.

teaching and practice of medicine and its related supporting sciences offered a rich field for early exploitation of SuperJANET and, at a conference held at Hammersmith Hospital on 9 November 1993 (using the pilot configuration of Figure 10.1), there were a number of reports and demonstrations of early experimental use, a few of which are mentioned here.[30]

In the field of histopathology, one project showed remote consultation using videoconferencing coupled with collaborative access to microscope slides and images; another demonstrated access to a microscope, shared between local and remote users. In both cases the emphasis was on enabling more rapid conferring when making a diagnosis in a field where expertise may be geographically dispersed.

Visualisation has been mentioned in many contexts. On this occasion, visualisation of the results of quantum mechanical calculations in the molecular modelling of medical drugs was shown on a local workstation accessing a remote supercomputer (a Cray Y-MP situated at RAL).

Teaching surgery requires students to be able to view operations, to hear what is going on, and possibly also to ask questions. Use of video was already developing in this context, particularly as it could give a better view than is possible, for example, from a viewing gallery. However, teaching of techniques in the more specialist areas may only be possible at a few locations in the UK. A demonstration was shown of a teaching session in which medical students at Edinburgh could view an operation in the Middlesex Hospital, in effect widening teaching access from a single institution to national scale. Further developments of this were also in progress at UCL, Cambridge and Manchester, building on experience already gained at UCL in using LIVENET, the University of London analogue-video teaching network, which was itself linked into SuperJANET video at UCL.

10.2 SuperJANET II: continued expansion and development 1994–1998

By the beginning of 1994, a good deal had been achieved in making new technology work – but there was plenty more in the queue, and realising anything approximating the ubiquitous and flexible SuperJANET conceived four years previously was still a long way off. When IP was formally declared the network protocol of choice for JANET in October 1993, this was in many ways a relief to everyone: users had wanted it for a considerable time; and now that JIPS was in operation, running two mainline network protocols, X.25 and IP, was proving increasingly onerous. In particular, although there had been investigations into operating X.25 over SMDS (which was possible in principle), the prospect of doing so was unattractive, especially considering the effort which would be required and that SMDS was no part of the strategic direction for SuperJANET. So, expanding the higher-performance commodity JANET service coverage was now simplified to extending IP over SMDS – and where X.25 was still needed for legacy reasons, it could be provided by X.25 over TCP (XOT).[31] On the other hand, ATM was part of SuperJANET strategic direction and it was now time to take the next steps towards its exploitation. And there was one further item to be

---

30 Reported in *Network News* 41, January 1994, where further details may be found.

31 *See* Chapter 6.

added to the list of technical things to do: BT was now ready to begin deploying the SDH links.

Deploying SDH turned out to be more of a challenge than originally anticipated. For example, a part of the standard defining how SDH could be used to carry PDH had effectively changed in deployment. The 155Mbps STM-1 rate had been carefully crafted to enable it to carry either three 45Mbps T3 circuits or four 34Mbps E3 circuits; while this was efficient, when it came to interconnecting the US and Europe, it was more convenient to be able to continue interfacing T3 to E3 and, if these were being carried by SDH/SONET, it was inconvenient if there were different numbers of E3 and T3 circuits being conveyed. So, for example, simply swapping JANET's 4×E3 in 140Mbps PDH circuits for 4×E3 in 155Mbps SDH ones was not possible. Moreover, full cross-connect SDH switches were very expensive and beyond the SuperJANET budget, so that a reconfigurable SDH network controlled by JANET was not economically feasible. But there was now a more important reason not to pursue this direction. Changing subscriber rate in a circuit network requires both the network and the end-equipment to make hardware changes. This may be acceptable in reconfiguring a growing, developing service network, where such changes are made on a timescale of years, but it is not economic for rapid or frequent, short-term reconfiguration. And it was the latter that JANET needed at this stage. ATM was the strategic universal or multiservice bearer of choice and it had matured to the point where 155Mbps could be supported directly. This offered the potential of being able to divide capacity among applications under software control. Thus the decision was made towards the end of 1993 to use SDH at the full STM-1, 155Mbps rate and concentrate on piloting IP over ATM, with the goal of carrying video and data simultaneously. In order to enable SDH/ATM trials in parallel with continued use of PDH for the IP backbone, BT agreed to leave the PDH links in service until the full set of SDH circuits had been provided.

In 1993, the newly constituted higher education funding councils had taken over from the UFC, which ceased to exist.[32] The contract with BT had been with the UFC. Following the achievements of SuperJANET during its first year and with the establishment of the plans for the next phase of development, the higher education funding councils had now ratified the continuation of SuperJANET funding, and consequently the BT contract, for the remaining three years. The first stage of SDH provision was achieved in the summer of 1994 with the establishment of five 155Mbps STM-1 links to Birmingham University, Imperial College, Manchester, RAL and UCL. As with the PDH circuits, these connections were terminated at BT's manual cross-connect in Birmingham to enable flexible reconfiguration. The eventual SDH provision was to be to nine sites, with twin 155Mbps links to five pilot sites.

Developing ATM as universal bearer

It had been apparent from the beginning that realising SuperJANET would take more resources and time than any single, initial phase. For example, in a paper to the UFC in the run up to obtaining the initial funding it was noted that:

---

32 *See* Chapter 9.

'The initial technology [ATM] is not expected to be mature enough for full service deployment in the initial phase of SuperJANET but it is expected to form the service platform for the network in subsequent phases.' (Universities Funding Council, 1991)

The initial ATM pilot had shown that switches were available which could provide circuit emulation of sufficient standard to support good quality constant bit rate video at 2Mbps. The next steps were to demonstrate IP over ATM and then to try both together. For wider deployment in an operational network, interworking between different manufacturers' equipment would need to be demonstrated; apart from this being desirable to avoid dependency upon a single source within one network, it would also be likely to occur at the boundary with another ATM network, whether campus, regional or international, academic or commercial. The remaining issue already apparent was that of connection setup. So far, a single application – the pilot videoconferencing service – was using the ATM network; connection setup was manual from Edinburgh, with connections being rearranged as necessary for each videoconference: hardly a scalable arrangement, especially if ultimately ATM connections might be needed on a dynamic basis for individual user applications.

Following the agreement towards the end of 1993 to expand the PDH ATM network, by July 1994, a 1413-site ATM network (*see* Figure 10.2) was in operation – EPSRC having funded the additional connection enabling Daresbury to participate. Initial trials of IP over ATM had been made and, with the first five SDH links imminent, plans were in hand to extend these. Some sites had now obtained their own equipment, both ATM switches and ATM interfaces for end-user host workstations. This was typically Fore equipment, originating in the US, which had already become successful in the local, campus ATM network market. The switch could support ATM links using 100Mbps TAXI, 34Mbps PDH and 155Mbps SDH; control was based on an on-board Sun workstation accessed via either a directly-connected terminal or Telnet over an Ethernet port. Host interfaces were available for Sun, HP and Silicon Graphics workstations as well as the PC. Cisco had also developed ATM interfaces for its routers, and JANET was one of its 'world test sites'. Such a router equipped with both ATM and FDDI interfaces could then be used to provide a high performance IP connection between campus LAN and ATM WAN.

Some initial testing of ATM over SDH was made by Imperial and UCL. However, equipping a further five sites with enough equipment to construct yet another pilot network was not an attractive option. At this point, BT offered JANET the opportunity for a joint ATM technology trial. In 1993, planning had begun for a Pan-European ATM Pilot with participation by 18 PNOs from 15 countries. This had been planned to begin carrying test traffic in late 1994. BT was a participant and wanted to conduct technology trials. There were also a number of research groups in the UK which had research projects funded by the European Commission and needed high-performance access to collaborators in Europe. The upshot was that, by early 1995, all five pilot SDH sites were configured to connect to BT's ATM cross-connect[33] switch in London, which not only created a

---

33 The term 'cross-connect' derives from the switching in the telephone trunk network of whole groups of circuits as a single multiplex circuit. ATM has a two-level multiplexing and switching architecture in which individual ATM virtual channels (or virtual circuits) are grouped together into virtual paths. By switching virtual paths, whole groups of virtual channels may be switched as one between physical transmission paths or links. An ATM switch performing this role is referred to as an ATM cross-connect.

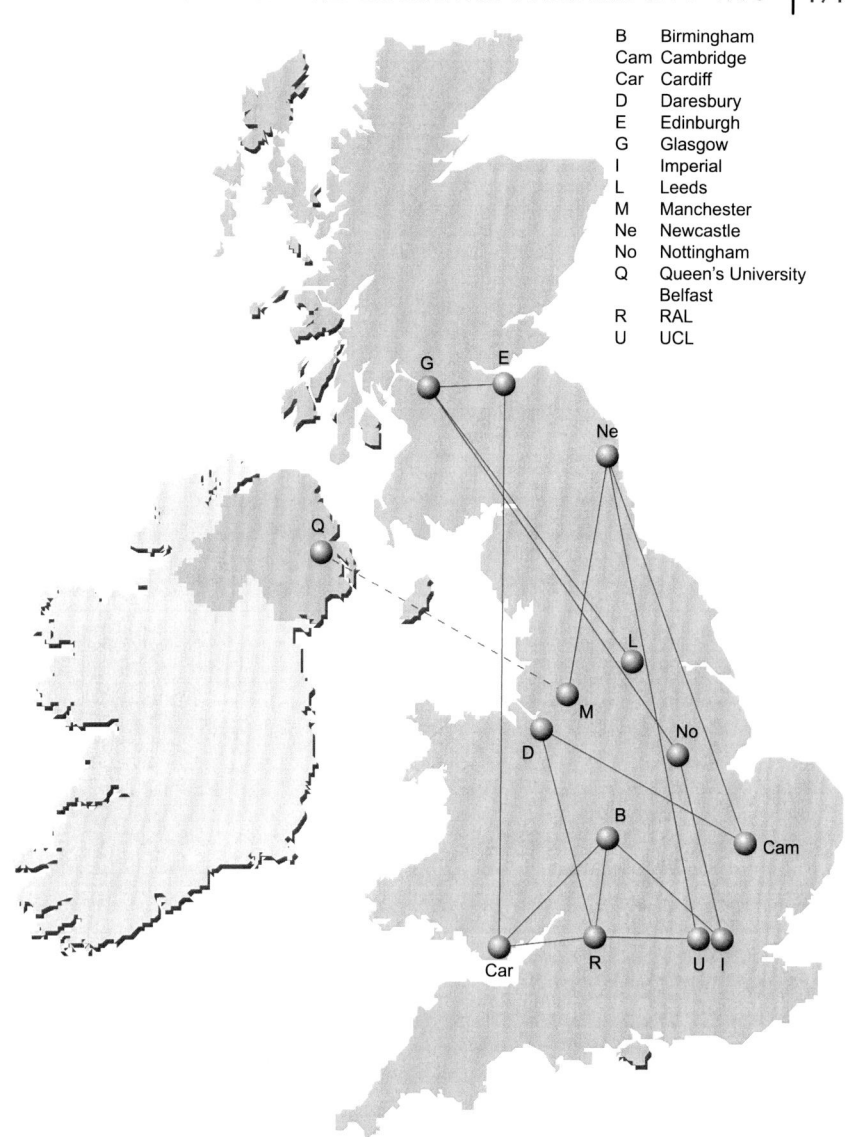

SUPERJANET II: CONTINUED EXPANSION AND DEVELOPMENT 1994–1998 | 191

B   Birmingham
Cam Cambridge
Car Cardiff
D   Daresbury
E   Edinburgh
G   Glasgow
I   Imperial
L   Leeds
M   Manchester
Ne Newcastle
No Nottingham
Q   Queen's University
    Belfast
R   RAL
U   UCL

Figure 10.2. 14-site ATM network, late summer 1994.

pilot ATM SDH network for JANET but also enabled access to the Pan-European ATM Pilot.

Towards the end of 1994, testing of mixed IP data and CBR video traffic over the ATM network was undertaken. This showed up some of the classic problems of implementing multiservice networks. As already suggested in Chapter 8, these two sorts of traffic need different handling in the network, something which neither the early switches nor early ATM host or router interfaces could provide – *see* panel 'Shaping and policing'. The issues were not peculiar to one manufacturer, though the effects certainly showed up in different ways. The situation was greatly improved as second generation equipment appeared; in the case of JANET, all the

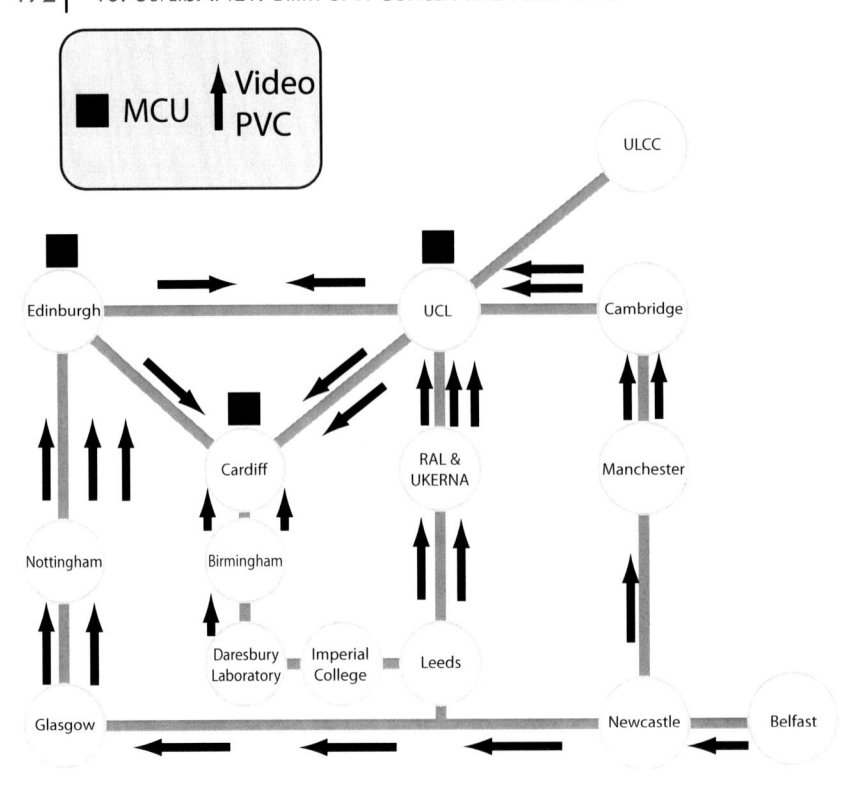

192 | 10. SuperJANET: Birth of a Concept and ATM Trials

Edinburgh
1. UCL
2. Cardiff
3. Glasgow
4. Nottingham
5. Leeds
6. Belfast
7. –
8. Edinburgh-Codec

Cardiff
1. Edinburgh
2. UCL
3. UKERNA
4. Birmingham
5. RAL
6. Daresbury
7. Cardiff Codec
8. WelshNet

UCL
1. UCL-1
2. Edinburgh
3. Cardiff
4. Cambridge
5. Newcastle
6. Manchester
7. Imperial
8. UCL-2

Figure 10.3. SuperJANET Video Network, showing MCU locations and PVC configuration.

DV2 switches were upgraded in March 1995 so that separate queues could be used for data and CBR video. Following these upgrades, successful trials were carried out of IP over ATM using a combination of 155Mbps and 34Mbps circuits; and from summer 1995, IP and video service began to be extended over the combined PDH and SDH service. By this time, the video service over ATM was using the MCU and PVC configuration shown in Figure 10.3.

SuperJANET II

At the end of 1994 approval was given to expand the coverage of SuperJANET as widely as possible by the end of the BT contract in 1997. SMDS was delivering a much improved service to a substantial part of the community and initial testing of all the components of the new technology of IP over ATM over SDH had demonstrated that although there were teething problems, ATM was beginning to deliver. For now the main deployment remained SMDS, but the strategic goal was confirmed to be ATM.

By the second half of 1995 the combination of ATM testing with newer equipment over the original PDH network and the initial five SDH sites using the BT ATM cross-connect in London had demonstrated that multiservice traffic could be carried, though careful configuration of interfaces and switches was necessary. However, the issue of ATM connection management remained and JANET was not yet at the stage where full deployment of an end-to-end ATM network was possible. But BT was completing the delivery of SDH links and the time was fast approaching when the PDH links would be withdrawn: it was now time to consider how to move to use of SDH alone.

In order to maintain the greatest flexibility for configuring links, it was decided that all the SDH links, other than those already installed for use in the technology pilot with BT, would be presented as multiple 34Mbps PDH circuits. As already described, only three such circuits can be provided in STM-1, which is wasteful of capacity and reduced the options for JANET network configuration, but it enabled some separation to be maintained between service, pilot and a small amount of research use. This plan was put into action during the second half of 1995. During 1996, ATM began to be used not only to deliver the video service but also to provide additional IP backbone capacity, particularly for international traffic, all of which went via London. The ATM topology and the IP VP trunk brought into operation during 1996 are shown in Figures 10.4 and 10.5.

Emergence of SuperJANET backbone

Towards the end of 1994 a different perspective began to take hold: the nodes established for PDH and SDH began to be regarded as 'points of presence' for SuperJANET. Partly, it was becoming clear that it would take a while yet for the new technology to mature, and until it did, resource would have to go into keeping the existing 'commodity' service operating. At the same time, the JANET community was now expanding and this also served to emphasise the amount of resources needed. The high performance LAN initiative had already enabled JANET service to dispersed campuses by means of a single access connection. Moreover, there had for some time been a growing number of sites which gained access to JANET by an evolving set of secondary connection schemes. Perhaps future access expansion could be accomplished by building on these, creating a series of separate access networks, developed as part of JANET to enable delivery of an increasing diversity of services. We shall return to the organisational stimulus for this (together with the influence that deregulation was now having) in Chapter 11.

The concept of a devolved set of regional access or delivery networks was now to begin to have its effect on the architecture of the network, and SuperJANET would begin to be referred to sometimes as a backbone. The name had been in

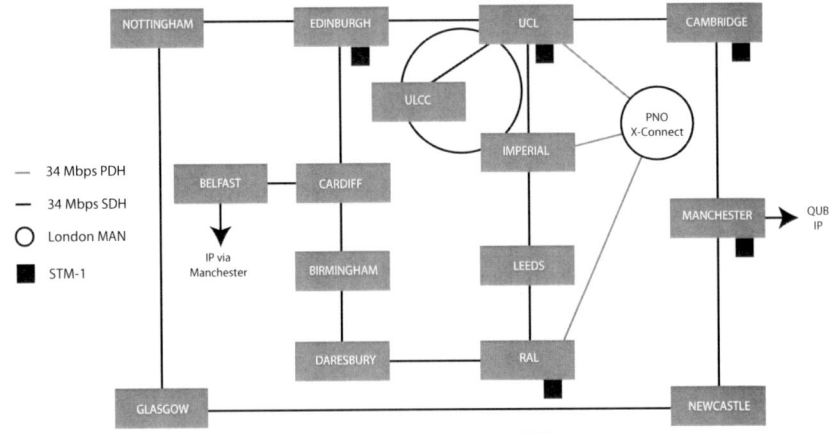

Figure 10.4. SuperJANET ATM topology from March 1996.

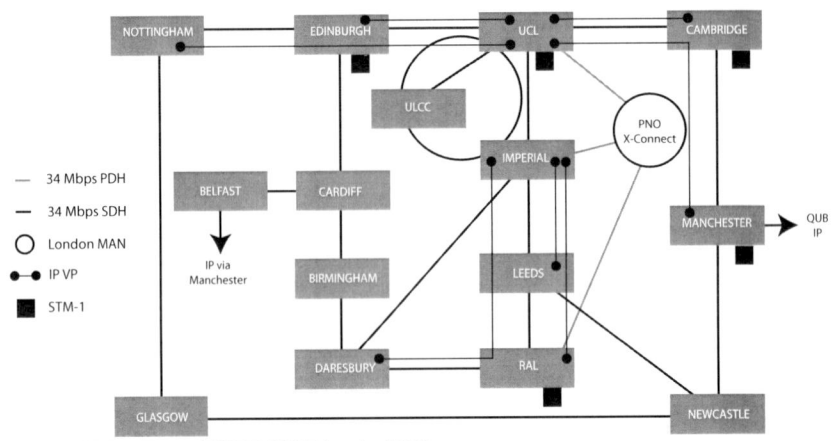

Figure 10.5. SuperJANET IP VP trunk, 1996.

the nature of a marketing ploy for the concept in raising awareness as part of the funding campaign – now SuperJANET, the name, would sometimes engender confusion: was it the backbone or the whole network?

Realisation of explicit regional access networks would begin from 1994 onwards, when the JISC announced the first stage of a programme to help in their creation. This was a natural follow-on to the high performance LAN initiative; and partly because the concept of metropolitan area networks was current at the time (with standardisation then still active in IEEE 802.6), partly because some of the networks were of urban extent, they came to be referred to as MANs.[34] The role of a MAN might range from campus backbone to a regional network interconnecting a number of institutions over a substantial fraction of the UK geographically, but they all provided the means of access to JANET for the institutions concerned. Here, they will generally be referred to as 'regional networks'; and, historically, their role has much in common with that of the original regional networks of the 1970s and early 1980s out of which JANET was initially constructed.

---

34 None ever used the DQDB technology, the subject of the IEEE 802.6 MAN standardisation effort.

Shaping and policing

If flows of packets are not to interfere with each other when traversing a network or link, then how much capacity each may take at times of congestion needs to be defined. Thus a given flow or ATM channel should not exceed a certain rate. Measuring that rate cannot in practice be done instantaneously but there remains the question of the length of time, number of packets or amount of data over which the measurement should be taken. If the objective is a constant flow then the average over any period should be the same as instantaneous and measuring the interval between the start of one packet and the next will be the nearest one can get to measuring this. Whatever technique is adopted, it can equally well be used either to measure and police incoming data or to regulate the flow of outgoing – called shaping. Moreover, the best that a shaper can do in terms of getting the most data through without any violating the policing at the other end of a link is to match the policing.

When transmitting data IP data packets over an ATM network, these need to be chopped up (segmented) into small pieces to fit into a sequence of ATM cells. Early host interfaces (such as the Fore interface) would send each set of cells for a packet at line rate. A series of packets would thus produce quite a bursty flow of cells. If the next link to be traversed was slower (as for example in the case of a host connected to a switch via a 100Mbps ATM TAXI link, followed by a 34Mbps PDH interswitch link), then either the switch would need sufficient memory to buffer whole packets-worth of cells or cells would be lost. Losing the odd cell doesn't sound too bad but, if a cell is part of an IP packet, then the receiving end cannot reassemble the packet and it is abandoned, usually triggering resending of the packet.

If a switch is being used for CBR traffic, where jitter and delay are both to be avoided, one way of proceeding is to police incoming cells very tightly, so that if a cell arrives after the previous cell sooner than it should do at the rate it has been allocated, it is dropped. Provided that all the sources obey the rates, this can be made to operate quite well, with very little delay or jitter, and no need to shape each channel on output. This was roughly how the DV2s operated initially. It worked quite well when all the traffic was generated by CBR H.261 codecs.

However, when the two forms of traffic met in the DV2s, packet data suffered serious packet loss, retransmission of packets occurred which made the situation worse, and packet data performance was killed. Equally, enough compressed video data was lost through late arrival or congestion at the switches caused by the packet data that the receiving codec lost synchronisation with the sender. Resynchronisation would typically take at least a couple of seconds, with a blank screen, a frozen picture, or a mess to look at meanwhile.

These were the particular problems experienced by JANET at the time, with the equipment in use. In general all early ATM equipment, from whatever manufacturer, experienced these sorts of teething problems. However, with the arrival of second generation equipment, most of these issues were overcome.

10. SUPERJANET: BIRTH OF A CONCEPT AND ATM TRIALS

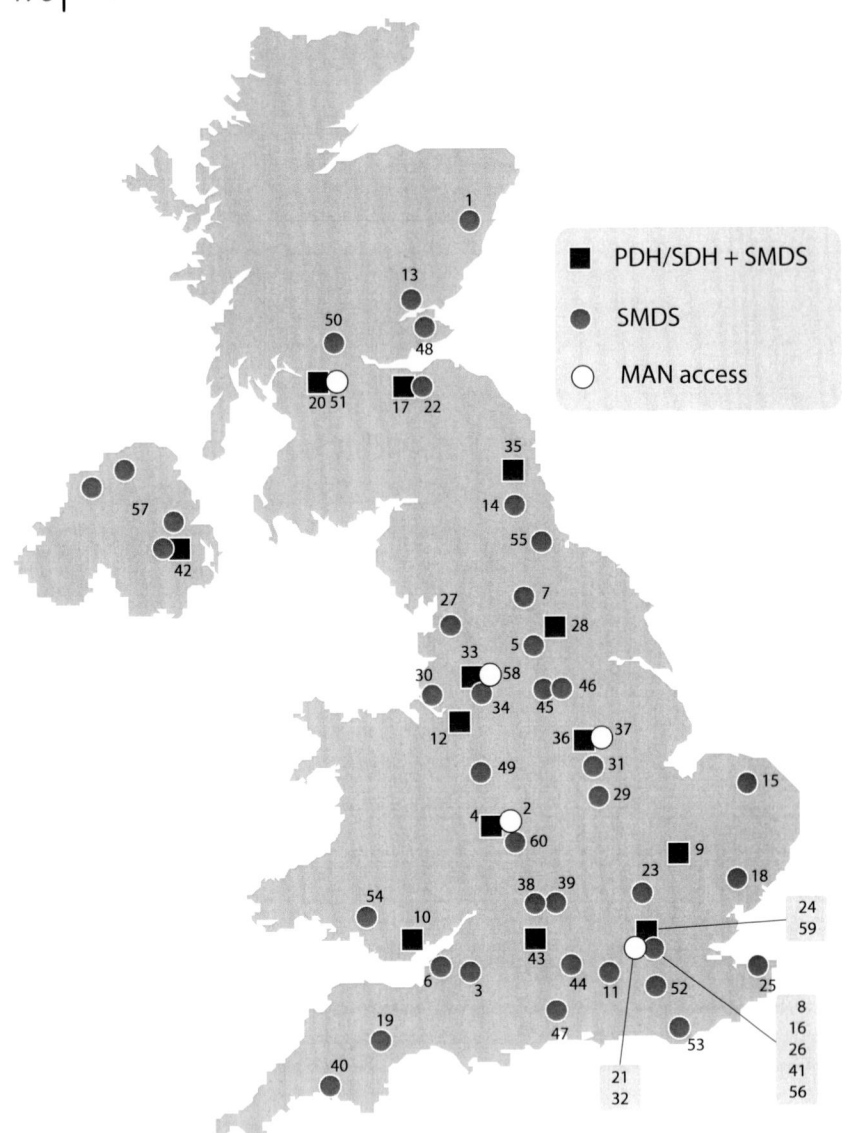

Figure 10.6. SuperJANET sites 1993/94; *see* Table 10.1 for key to sites.

To some extent, JANET had always been constructed like this, along the lines of a backbone interconnecting major switches or routers (sited at the NOCs) to which other sites connected, star-fashion. The difference now was that this became much more explicit and, in IP architecture terms and network operational terms, they reverted to being independent networks, responsible for their own internal (interior) routing, distinguished in principle from whatever exterior routing might be in use in the JANET backbone.

During the first year of the SuperJANET project, the network was expanded to serve nearly 60 sites (*see* Figure 10.6). As part of SuperJANET II, a further 27 organisations were connected using either 4Mbps SMDS or 8Mbps access lines

Table 10.1. SuperJANET sites 1993/94. (See Figure 10.6)

Site	Access	Site	Access
1. Aberdeen	SMDS	31. Loughborough	SMDS
2. Aston	MAN	32. London School of Economics	MAN
3. Bath	SMDS	33. Manchester	PDH/SDH, SMDS
4. Birmingham	PDH/SDH, SMDS	34. Manchester Metropolitan	SMDS
5. Bradford	SMDS	35. Newcastle	PDH/SDH, SMDS
6. Bristol	SMDS	36. Nottingham	PDH/SDH, SMDS
7. British Library Doc Suppy Centre	SMDS	37. Nottingham Trent	MAN
8. Brunel	SMDS	38. Oxford	SMDS
9. Cambridge	PDH/SDH, SMDS	39. Oxford Brookes	SMDS
10. Cardiff	PDH/SDH, SMDS	40. Plymouth	SMDS
11. Cranfield	SMDS	41. Queen Mary & Westfield	SMDS
12. Daresbury	PDH/SDH, SMDS	42. Queen's University Belfast	PDH/SDH
13. Dundee	SMDS	43. RAL	PDH/SDH, SMDS
14. Durham	SMDS	44. Reading	SMDS
15. East Anglia	SMDS	45. Sheffield	SMDS
16. East London	MAN	46. Sheffield Hallam	SMDS
17. Edinburgh	PDH/SDH, SMDS	47. Southampton	SMDS
18. Essex	SMDS	48. St. Andrews	SMDS
19. Exeter	SMDS	49. Staffordshire	SMDS
20. Glasgow	PDH/SDH, SMDS	50. Stirling	SMDS
21. Hammersmith Hspl	MAN	51. Strathclyde	MAN
22. Heriot-Watt	SMDS	52. Surrey	SMDS
23. Hertfordshire	SMDS	53. Sussex	SMDS
24. Imperial College	PDH/SDH, SMDS	54. Swansea	SMDS
25. Kent	SMDS	55. Teeside	SMDS
26. Kings College London	SMDS	56. ULCC	MAN, SMDS
27. Lancaster	SMDS	57. Ulster *(4 sites)*	4 × SMDS
28. Leeds	PDH/SDH, SMDS	58. UMIST	MAN
29. Leicester	SMDS	59. UCL	PDH/SDH, SMDS
30. Liverpool	SMDS	60. Warwick	SMDS

as part of the MANs initiative. Over the next few years, around 12 regional networks were created and by the end of the BT contract, approaching 100 sites were connected. SuperJANET had introduced advanced networking technologies into the networking programme and opened opportunities for multimedia applications and services. It had yet to realise all of the initial concept, but it had laid sufficient service foundations to enable the development over the next decade of such concepts as, for example, what would later be termed 'e-learning': a substantial achievement for the time.

In recognition of these achievements, in 1994 SuperJANET was awarded the British Computer Society Award,[35] one of three that year.

International access

At this point, with the SuperJANET architecture beginning to provide a separate, autonomous backbone for JANET, it became clear that peering with other networks, including particularly international access, should now in principle be a backbone function: connection should be specifically to the backbone and its configuration and provisioning should reflect this. Although not really new, it was not exactly how the interconnections were arranged in practice; nevertheless, in the next few years, it would become more explicit.

The first step has already been mentioned in Chapter 7: the beginning of rationalisation of peering arrangements among UK Internet Service Providers (ISPs) in London which resulted in 1994 in the creation of the LINX (the London Internet exchange), initially located in Telehouse in London's Docklands. The LINX would later expand beyond a single location, in particular Telecity, also in Docklands.

Though organisationally convenient, this still left the problems of providing peering capacity and peering resilience for JANET. Indeed, prima facie, resilience was if anything worse: both ULCC and Telehouse presented single points of failure for the interconnection of JANET with external networks. Part of the approach to addressing this took the form of what came to be referred to as Dual Interconnect Across London, DIAL. By 1996, the London MAN[36] was in operation, providing a 155Mbps redundant interconnection for JANET between ULCC and Telehouse. This still left ULCC and Telehouse as single points of failure. Subsequently, a further international point of interconnection was established at UCL and independent connections were also made via Telecity as the LINX itself expanded. Needless to say, this did not all happen at once but rather over a period of years during the latter part of the 1990s, each upgrade or addition typically being a combination of improved resilience and increased capacity.

It will not have escaped notice that the international interconnections are all in London,[37] an inconvenience for both resilience and traffic patterns on JANET. International traffic forms a substantial fraction of the traffic generated by everybody on JANET which makes London a hotspot. At this time, during the mid-1990s, about half of all JANET traffic was to (or from) the US. The DIAL concept was subsequently exploited to help partly alleviate this, for example, by adding a link in 2000 from the UCL backbone node to the Leeds node, effectively adding internal international capacity, resilient against access failure via ULCC.

We've seen already (Chapter 7) how increased international capacity was in continual demand from the moment it was achieved: growth in network capacity internationally, as elsewhere, was a facet of its success, the only topic of discussion being its rate. Transatlantic capacity had reached 4Mbps by the summer of 1995; and by 1996 it was 20Mbps, then being provided by a combination of Sprint and

---

35　See *Network News*, No.43, 1, November 1994 (http://www.webarchive.ja.net/services/publications/archive/newsletters/network-news/issue43/bcs.html, accessed March 2010).

36　Subsequently the London Metropolitan Network, LMN.

37　Actually, most submarine cables tend to come ashore in Cornwall but access for the likes of JANET is in London.

ANS. Following further deregulation in the international telecommunications market, a more substantial upgrade in capacity was brought into operation in June 1997 with the commissioning of a 45Mbps T3 link, which more than doubled the capacity for JANET users to the US – though, even as this was being announced, it was expected that it would be saturated by the autumn. A second T3 link was added in 1998; by January 1999, both would be saturated with traffic coming into JANET; by July of the same year this 90Mbps had been replaced by 310Mbps, in the form of two 155Mbps SDH/SONET circuits; and by March 2001 this had increased to just under 1Gbps (930Mbps, provided by six 155Mbps circuits). By then it was evidently time to contemplate the next change in approach to provisioning international network capacity.

The pattern was the same for European interconnection but with the added complication of involvement by many countries. The Commission of the European Union in its series of Framework programmes for development and research had taken the stance that all national infrastructure was a matter for national governments. However, it was persuaded that international interconnection infrastructure, at least in its development phase to underpin collaborative information technology research and development, was something it could help to fund – indeed, an initial example of this was the European funding of COSINE IXI in 1989. As we've already seen (Chapter 7), by 1994 it was time for Europe to seek a substantial upgrade in interconnection capacity.

The EU was launching the fourth in its series of programmes, Framework IV. Overall direction of the programme was split amongst a number of directorates. Two of these had interests in IT: DG-III, which was responsible for Esprit and ACTS (Advanced Communications Technologies and Services) programmes, and DG-XIII, which was responsible for the Telematics for Research programme. David Hartley, CEO of the newly created UKERNA, who in 1981 had chaired the Network Management Committee which oversaw the creation of JANET, then in 1990 the Network Association Steering Group which oversaw the creation of UKERNA, now turned his attention to achieving a European interconnect facility of a capacity commensurate with SuperJANET. Encouraged by a specific invitation from the EU to create such a facility as part of Framework IV, in February 1995 a group of representatives of European NRENs met in London to consider making a proposal in response. The suggestion was that the interconnect should operate at transmission speeds in the range 34–155Mbps. A consortium of all the NRENs in Western Europe submitted a proposal called TEN-34 (Trans-European Network at 34Mbps) in early summer, with DANTE as the coordinating partner. Subsequently several more countries joined from Eastern Europe. The proposal was accepted. The project was overseen by a management committee chaired by David Hartley. Led by DANTE, by August 1996 planning had been completed and the service began in March 1997, with UKERNA and the NOSC at ULCC having won the TEN-34 operations contract in February. The links comprising the service varied in capacity from 6 to 24Mbps; the overall achievement was of a substantial upgrade. Included in the service was a T3, 45Mbps transatlantic link for Europe. This was independent of the links JANET already had but it now also meant that there was the possibility of additional backup for JANET transatlantic traffic.

TEN-34 was significant for providing the first integrated, pan-European NREN backbone – the successors of which have continued to provide European NREN backbone services to the time of writing. The promise of substantial improvement in capacity coupled with EU funding was sufficient to persuade all of Europe to participate. However, it was not the only pan-European network activity taking place at this time. It will be recalled that in 1993 BT had announced its intention of participating in the Pan-European ATM trial (see under 'Developing ATM as universal bearer'). This was a European PNO trial, for which planning had begun in 1993. As described above, the ATM segment of SuperJANET made use of the BT Pan-European cross-connect in London as part of the construction of the ATM-over-SDH trial. Since the Pan-European trial itself needed test traffic, it was agreed that it could be used to carry research and development traffic generated by EU Framework projects. Access to this was available to JANET users for nearly a year from March 1995 and a number of projects with UK participation took advantage of this. (This was available a year before the main ATM segment of SuperJANET migrated from PDH to SDH in March 1996.)

The Pan-European ATM pilot ended in December 1995 and with it the BT cross-connect facility in London. A proposal was put to JISC to extend collaborative technology trials with BT for a further year, until the prospective end of the BT SuperJANET contract; however, this was turned down because it did not seem cost-effective use of funds for the short time remaining. This left some 155Mbps SDH links unused. Meanwhile, European PNO ATM activity had not ended. Having completed the European technology trial, the PNOs moved to the next stage of a pre-competitive market trial called JAMES (Joint ATM Experiment on European Services), which began in April 1996, lasting for two years until March 1998. Some of the spare SuperJANET 155Mbps capacity was used to link Manchester and RAL to this, which again was used by several EU projects having UK participation.

10.3 SuperJANET III: ATM 1997–2001

With the additional resources provided under the auspices of SuperJANET II, by 1996 SuperJANET was now providing an upgraded and reasonably stable service over a combined infrastructure of SMDS, ATM and PDH. ATM over SDH and IP over ATM had both been demonstrated to work, and the next step was to move to a service provided by an ATM bearer over SDH. The extent to which ATM might become the end-to-end bearer of choice was still not yet clear but, at least within the backbone, ATM seemed to offer the potential for provisioning individual (virtual) channels with a greater degree of flexibility than SDH. It had also enabled the provision of a videoconferencing service which might be able to coexist on the same bearer network as the packet data service, though it was restricted to sites with end-system ATM access. With broadband ISDN still predicated on use of ATM, coupled with the known development of ATM switches capable of handling link rates of 622Mbps and the availability of routers which could support IP over ATM, JANET now confirmed its decision to exploit its investment in this apparently strategic technology.

However, parts of the SuperJANET development programme had taken longer than anticipated five years previously. In particular, the transition to SDH was taking much longer than had been hoped; and the development and evolution of the regional network (MAN) and backbone architecture, with its origins in a combination of administrative reorganisation and the High Performance LAN initiative, was still in full swing, partially tied as it was to regional and local funding cycles, even with the help of the JISC funding for the MAN initiative. In view of all this, in mid-1996 the BT SuperJANET contract, which had been due to end in March 1997, was initially extended to September 1997; and subsequently, it was further extended to the end of March 1998.

Procurement of the replacement infrastructure took place in 1997 and resulted in the award in October 1997 of a three-year contract to Cable & Wireless.[38] This was a pretty tight timescale with Christmas looming. In the event the replacement came in two parts, in January and February 1998, and the switchover to the new network was achieved during March.

The architecture of SuperJANET III confirmed that JANET was now fully committed to a network composed of autonomous campus networks, regional access networks and a central backbone network: the SuperJANET III replacement was of the backbone, accompanied by continued support for regional access network development. However, it also continued the pattern begun with the BT contract in which management and operation of the backbone was split between the supplier on one hand and UKERNA and the NOSC on the other.[39] The network, shown in Figure 10.7, was composed of an interior core, owned, managed and operated by Cable & Wireless; this was formed of a ring of nodes situated at Bristol, London, Leeds and Manchester. The exterior part, managed by UKERNA, consisted of 13 regional network nodes (termed Backbone Edge Nodes or BENs for the lifetime of SuperJANET III) connected to the interior core nodes by 34- or 155Mbps circuits (two 155Mbps circuits in the case of ULCC). Sites would access JANET directly via the regional network that they were already a part of, or else by leased line to the nearest BEN (or regional network), or by continued use of BT's SMDS. This last was soon to cause difficulty because it effectively partitioned JANET, complicating the provision of both routing and capacity.

Expanding community and new services

For some years during the early and mid-1990s there was protracted discussion about the need for the Further Education sector to gain access to JANET and its associated services. Actually, there was essentially 'violent agreement' in all quarters that it was desirable and necessary, and colleges had been pressing for it for some considerable time – indeed some had of necessity found ways to connect either to the Internet using public services or via JANET 'sponsored' connections, a mechanism originally conceived to enable non-educational organisations working with the community to gain access. There was just one problem: where to find the money. Eventually (as we shall see in Chapter 11), in 1997, HEFCE began a programme managed by UKERNA of funding and rationalising connections for the English FE colleges, beginning with an initial 72 colleges to be connected during 1997/98. Wales had already begun an independent programme to provide colleges with network access, one aspect of which had been WelshNet and the beginnings of provision of videoconferencing for colleges; Scotland would initiate a programme a little later.

The university community had evolved network expertise as an extension of managing institutional computer services over the previous two decades and more, alongside the development of JANET itself. Though some FE colleges had

---

38   Cable and Wireless Communications Services Ltd.

39   Actually there were more participants in managing the network on the UKERNA side, since Edinburgh had operated the pilot ATM network together with the ATM-based pilot videoconferencing service (*see* Section 10.1).

10. SuperJANET: Birth of a Concept and ATM Trials

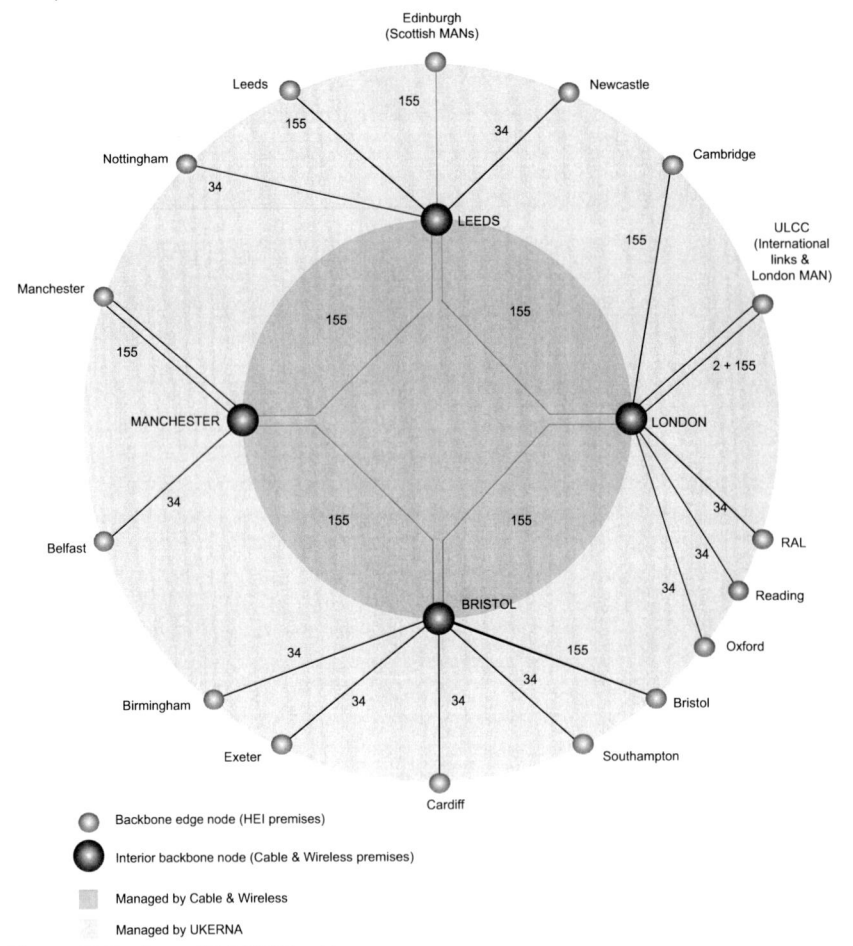

Figure 10.7. SuperJANET III backbone.

similar such in-house service expertise, it was by no means general, particularly in the case of smaller colleges. Accordingly, UKERNA and the NOSC began to offer a managed router service whereby the NOSC would manage the JANET access router for a college (provided certain conditions were met, such as the router being an approved model, etc.).

The decision made in SuperJANET III to use ATM has been cast in terms of its still being regarded as a 'strategic' technology. Strategic for what? Recalling that one of the strands driving the original SuperJANET concept was to be able to deliver continuous media as well as packet data, an aspiration shared with broadband ISDN, perhaps the more significant, long term underlying commitment partially signalled by this choice of technology was that JANET would henceforth be a multiservice network. Unfortunately, although it would be confirmed that continuous media delivery was now a requirement here to stay, several circumstances would conspire to deny JANET achieving unified delivery with SuperJANET III.

The ATM-based pilot could not become a service unless ATM delivery became ubiquitous. The SuperJANET III backbone was based on ATM but most sites connected to the IP service, not the underlying ATM one, so that end-to-end ATM delivery was still hardly more prevalent than in SuperJANET II. This presented a serious obstacle to deployment of a video service accessible to all, but it was not the only one. Public videoconferencing was now widely available, including internationally, over ISDN and there was now pressure to be able to interwork with this; although this service and the SuperJANET pilot used the same (H.261) compression scheme, interworking was not readily available because of the different transmission rates. There were other technical issues. The pilot video service depended on end-to-end emulation of an E1 circuit. As has already been indicated, owing to the absence of standards emerging for ATM cell policing and scheduling behaviour, difficulties had already been experienced in operating mixed data and circuit emulation services in a network with equipment from a variety of manufacturers. To the extent that these had been overcome in pilot operation, hardware and software equipment modifications had been required. In SuperJANET III, yet another variety of switch[40] was introduced in the core which had further different properties: it became evident that attempting to tune individual channels to provide satisfactory E1 emulation without interfering with the data service (or vice versa) was not a practical proposition on a service basis. Furthermore, the E1 interfaces needed for this service had become so expensive as to be unviable in any numbers. In view of all this, it was decided to take a different approach, based on ISDN, and for a while the studio-based videoconferencing service became separated from the JANET service network. The way in which videoconferencing services developed thereafter is described in Chapter 12 (Section 12.3).

Although basing a national videoconferencing service on ATM was not practical at this time, as the SuperJANET III service stabilised it was decided to pilot a limited form of 'managed bandwidth service' (MBS), a term coined to describe generally the provision of defined capacity across a network. There was some demand for such a service in support of research projects; it also enabled the continuation of trials of ATM features which would be needed in case of any return to multiservice operation. As a result, in 1999 a pilot Managed Bandwidth Service was provided, based on manually configured ATM channels, using such facilities for scheduling and policing transmission capacity as were available. The service was restricted to no more than ten percent of the link capacity on the designated route, in order to protect the main JANET service. The service was used by a few projects, including European Framework V projects, which needed either dedicated capacity or to support projects which required end-to-end ATM connections (using the corresponding facilities of TEN-155). Although this was a requirement of only a small minority of the community, it was one which harked back to the origins of SuperJANET: that the network should be capable of configuration at a low level to serve special transmission requirements not capable of satisfaction through the 'commodity' service. The requirement was one not peculiar to JANET and would become sufficiently important over the next few years as to contribute to the way the network would develop.

---

40   The Cable & Wireless core was based on Cisco Lightstream switches.

The millennium bug

Much could be written about the design, development, programming and effort which has gone into dealing with time in computers and networks – to say nothing of its nature and role in physics and philosophy – but here is not the place! Many activities in automated systems are tied to time. Briefly, there are typically three aspects which have to be addressed in computer-based systems: obtaining a current date and suitably accurate time of day; measuring the passage of time; and converting the end of a time interval to a date and time of day by reference to a known date and time at the start of the interval. This last is a calculation routinely carried out in any computer, for example, when updating its date from its internal clock, in the intervals between obtaining a new reference date and time from an external source.

The so-called millennium bug was concerned with the last – and apparently simplest – of these activities. There are two aspects associated with this: representation of the date and calculation of the passage of time from a chosen reference date. The simpler issue is associated with the representation in storage of the date in terms of the day, month and year. The first two represent no particular problem but, dating from the early days of computers, when memory was expensive and small, space was frequently saved by storing only the last two digits of the year and choosing a reference year in the 1960s or early 1970s (sometimes referred to as the 'era' or 'epoch' in this context). The specific issue was then whether, as the millennium dawned, a date represented by '00' would be correctly interpreted as 2000 or incorrectly as 1900 – and similarly for the succeeding years. (If the two-digit representation continues to be retained, there will then be another problem when a century has elapsed after the chosen 'epoch'.)

The second issue has to do with leap years. In everyday life, people are in the habit of associating the passage of time with astronomical events, days and years in particular. Inconveniently, whichever way is adopted, there is not an integral number of days in a year. To prevent the seasons gradually moving through the calendar, while maintaining the convention of having a whole number of days in a year, the mechanism of the leap year with an extra day was introduced by Julius Caesar (46 BC) and then modified by Pope Gregory XIII (1582). The *de facto* international standard Gregorian calendar specifies leap years whenever the year is divisible by 4, except centennial years *unless* they are divisible by 400. The potential issue then arises from the many conversion programs having been written all over the world in the previous few decades but the last part of the Gregorian rule being relatively obscure. Would the code everywhere correctly conclude that the year 2000 was a leap year?

Nobody being prepared *a priori* to assert categorically that nothing nasty would happen in case something, somewhere got either of these wrong, for the few years leading up to 2000, effectively a world-wide code audit was instituted.

JANET and its community, along with everyone else, was caught up in this one-off event of a lifetime, and contributed its share of world-wide effort in planning for all conceivable eventualities. In the event, JANET did not experience any problems, either as the year changed to 2000 or at the end of February. After the events, there was a rash of media reports from those who were prepared *a posteriori* to assert that the potential problem had been exaggerated; there were also reports of babies registered as being 100 years of age at birth, while in some places gaol sentences were increased or reduced by a century!

International access

In Europe, the original TEN-34 project had targeted transmission rates in the range 34–155Mbps. Having achieved initial service in March 1997, DANTE and the European NREN consortium submitted a proposal under Framework V[41] to develop and upgrade the network further. The project was called QUANTUM (QUAlity Network Technology for User-oriented Multimedia) and was effectively the test and evaluation programme which preceded the deployment of TEN-155. TEN-34 already exploited ATM technology; for TEN-155 it was planned to offer options for ATM MBS as well as the main IP service. The alpha testing of the MBS network technology (Perkins and Cooper, 1999; de Arce *et al*, 1999) was undertaken (under a Byzantine administrative arrangement involving ERCIM[42] and RAL) by the Framework V multimedia project, MECCANO (Multimedia Education and Conferencing Collaboration over ATM Networks and Others), the third and final project in the UCL-led MICE trilogy. During 1998 the network was upgraded to a transmission rate of 155Mbps and TEN-155 became operational in December 1998. Subsequently, in 2000, 622Mbps links were installed.

Apart from the steady addition of transatlantic capacity already mentioned previously, there were also developments in the US networking scene which interacted with JANET at this time. In 1995, in partnership with MCI, NSF had upgraded the academic US backbone to 155Mbps (known as the very-high-performance Backbone Network Service, vBNS); subsequently, it would be upgraded to 622Mbps in 1998. However, the MCI-NSF partnership was for a limited period and in 1996 UCAID, the University Corporation for Advanced Internet Development, was formed and took over responsibility for the provision of the academic backbone to the US in 1998. The title Internet2 was adopted for the project to develop the next generation of academic and research network service (and subsequently came to refer to UCAID also). In 1999, UCAID and TERENA agreed to promote collaboration between Europe and the US. UKERNA also made such an agreement with UCAID and began a series of open meetings with the Internet2 project. Plans were developed for peering SuperJANET and Abilene, the next generation US backbone, and this peering began service at

---

41 Strictly speaking, the project was submitted in the closing stages of Framework IV; however, it was intended for use in support of Framework V, and funds from the latter took over once it was established.

42 The European Research Consortium for Informatics and Mathematics; *see* Glossary. MECCANO, an existing Framework V project, was in need of European network transmission infrastructure and well-placed to carry out alpha testing of that proposed for TEN-155, essentially on a *quid pro quo* basis. DANTE had subcontracted ERCIM to participate in the testing, on the basis that several of its members were already part of MECCANO. However, UCL was not a member of ERCIM so, as UK ERCIM member, RAL participated with UCL in the MBS alpha test part of QUANTUM.

34Mbps in January 2000. Later in the same year eight projects were funded to enable transatlantic collaboration. As with all other network links, as the projects began operation so more capacity was needed and, in January 2001, the peering was increased to 155Mbps, while a direct connection at 622Mbps was sought.

As was to be expected, the increased international capacity then had a knock-on effect and more capacity was needed across London and within JANET. Accordingly, further 155Mbps links were installed between London and Manchester and between Telehouse and Leeds. Internally, in London, all the existing links in and out of Telehouse were upgraded from 100Mbps to 1Gbps.

Technology developments

While SuperJANET was undergoing its first three phases of development, further advances had of course also taken place in basic network technologies. The most influential of these has already been mentioned: the development of 100Mbps, fast Ethernet (Chapters 5 and 8). Although early products began to become available in 1993, before standardisation in 1995, it was not really until the second half of the 1990s that the full impact of this began to be appreciated. IP had become established as the network 'platform' of choice: every operating system for computers large or small was now developing network applications and support on the basis of IP. Within the local area, IP had been operating primarily over Ethernet for around a decade; and now that switched, full-duplex technology had released Ethernet from all of its scaling issues, it was confirmed as the technology of choice – to the desktop or any other end-system. This put an end to any further development of FDDI (specifically, FDDI-II, which never became a product, and FFOL, the FDDI follow-on, which was never developed even as a paper design). In the presence of the success of Ethernet, a LAN emulation (LANE) technology which operated over ATM was developed in the early 1990s. The trouble was that quite a lot of additional mechanism was needed: a server process was required to emulate the group communication facility of Ethernet; another to map IP addresses to ATM addresses; and even though LANE enabled the use of existing IP support in a host, ATM host hardware and basic software support were still needed. When, on top of this, gigabit Ethernet appeared in 1998, the fate of ATM was sealed because now the effect of the small cell size came home to roost: ATM switches could no longer match the performance of Ethernet switches. Moreover, gigabit routers now began to appear, making ATM no longer the technology of choice for either campus or wide area.

However, ATM had bequeathed a legacy. At the time when taking individual routing decisions for every IP packet was still a factor holding back router performance, the simplicity of being able to switch on a label like the virtual channel identifier in ATM[43] was exploited for IP packets by inserting a four-byte header before the IP header, and then having a router use this label to look up which output port to use. In this new context the technique was called multiprotocol label switching (MPLS).[44] It needed a signalling system to manage and distribute the labels, for which simplicity favoured a dedicated protocol called Label Distribution Protocol (LDP), while protocol reuse favoured RSVP.

---

43 The idea is much older, going back at least as far as the beginning of X.25 in the early 1970s.

44 Deriving from the original form in which the label might be partially constructed from an ATM label, enabling an ATM switch to assist with the necessary switching.

Soon after its development, as router design advanced, the technique no longer offered a performance advantage and MPLS would have been discarded were it not that label-switched paths can be set up which differ from regular routes, a possibility which can offer an advantage for routing particular flows along especially provisioned paths: a topic generally referred to as traffic engineering. Subsequently, a generalised version of MPLS (GMPLS) was partially developed on the basis of the observation that although the header in MPLS associates a packet with a particular virtual path or flow in a packet network, it might also be used to associate the packet with a TDM or WDM channel in an underlying transmission network.

Two other developments also appeared, both associated with LANs. Given the ubiquity of 802-style LANs, particularly Ethernet, PNOs developed a proprietary technology, originally called LAN Extension Service (LES),[45] whereby LAN services could be extended over the wide area on a point-to-point basis using a leased line. From the late 1990s onwards, JANET regional networks took advantage of this service as an easy and convenient (and cheap) way to extend coverage of a LAN to MAN proportions (though the behaviour under line error conditions was typically ill-defined).

The second development was the VLAN, the virtual LAN. This enabled a single physical, switched LAN infrastructure to be used to implement a number of logical LANs, effectively partitioning the interconnected hosts into a mutually exclusive number of groups on the basis of switch and port number, MAC address or IP address. Since each VLAN acts like a separate LAN, a broadcast is restricted to the VLAN in which it originated rather than interrupting every host on the physical network. This partitioning of the broadcast domain can help performance in a large network. Moreover, since each VLAN consists of a mutually exclusive set of hosts which cannot communicate at layer 2, VLANs may be used to implement a basic level of security amongst different groups of users. These factors, coupled with the convenience of the separation of logical host or workstation groupings from details of physical network topology and port location, ensured the rapid growth in popularity of VLANs in large organisations, where regular reorganisation with associated office relocation is endemic.[46]

Grid computing and e-Science

In the latter part of the 1990s, events in the scientific world were moving towards the adoption of an enhanced approach to the exploitation of computation and information processing in scientific research. The instruments of big science were so expensive that, like the James Clerk Maxwell telescope in Hawaii or the Large Hadron Collider at CERN, they had to be shared internationally. Likewise, there were valuable databases scattered around the world, enhanced access to which was increasingly widely needed. Moreover, in some instances a number of facilities (be they telescopes or computers, to pick just two examples) or institutions collaborating in data processing might need to co-operate to undertake advanced

---

45 Subsequently rebranded as part of 'Short-Haul Data Services', SHDS.

46 To correct any impression of its being an instant universal panacea, it should be noted that VLAN configuration requires careful management in a large organisation. Moreover, until the complications induced in the spanning tree algorithm by the introduction of VLANs had been overcome, which took some years, the operation of the algorithm could take many minutes to converge and restore the network after a break, during which it would be unusable throughout an organisation.

investigations – notable examples being in astronomy and particle physics (*see* panel 'Communications for radio astronomy and particle physics'). Everything was pointing to a need for communications and processing technologies to facilitate co-operation not only amongst teams of investigators but also amongst information and processing resources.

One of the activities undertaken as part of the US Gigabit Testbed Initiative, as well as over early stages of SuperJANET, was the partitioning of large computing tasks into pieces which might with advantage be processed on supercomputers of differing architecture, either serially or in parallel. An older example of splitting up processing amongst many processors is provided by particle physics processing: experiments consist of recording huge numbers of events, each of which is independent and is processed independently, and none of which requires any great processing capacity. Since the mid-1970s, ways have been devised for exploiting this to enable parallel processing by numbers of small processors to replace use of large, expensive computers. The advent of the networked PC made this practical and rapidly gave rise to early examples of processor farms, found worldwide at the time of writing in many contexts. In the 1990s the notion that this concept could be further automated took hold and led to work which originated out of Argonne National Laboratory (ANL) in the US, described in the book by Foster and Kesselman (1999) and referred to as 'Grid computing' or sometimes just 'The Grid' if the context is unambiguous – the term originating from the oft-used analogy with electricity distribution to denote a ubiquitously available computing utility.

The concept of distributing a task was quite intensively studied in the 1970s, perhaps most notably at Xerox PARC and the University of Cambridge, the latter building a system in which a number of personal systems could be dynamically allocated and assembled from components scattered around a LAN following user authentication.[47] Indeed, the example of serial processing of a task by a number of remote systems was also catered for in the JANET Red Book protocol for handling job transfer (Chapter 3). During the 1980s, distributed systems engineering continued with much of the focus latterly (early 1990s onwards) being on the software engineering constructs to enable dynamic binding. The Grid concept was of a potentially world-wide set of computation, data and information resources embedded in a universal communications environment which users could access individually or in concert. Towards the end of the 1990s the idea was taking hold as a major strategic component of how 'big science' could be tackled in future. In this context it played a major role in stimulating 'e-Science' research programmes world-wide, not least in the UK. In terms of realising Grid-based e-Science, one way of viewing the Grid is as a set of dynamically assembled virtual supercomputers, together with the data and information sources relevant to whatever computational problem is under study. Apart from the information repositories and computing elements required, it is evident that a ubiquitous network of substantial capacity is needed to which all users and resources are connected.

The term 'e-Science' was coined to refer to 'electronically' mediated scientific research, 'electronic' here being taken to encompass every aspect of information and communication technology. From a communication perspective, the increasing volumes of data to be transported on timescales dictated by the need for effective co-operation renewed attention in the scientific community toward achieving high-performance networking. Two scientific activities at this time achieved

---

[47] Described in the book by Needham and Herbert (1982).

Communications for radio astronomy and particle physics

Big telescopes need large apertures, where 'large' is measured in terms of the wavelength of the light or radio waves being used in the observation. Big optical telescopes have apertures up to ten metres: a big radio telescope might need an aperture of several thousand kilometres. Since this is impractical, observations from a collection of telescopes are assembled – synthesised – to create a sort of patchwork of samples of what would have been observed with a telescope having an aperture diameter of roughly the largest separation amongst the collection of observations. Sometimes referred to as 'synthesised aperture' or 'aperture synthesis', the technique is also more commonly called 'very long baseline interferometry' or VLBI. The technique has been in use for decades, traditionally with many large tapes of observation data all being collected together for synthesis and analysis. This might take weeks or months and only after that would it be known what the results were – or, worse, that something needed adjusting and the observation repeating. Once wide-area communication reached transmission rates in the region of gigabits per second, replacing tapes by direct communication between telescopes and processing centre became a possibility. A major processing centre, in Europe and worldwide, is JIVE, the Joint Institute for Very Long Baseline Interferometry in Europe, situated in Dwingeloo, The Netherlands. By the end of the 1990s communication speeds were such that achieving direct transfer of observation data from telescopes to JIVE began to be a real possibility, and high-speed networks throughout Europe and around the world started to play their parts in achieving this.

At the other end of the physical scale, particle physics was accumulating ever-increasing volumes of data for analysis with each succeeding generation of accelerators. By the second half of the 1990s planning was under way for handling the projected requirements of the international particle physics programme which would begin in earnest a decade later with the commissioning of the Large Hadron Collider (LHC) at CERN.[1] The predicted requirement was for transport of at least three orders of magnitude more data than in the past. Equally important was the decision that instead of concentrating a major part of the processing at CERN, it would be distributed among participants; consequently, it would be important that data flowed steadily from CERN to the home institutions of experimental groups. The general idea was to disseminate the data to participating countries, thence to participating institutions, via a hierarchical set of 'tiers'. The tier 1 centre in the UK is RAL; tier 2 centres are typically composed of universities. Apart from considerable development having been put in by those responsible for data handling and storage systems in the particle physics community, a crucial component would be the ability to achieve the required data transport capacity from CERN to RAL.

---

1 As of 2009, the first physics experiments are scheduled for 2010.

particular prominence in respect of communications: both served as drivers and are outlined in the accompanying panel 'Communications for radio astronomy and particle physics'. In the UK the e-Science Programme, led by the Research Councils, became the focus for enhancing network support for the sciences. Planning for the e-Science Programme began in the latter part of the 1990s; the major part of what became a £¼ billion programme was funded during the years 2001 to 2006, towards the end of which emphasis altered from pioneering e-Science techniques to engineering their adoption in the mainstream research programme. At one time there was discussion in some quarters about whether parts of the research programme might even need their own communications infrastructure as a part of a shared infrastructure might not suffice. Considering the sheer cost of implementing and operating such an infrastructure, such a course was not seen as cost-effective, but it is to be noted that it was against this background that the Follet Review was carried out in 2000 (*see* Chapter 11) at a time when the JISC – and hence JANET – community had already broadened from HE to include FE; JANET was soon to become the schools interconnect; and there was real concern that, with all this expansion happening, network support for science might be inadequate. It was in these circumstances that one result of the Review was the creation of the JISC Committee for Support of Research (JCSR).

As traffic levels continued to rise on SuperJANET, the management of the interconnects between the IP-over-SMDS and the main IP-over-ATM segments became increasingly difficult. Approval to phase out the use of SMDS within JANET was obtained and put into effect by the end of March 1999. Existing SMDS site connections were brought to their nearest regional networks or BENs via either 8- or 34Mbps leased lines; and two new BENs were established at Reading and Southampton. This helped the management of the network but now another, more fundamental problem appeared: by 1999, only a year after it came into service, the network was running out of capacity. Although vBNS in the US had a 622Mbps infrastructure, this was not available as an upgrade path for SuperJANET III, even had it been affordable. The only solution was to identify particular routes across the network between BENs that UKERNA managed and to add external links that bypassed the Cable & Wireless 155Mbps core. The advent of gigabit Ethernet, by now being installed in campus and regional networks, not only exacerbated this problem – because one of the points of SuperJANET had been to enable in the wide area what had been achieved in the local area – but the small ATM cell size was now making it difficult to achieve the switch performance needed to support gigabit links. On the other hand, not only could Ethernet switches support these link speeds but now so could routers. As well as the SuperJANET goal of supporting newer applications requiring greater network performance, the connection of the Further Education sector was now well under way, which would more than double the number of connected institutions. Even though the access links were typically slower than for Higher Education institutions, the number of connections would still require substantial additional capacity in the core. It was now time to plan the next phase of SuperJANET, particularly as the Cable & Wireless contract was for only three years and would finish at the end of March 2001. We shall return to the next stage in the development of the network in Chapter 12, after a brief look, in the next chapter, at the way network devolution evolved to encompass the whole of education.

11

Devolved Networks

The reorganisation of higher education and research funding in general and of JANET in particular have been outlined in Chapter 9, together with a more detailed look at the formation of the new organisation, UKERNA, which took over responsibility for JANET in 1994. The expansion of the community and broadening of JANET's remit would also provoke organisational changes in the management of JANET at UK country and regional level. In this case the key development would be devolved organisation, perceived as an integrated means of coping with growth; enabling participation in provision of regional services compatible with JANET but not within its immediate remit; being eligible to benefit from regional sources of funding; and exploiting regional communications opportunities available as a consequence of the further telecommunications deregulation begun in 1990. In pursuing this devolved model it may be observed that JISC was simply recognising all of these aspects, as well as effectively recruiting more effort into managing JANET as a whole.

Exploiting hierarchical structures is a classic approach to coping with scale, particularly if what one is dealing with is in some way naturally composed of many parts, some of which are supposedly subordinate to others. Splitting a larger entity into smaller parts can provide a natural way of reducing a system to constituents of tractable proportions. If the parts are still a bit big, one can repeat the process. The approach applies as much to the organisation of funding and management as it does to technical design of systems. Government (where the process is sometimes called devolution) and business (delegation) provide endless examples of the hierarchical funding and devolved management paradigm.

11.1 Regional networks (MANs)

The original topology of JANET was a backbone linking a series of NOCs (together with gateways providing connectivity to other national and international networks). A region was initially served by a natural star topology of links centred on a NOC. Organisationally, a small amount of effort at each NOC was co-ordinated by the Network Executive to provide the operation and management of JANET. Both organisationally and architecturally, these arrangements could be viewed as a simple hierarchy. However there was little exploitation of this, technically or organisationally, and JANET, regarded as a network which interconnected a set of higher education institutions, was operated as a single, homogeneous whole.

The re-emergence of regional networks during the 1990s was both organisational and architectural. Architecturally, the High Performance LAN Initiative (1990/93) occurred in time to enable institutions to install campus (or, in some cases, distributed campus) network infrastructure of a capacity commensurate with SuperJANET (as we have seen in Chapter 10). And although none of these networks was technically an IEEE 802.6 MAN, the fact that the coverage was larger than that provided by a LAN – and some were of an extent comparable to an urban or metropolitan area – was sufficient for these new networks to be generically dubbed 'MANs' at the time. However, from about 1994, after UKERNA had come into being, there began a gradual evolution as the larger of these so-called MANs, including where the JANET NOCs were situated, took on a somewhat wider role as more general access networks. There had always been institutions which were connected to JANET via a directly-connected organisation; now the process of gradual reorganisation of all JANET access being via an access network attached to a separate, distinct backbone began in earnest. These new access networks would cover a region substantially larger than a city – comparable, indeed, to the original regional networks of the 1970s (*see* Appendix B, where the originals and all those extant in 2008 are listed). Latterly they have increasingly been referred to as regional networks – the term adopted here. Although campus networks were often based on FDDI at this time, many regional networks were initially based on ATM. Regional networks began to be incorporated as essential components of the provision of network services as the ATM part of SuperJANET began to come into operation in 1994/5. During SuperJANET III (1997–2001) the majority of site connections evolved toward being via a regional network, so that by SuperJANET4 it was assumed (both architecturally and administratively).

However, the revival of regional networks was primarily of organisational significance: devolving administration and management of JANET to regional level not only enabled a more scalable approach to network organisation, it was also propitious in several other ways. The further deregulation of the telecommunications industry mentioned briefly in earlier chapters led to the appearance of a great many incipient[1] local and regional broadband communications companies, most seeking to purvey cable TV in combination with telephone services in the period when BT was still barred from delivering TV services over its infrastructure. While all but a few of these cable companies disappeared after a few years, the lowest level of infrastructure remained and localised competitive communications services became a reality. For a centrally administered network to exploit such opportunities would have been an administrative nightmare: devolved regional networks could do so with advantage.

There were two other respects in which a regional network was better positioned than a national network. For some developments, there were sources of funding specific to a region. The actual source might be regional, for support of the particular region in some way; or it might be national, even international – typically European – but for the specific benefit of a region because it was recognised as being either at some disadvantage or otherwise worthy of extra assistance. A regional organisation was potentially well-placed to take advantage of such funds whereas a national one was not. And, finally, there was the provision of network facilities to communities or for purposes not within the remit of JANET but which might be amenable locally to combination with JANET. Examples, at the time, of regions where these circumstances might obtain were the Highlands

---

[1] As many as 3000 to 4000 at the height of the 'rush', according to some observers at the time. Most rapidly went into liquidation as the reality of obtaining an operator's licence took hold, leaving what there was of the ducting infrastructure to others.

and Islands of Scotland and the North West of England; subsequently, Wales and Northern Ireland have also provided examples.

The JISC provided funding and assistance to organisations collaborating to set up a regional network. It was recognised from the outset that a regional network would operate as an independent network; as such, it was free to carry traffic for other customers in addition to JANET traffic. A regional network would be a separate organisation, independent of UKERNA, possibly a university or, increasingly often, a company. Any funding or revenue streams associated with non-JANET services would effectively be combined with that from the JISC.[2]

Administratively, a regional network provided the JANET services via a contract with UKERNA, the contractual framework including a Service Level Description specifying the JANET services in detail. Funds generally flowed from the relevant higher education funding council or DELNI via UKERNA to the regional network.[3] A regional network, being an independent network and organisation, had its own policies. Just as the JANET AUP refines the basic national telecommunications regulatory framework to suit the purposes of its community, so a regional network AUP defined policy consistent with its own purposes and the JANET services it provided. JISC placed no restrictions on the type of usage or the type of organisation that may connect to a regional network: that is, connection policy is a matter for the regional network. However, since a regional network carried JANET traffic, it was required to reach a minimum level of network service. Also, as would be expected, a regional network could only route eligible traffic onto the JANET backbone; transit traffic from a non-JANET organisation intended for the wider Internet had to leave the regional network via a non-JANET link.[4]

There were, of course, a number of variations on the basic scheme. Two examples were the Thames Valley Network, TVN, and the East of England Network, EastNet. The former came into being simultaneously with SuperJANET4 (2001) with three (Oxford, RAL and Reading) of the four sites being essentially part of the backbone (the fourth site, Oxford Brookes University, was connected to Oxford via private fibre). Both TVN and EastNet, centred on Cambridge, chose to arrange for operation to be undertaken by the JANET NOSC (TVN from the beginning, EastNet subsequently); both were examples of what came to be termed a 'regional delivery network'. By contrast, in the North West, C&NLMAN would make common cause with other interests in the area, particularly schools, as a result of which it would develop a very close relationship with the schools regional network for that area, CLEO. The re-emergence of regional networks did not all happen at once: while the beginning can be traced to 1994, universal adoption of the model of JANET delivery for HE being via regional networks was only formally completed with the creation of the Northern Ireland Regional Area Network (NIRAN) in 2004 – at which point the geographical areas covered by the regional networks as shown in Figure 11.1.

---

2 Technically, particularly in the case of a company, it also meant that the financial consequences of a regional network failure would, *prima facie*, be a matter for the regional network, not UKERNA; however, this was not a *raison d'être* nor would it remove ultimate responsibility for the regional service from UKERNA.

3 Originally it was not quite so simple or uniform, DELNI and the higher education funding councils preferring to pass capital funds directly to the regional networks. This practice largely ceased in 2007 with SuperJANET5, partly as more regional networks became companies (which presented technical obstacles for the educational bodies in transferring funds), partly in the interests of improved accountability, and partly to enable better control and integration with recurrent funding.

4 In practice, regional network AUPs and connection policies are consistent with the JANET AUP.

214 | 11. Devolved Networks

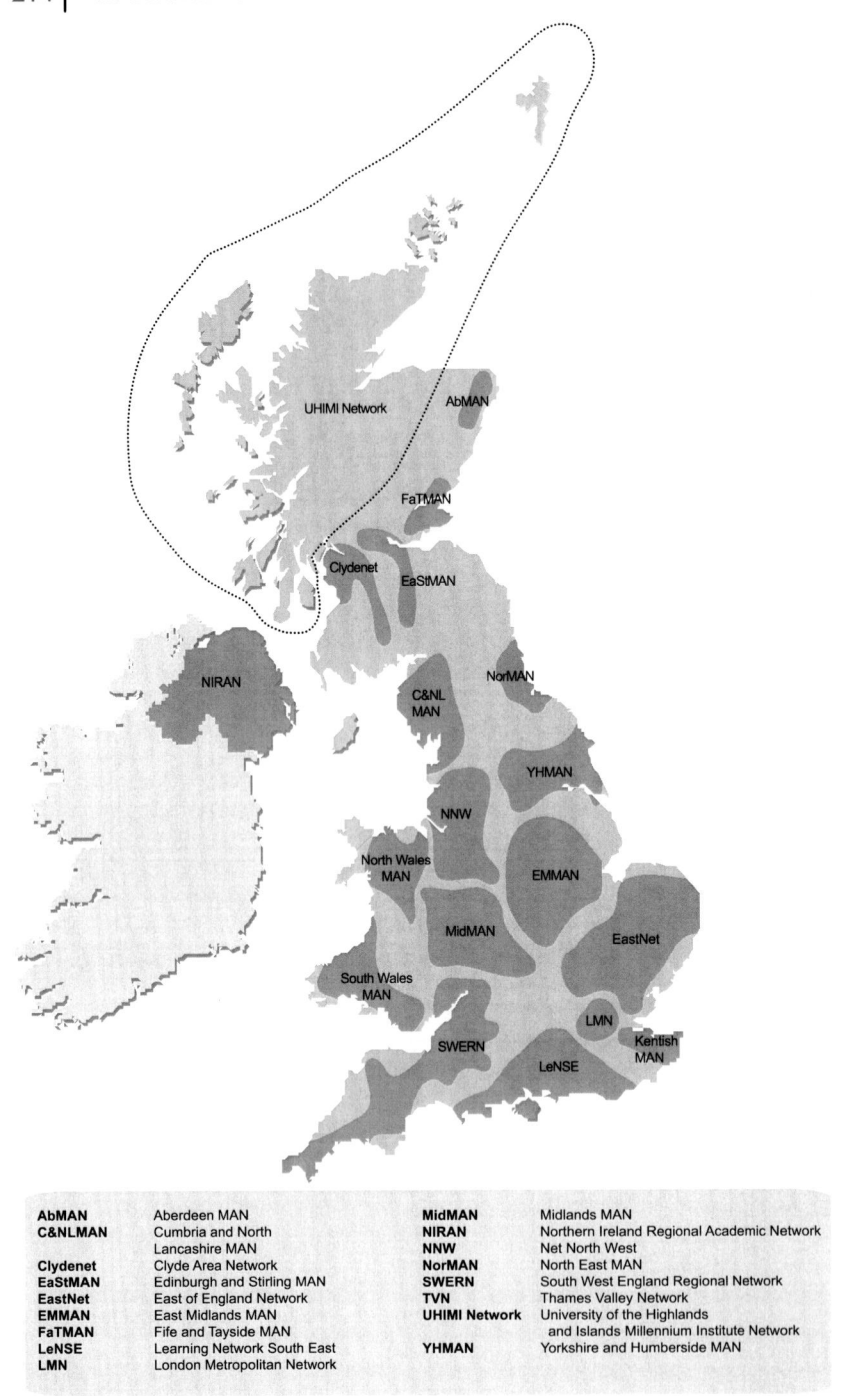

Figure 11.1. Regional Networks (MANs) in 2004.

AbMAN	Aberdeen MAN	**MidMAN**	Midlands MAN
C&NLMAN	Cumbria and North Lancashire MAN	**NIRAN**	Northern Ireland Regional Academic Network
		NNW	Net North West
Clydenet	Clyde Area Network	**NorMAN**	North East MAN
EaStMAN	Edinburgh and Stirling MAN	**SWERN**	South West England Regional Network
EastNet	East of England Network	**TVN**	Thames Valley Network
EMMAN	East Midlands MAN	**UHIMI Network**	University of the Highlands and Islands Millennium Institute Network
FaTMAN	Fife and Tayside MAN		
LeNSE	Learning Network South East	**YHMAN**	Yorkshire and Humberside MAN
LMN	London Metropolitan Network		

During the early 1990s there had been protracted discussions about connecting FE colleges, primarily revolving about the issues of funding and governance. In principle, there was little doubt in anybody's mind that the FE sector should be provided with network services like those available to HE – indeed, it could be seen as a natural progression from the JANET concept going back at least to the early 1980s. However, there were concerns from HE that connecting the much larger FE sector would dilute the service provided to HE; and within FE, the corresponding concerns were that the JANET service was focused too much toward HE and would be unable to accommodate the large number of additional organisations.

In the event, it was the Welsh FE colleges which were the first to be connected as part of a single programme centrally funded by the Further Education Funding Council for Wales. The project to do this, FEnet96, was undertaken by the University of Wales College of Swansea[5] and it was this programme which developed the concept of providing support and training, usually by staff based at HE institutions, for FE colleges not only in use of JANET but also for internal networking requirements. The model would subsequently be adopted by JISC as the model for its Regional Support Centres (RSCs).

Elsewhere, although a growing number of colleges had been connecting to JANET via 'sponsored connections',[6] the next centrally funded initiative was when, in agreement with the Further Education Funding Council, HEFCE funded an initiative in 1997[7] to connect 72 English FE colleges, approximately one sixth of the total, which received direct HEFCE funding for parts of their programmes. Although, technically, the purpose of the funding was to support the HE programmes, no restriction on usage was placed on the use of the JANET connections. The programme to connect the bulk of the English colleges only took place when funding was subsequently provided through the National Learning Network initiative which covered a range of ICT support, including connection to JANET. At the about same time the Scottish FE Funding Council (SFEFC)[8] and the responsible government department in Northern Ireland asked UKERNA to establish programmes to connect their FE colleges onto JANET. In all cases, the RSC model of college support that had been developed in Wales was an essential element in the provision of the JANET service. During this period, the number of FE colleges was gradually reducing as colleges merged with their neighbours. This created significant problems as the resulting colleges then typically had a number of campuses spread over a wide area and the JANET 'rule' was 'one organisation, one funded connection'. However, since the merger policy was strongly supported by the funding councils, a one-year 'transition policy' was agreed for multiple connections. At the same time, the development of regional networks also enabled greater flexibility in the provision of multiple connections.

The next development was that in recognition of the widening community, in 1999, the further education funding bodies became partners in JISC. Then, in 2000, Professor Sir Brian Follett conducted a review of JISC, in the light of the broadening of its remit since its creation in 1993, which resulted in some restructuring of both JISC itself and its sub-committees, including the creation of the JISC Committee

---

5   Established as Swansea University in 2007.

6   *See* Chapter 13.

7   *See UKERNA News*, No.4, p.2.

8   *See* Glossary.

for the Support of Research (JCSR) which came into being at the time of the UK e-Science programme (Chapter 10). All this had a knock-on effect on UKERNA, which was now serving a wider community but, in terms of its own governance, only had representation from its original constituency. There had already been an effective shift in stance in early JANET policy to accommodate connection to JANET being open to all bodies with an interest in education or research on a non-commercial basis; subsequently, in some regional networks there had also come about a sharing of network infrastructure with other public services – an example being the Lifelong Learning Network in Wales (LLNW) (see below). All this evolutionary broadening of the JANET community led to alterations in the governance of UKERNA. Indeed, a review of this had been mooted around the time of the Follett review. The Follett Report recommended that UKERNA remain the network provider for the sector now represented by the JISC and, furthermore, a review of its governance should be completed. Fred McCrindle[9] undertook the review and as a result in 2002 the governance of UKERNA was amended by restructuring membership of The JNT Association into classes capable of incorporating various categories of funding organisation. The Articles of Association were correspondingly amended to enable additional representation on the Board by the election of additional directors. The arrangements also incorporated provision of mechanisms to assist any future similar widening of the Association's remit.

11.2 Schools networks

The FE sector includes sixth-form colleges and, beginning in 1998, the schools sector began to look toward JANET for network interconnect and some application-level support: in effect, this was an early sign of recognition that exploiting the intrinsic synergy of the whole education sector would prove fruitful. Schools in the UK are generally the responsibility of local authorities and school funding forms a major part of local authority funding, which is via the devolved government departments. Correspondingly, the detailed arrangements differ in England, Scotland, Wales and Northern Ireland. The main contribution of JANET in bringing educational network services to schools may perhaps be summarised as being, at least initially, in providing network interconnect services, videoconferencing, content and, particularly in more recent times, security. In general the extension of JANET to the schools sector took place as SuperJANET4 became operational in 2001, and in parallel with the continued adoption and upgrading of regional networks for delivery. Appendix H lists the organisations associated with the provision of school networks throughout the UK (as of 2009).

Wales

Early development of the Welsh Video Network (WVN) in the mid-1990s helped to provide the stimulus for extension of networking services to all sectors of education in Wales. The Lifelong Learning Network for Wales (LLNW) was created to provide high-capacity services to schools, libraries and lifelong learning

---

9   Then Principal of Reading College and School of Arts & Design.

centres. This built on existing experience and provision already established in the HE and FE sectors, including the WVN. The first phase of LLNW was the provision of a core network interconnecting the 22 local authorities. This was procured by the Welsh Assembly Government, which contracted as procurement agent Welsh Networking Ltd, the consortium that originally operated the South Wales MAN (SWMAN) and which, in turn, subcontracted UKERNA to administer the procurement. The core network came into operation in 2002 and effectively provided points of presence for local connection and delivery of services. Connection to JANET provided the interconnect with the rest of the UK education sector.

Since 2005, in an effort to achieve more efficient use of public resources, a policy of aggregation of public networks has been followed in Wales, with increased sharing of basic communications infrastructure within education and also with other local authority and government services. In particular, in 2006 both the Welsh regional networks, the North Wales MAN (NWMAN) and SWMAN, moved over to use of the LLNW infrastructure.

In exploiting aggregation in this way, LLNW was paving the way for what has since been termed 'shared services' in the public sector. The next step was the Public Sector Broadband Aggregation (PSBA) Network. The Welsh Assembly Government announced funding for this in 2007 and following open procurement the PSBA Network was constructed on the basis of LLNW. Its initial phase provided a resilient core interconnect operating at 2.5Gbps in 2008. The provision of educational applications and content has been undertaken by the National Grid for Learning in Wales, NGfL Cymru, which since 2007 has had an expanded remit, covering all of school-level education and colleges. Since coming into operation, the PSBA Network has been gradually developed to provide support for local and Welsh government and the NHS in Wales.

England

In 1999 the Local Education Authorities (LEAs) began forming regional consortia for the purpose of procuring network services, both within the region and to the Internet. Each Regional Broadband Consortium (RBC) also provided educational application services. Of course, by virtue of being connected to the Internet, the RBCs were also connected to each other. However, it was recognised from the outset that the RBC interconnect would need to provide substantial capacity and a high standard of end-to-end service for the sector to obtain full benefit. Moreover, as elsewhere in the education sector, service free at the point of use was required. As planning for the interconnect began in 2001, at the time when SuperJANET4 was being deployed, it was evident that JANET would provide a satisfactory and relatively simple way of achieving the interconnect. Moreover, it had the added advantage that the whole of the education sector would be interconnected via the same high performance backbone, facilitating and encouraging co-operation amongst primary, secondary and tertiary levels. Confirmation of funding for JANET to provide the RBC interconnect came from Charles Clarke, Secretary of State for Education and Skills in January 2003.[10]

Initial exploitation of the JANET backbone interconnect, which came into operation in 2003, focused on videoconferencing and educational content

---

10   See *UKERNA News*, No.22.

distribution, an example of the latter being the Pathé news archive which became available in 2004. In the same year, as a result of work by the RBCs, the British Education Communications Technology Agency (Becta) and UKERNA, three documents were published defining network service standards for schools, relating to network design, network security and videoconferencing. However, unlike basic network protocols which are needed as a prerequisite to communication, no such irresistible lever exists for these standards and take-up appears to have stalled for the time being.

In 2006 the Department of Education and Skills (at that time the department responsible for funding schools in England) agreed a regime for funding continued use of the JANET backbone as the school interconnect – effectively the schools sector in England became a JANET funding partner. A significant effect of this is that the RBCs (and the ten LEAs which chose not to join an RBC – *see* Appendix H) now had access to all the services in the JANET SLA.

Scotland

As elsewhere, 2002 was marked by the upgrading of most of the Scottish regional networks. At the same time, the beginnings of extension of similar provision to schools began to be planned, led by the Scottish Schools Broadband Project, known as Spark. The means adopted was to link all 32 Local Education Authorities to JANET, via the regional networks, to provide the basis for hosting school community learning, administration and communication services, including videoconferencing and content services, as in the HE/FE sector.

The basic Spark interconnect came into service in autumn 2003 as the Scottish Schools Digital Network[11] (SSDN) Interconnect and included five additional educational organisations directly connected to the SSDN Interconnect. Initial use, as in other schools networks, focused on videoconferencing together with content access and distribution.

During the following year an initial educational application infrastructure was added in time for the start of the school year in 2004. The initial experience gained during the next two years was then used in 2006 to begin the design of an upgraded version of the facility, now called Glow,[12] which began to be deployed in 2008 by Research Machines on behalf of Learning and Teaching Scotland (LTS). Glow really refers to the application overlay which – like the World Wide Web on the Internet – is what is most visible, even significant. More recently, upgrading of the interconnect arrangements (Interconnect 2.0) between JANET and local authorities (via the regional networks) has begun; and at the time of writing, roll-out of Glow is still in progress.

Northern Ireland

By contrast, a very different approach was taken in Northern Ireland. Online learning infrastructure is provided by the Classroom 2000 Project (C2k) to all the schools and libraries. Provision of content, network and operation is contracted

---

11   Renamed from Spark because that name was already in use.

12   The design of Glow drew upon that of Kennisnet, The Netherlands school network, which members of the SSDN team visited in 2003.

out and managed through a lead contractor (Hewlett Packard as of 2009). C2k is primarily funded by the Department of Education, Northern Ireland (DENI), with additional funding from the European Union. Responsibility for provision of school education is organised through five regional Education and Library Boards (ELBs). Correspondingly, C2k is managed as a collaborative project by the ELBs. Deployment was originally in two phases: primary schools in 2001 and post-primary in 2003, with contracts typically being of 3 to 5 years duration. Connection to JANET in Belfast, as for NIRAN, provides the interconnect with the rest of the UK education community.

From an organisational perspective, perhaps the most significant achievement in recent years has been the management of scale, culminating with the creation in 2003 of a network infrastructure which covers the whole of the UK education and research sector and has enabled the subsequent construction of ubiquitous services, particularly in the schools sector. In terms of network implementation, the whole network is hierarchical, typically of three or four levels. Organisationally, it is now much more complex, particularly in the schools sector, there being both hierarchical funding and separation of the organisation of content provision, network provision and educational application infrastructure provision. It is estimated that, as of 2009, the national online educational and research infrastructure serves in excess of 18 million users, to be compared with perhaps less than a hundred thousand[13] in the early 1980s.

---

13   Estimates are hard: now, all use the network; then, probably only 10% used computers (the only reason to use the network), primarily for calculation.

12

SuperJANET: Realisation with WDM

At the end of Chapter 10 we left the development of the network in 1999 at the point where use of SMDS had been eliminated; the connection of the Further Education sector was in full swing; and SuperJANET III, a three-year contract with two years left to run, was already short of capacity. Internally, sites were now beginning to install gigabit Ethernet and suddenly ATM was no longer strategic for anyone except in specific niche areas. Band aids were being applied to the core network in the form of extra links between BENs (Backbone Edge Nodes), bypassing the Cable & Wireless core, and two factors were looming which could only serve to increase the load on the network. Firstly, video needed to be extended much more widely, preferably bringing it up to the standard already being experienced for teaching in Scottish universities using M-JPEG over high-capacity ATM links; secondly, extension to schools was being mooted, which implied a likely need for even more capacity somewhere in the lifetime of the next phase of the core network. This all had the feeling of *déjà vu* – JANET was again chasing capacity, as a decade before; but now, fortunately, technology had caught up.

12.1 SuperJANET4: gigabit routers
2001–2006

The original SuperJANET and its upgrade phase, SuperJANET II, had effectively been development and trial of new technologies: SDH, ATM and, incidentally, SMDS. Though the latter had had no strategic place in SuperJANET's development it had been Hobson's choice to upgrade the network in the interim, as well as reaching places not covered by the technology pilots. SuperJANET III had originally been seen as a natural successor in terms of consolidating and building on B-ISDN ATM technology.

However, in using SMDS provided by BT and the SuperJANET III ATM core provided by Cable & Wireless, JANET experienced twice the dual problems of not having control of its own infrastructure, which meant that reconfiguration was a cumbersome, administrative process; and there being no built-in upgrade path for capacity, both of which made service evolution harder – whereas SuperJANET had been intended to encourage evolution. In each case, SMDS and ATM had run out of capacity and the only option open to JANET was to augment the capacity by bypassing the main service (in SuperJANET II, by evolving the IP and IP-over-ATM pilots into a parallel backbones; and in SuperJANET III, by providing

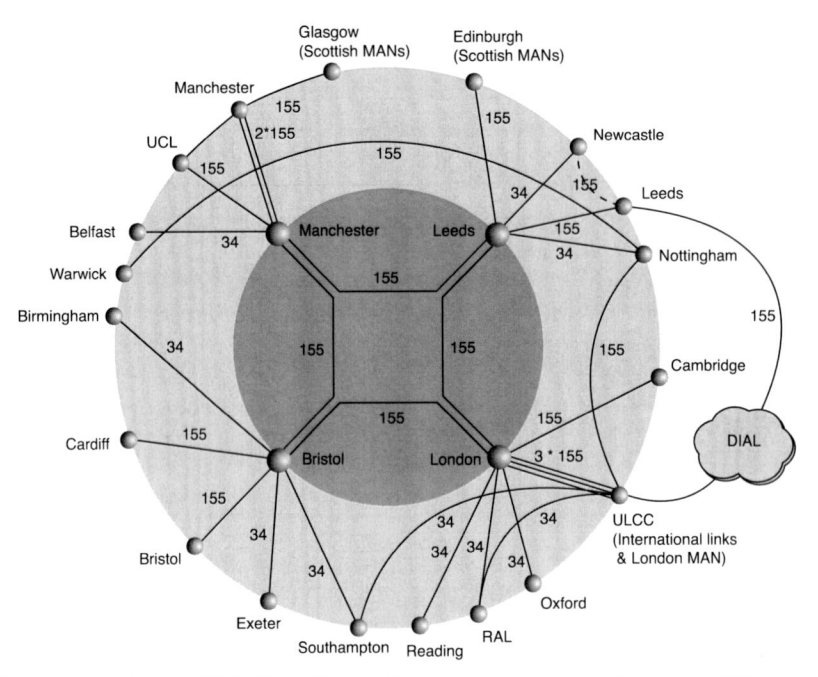

Figure 12.1. SuperJANET III backbone with 'band-aid' upgrades, September 2000.

external bypass links on congested routes – *see* Figure 12.1).[1] Both aspects would influence SuperJANET4 in fundamental ways.

Bearing in mind the foregoing remarks, it will come as no surprise that SuperJANET4 would be an IP network. There was no debate about this: ATM was no longer strategic in any enterprise network, though it remained a bearer service in some niches: for example, it has been built into domestic ADSL-based IP service for around a decade. Another aspect of SuperJANET4 design was that it would be explicitly based on the autonomous backbone and regional access network model. Originally, in SuperJANET, the limitation of the SDH-based network only being present in a few major locations around the UK had arisen primarily through cost, rather than a conscious decision to construct a backbone. Then, as the High Performance LAN Initiative had evolved into the MAN Initiative and the regional network model had been adopted administratively for the reasons describe in the previous chapter, so the concept was gradually confirmed in the overall architecture of JANET. SuperJANET III had more or less implemented this architecture, though the backbone was in two parts and its management was split. As the planning began in 1999 for SuperJANET4, this backbone and regional access architecture was taken as intrinsic.

As pre-procurement discussions with potential suppliers began, another element re-entered consideration: at what communication layer was the underlying infrastructure to be procured? UKERNA would be responsible for managing the IP routers, which would be the subject of a separate procurement. But what of

---

1   Interestingly, this technique of providing parallel paths to relieve congestion has its analogue in packet switch design, where it is called 'dilation', though in that case, typically, the whole fabric is duplicated, not parts of it in piece-meal, band-aid fashion!

Boring under the Cam

The connection of EastNet to the C-PoP in London presented unexpected problems in Cambridge. For the preferred route into the University, a fibre cable needed to cross the Cam. The Cam is unusual in that there is an underground river flowing underneath that visible on the surface. After considering alternative routes, a robotic boring machine was deployed to dig a 15-metre deep tunnel under both rivers – *see* accompanying illustration, Figure 12.2 – evocative of Project Mohole[1] or perhaps Jules Verne's *Journey to the Centre of the Earth*!

---

[1] Project begun in 1961 to drill through the earth's crust; the name derives from the Mohorovičić ('Moho') discontinuity which separates the (continental and oceanic) crust from the mantle.

---

the transmission infrastructure? Was a fibre network now conceivable? How far was it practical to go in this direction at this time? It will be recognised that this harked back ten years to part of the original SuperJANET concept. Now, wave-division multiplexing equipment was becoming more readily available: in fact it had reached much the same stage in 1999 as SDH equipment had almost a decade before and a number of providers were in the process of deploying it on a trial basis. If JANET could obtain sufficient contractual interest in the underlying optical and fibre infrastructure, maybe some sort of test and development network might be possible; and, perhaps even more attractive in view of previous

Figure 12.2. SuperJANET4: boring under the river under the Cam. *(Photo: Rolly Trice.)*

12. SuperJANET: Realisation with WDM

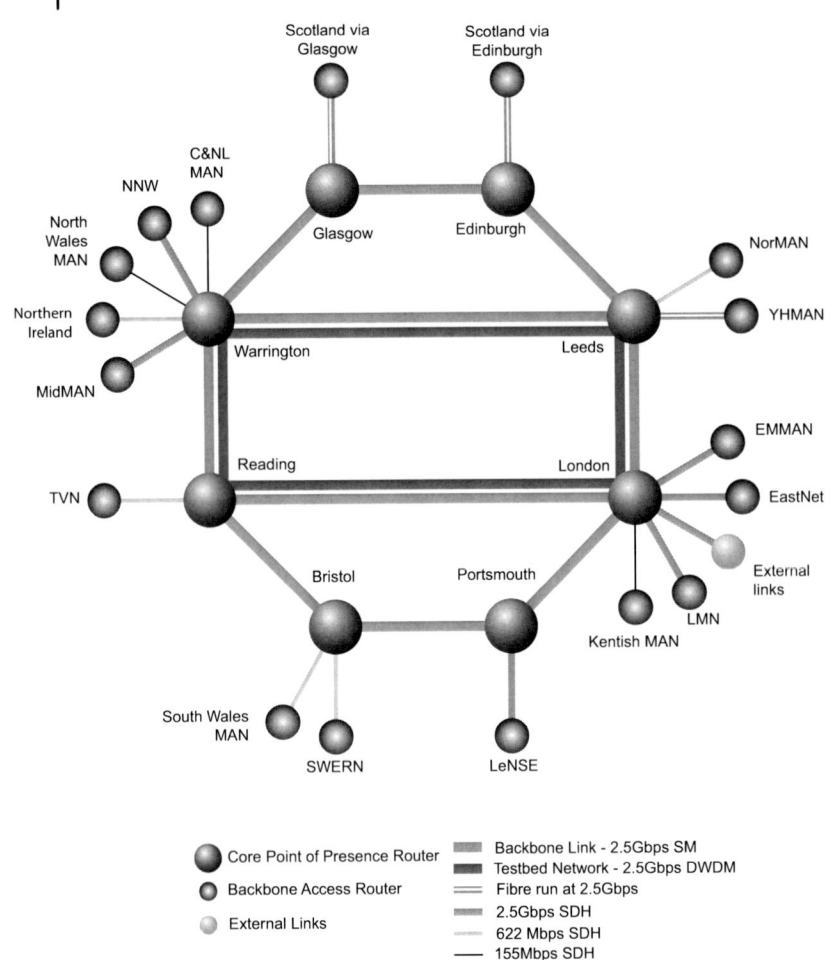

Figure 12.3. SuperJANET4, June 2001.

experience, maybe even the means to a substantial capacity upgrade at controlled cost could be built in.

In the event it turned out still to be premature for JANET to attempt to invest directly in the fibre part of the infrastructure to support what was now a mature network service but it was possible to obtain a 2.5Gbps fibre-based SDH infrastructure, with both a test network facility incorporated and an upgrade to 10Gbps built-in part-way through the contract. MCI WorldCom was chosen as supplier of this transmission infrastructure in June 2000 and most of it was delivered before the end of the year – though not without incident: see panel 'Boring under the Cam'.

The initial topology of the SuperJANET4 backbone was that of a ring of eight nodes: Glasgow, Edinburgh, Leeds, London, Portsmouth, Bristol, Reading and Warrington, as shown in Figure 12.3. These nodes were termed Core Points of Presence, or C-PoPs, and they were located in WorldCom premises. Two further links, between Leeds and Warrington, and London and Reading, respectively, provided a combination of some additional capacity and resilience. All 10 links provided SDH presentation, initially operating at 2.5Gbps. Cisco was chosen

Figure 12.4. GSR 12016 routers in a SuperJANET4 PoP. *(Photo: Rolly Trice.)*

as the router supplier and eight GSR 12016 routers provided the C-PoP nodes (Figure 12.4). Connections to regional network points of presence were made at speeds ranging from 155Mbps to 2.5Gbps. SuperJANET4 came into service on 1 April 2001, on a five-year contract ending in December 2005.

Providing a high-performance network based on the latest technology sometimes has its unexpected features – unusual civil engineering has already been mentioned. A property of any generation of electronics pushed to the limits of performance is that it is hot, large and heavy. The first implies large consumption of energy, both to produce the heat and then to get rid of it again. The other two, in the case of SuperJANET4, meant that the router equipment racks were so large that special arrangements had to be made to house them and installation could only be achieved with the help of a forklift truck. Moreover, when after a couple of years the installations were examined, some of the false flooring was found to be so bowed under the weight that structural assessment was needed to see if it would collapse during the lifetime of the routers!

In 2002 the SuperJANET4 backbone was upgraded to 10Gbps (Figure 12.5) – as fast as an IP network could operate at the time outside a laboratory. WorldCom had recently brought into service its new UK DWDM[2] infrastructure and the SuperJANET transmission infrastructure was now using this. Corresponding upgrades were made to the core IP routers, and an on-demand programme of upgrading the access links from the core to the regional network was begun.

In Scotland, during the final days of SuperJANET II and the whole of SuperJANET III, the video service for teaching had been based on M-JPEG over ATM (*see* Section 12.3). The Scottish regional networks, the MANs, had been

---

2   Dense WDM: form of WDM in which many wavelengths are transmitted on a fibre; harder than just a few, so took longer to develop and costs more. *See* Glossary.

226 | 12. SuperJANET: Realisation with WDM

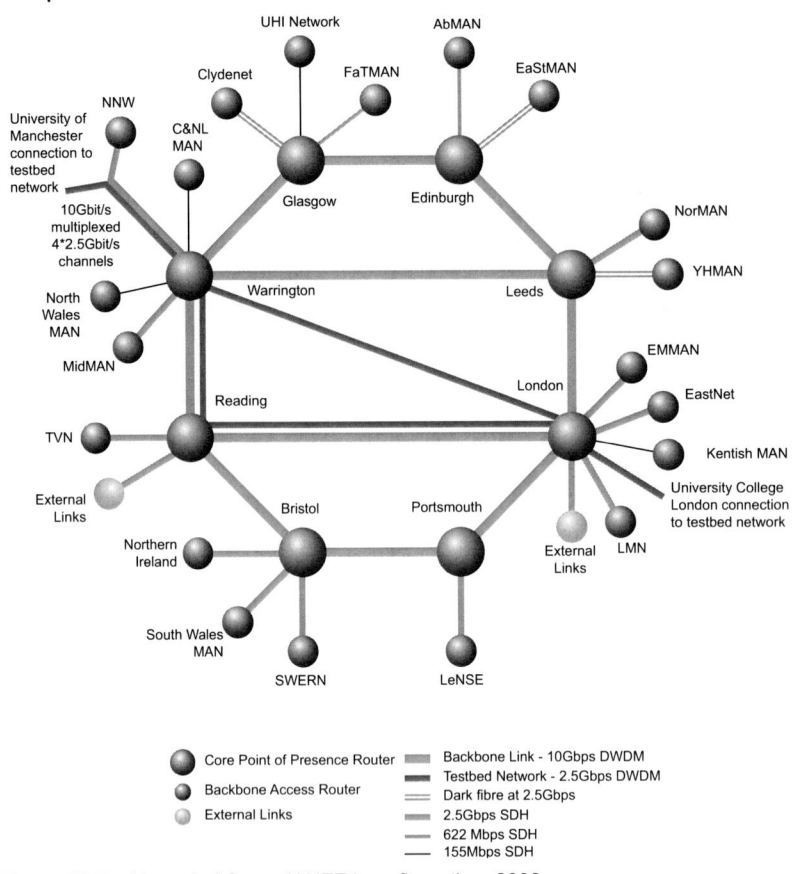

Figure 12.5. Upgraded SuperJANET4 configuration, 2002.

connected together independently of either SuperJANET III or SuperJANET4 backbones by what was termed the Scottish MANs Interconnect. In the interim, since SuperJANET4 had come into service, this Interconnect had become due for replacement. Now, in 2002, following the introduction of IP-based videoconferencing and the SuperJANET4 backbone upgrade, the latter took over much of the interconnect role, and the five Scottish regional networks now connected directly to the backbone: ClydeNET, FaTMAN and the UHI network to the Glasgow C-PoP, and AbMAN and EaStMAN to Edinburgh to Edinburgh.[3]

With SuperJANET4 the UK academic network had regained several important attributes. Overall responsibility for management, operation and development of the packet-level infrastructure was back in the hands of the community, represented by UKERNA: it was no longer split between UKERNA and the supplier. And the cost of transmission capacity upgrades was beginning to be brought under some semblance of control: they were still not cheap but, by building costed options into the contract which could be exercised at the discretion of the community, budgeting for growth had taken the first steps in the right direction. Why were (and are) these aspects so important? One clue may found in the difficulty the original Wells working party in 1973 had in making the case for an academic

---

3 See Appendix B for a list of the regional networks as of 2008, and their full names.

network when it was impossible to specify what it might be used for: the natures of the applications which would emerge were unknown and unpredictable – and this was to be expected of academic use of an open network. By the time of the World Wide Web and the beginnings of SuperJANET, this aspect of any NREN had become accepted but it didn't make constructing and engineering it any easier: if you can't specify the use, then nor can you engineer it 'optimally'! Under these circumstances it becomes crucial to retain as much control as possible over the development, configuration and capacity upgrading of the network – and SuperJANET4 made advances in these respects, compared to the earlier phases of SuperJANET. In the same vein, incorporated in the SuperJANET4 contract was a test network facility, consisting of an additional set of links forming a ring connecting the four C-PoPs at London, Reading, Warrington and Leeds. Following the 2002 performance upgrade, this was configured to enable support for pre-service work on IPv6 (the next version of IP) and managed bandwidth in an IP network. In 2003, a further enhancement was made to SuperJANET4 to provide resilience for the Northern Ireland connection which was provided by a link connecting Belfast and Glasgow. In collaboration with HEAnet, Dublin was linked to Belfast and London, providing a fall-back connection for Belfast and, under normal operation, improved access between HEAnet and JANET. Figure 12.6 shows a geographic view of SuperJANET4, including these links.

Quality of Service and Managed Bandwidth

ATM and SuperJANET III had failed to deliver the multiservice network which JANET increasingly needed. Now that JANET, along with everybody else, was fully committed to the universal packet network using IP, and with SuperJANET4 in prospect, it was time to review the status of the technical developments in support of multiservice networking using IP. The topic of QoS has been briefly described in Chapter 8, as it applies to both ATM and IP networks. Towards the end of the 1990s, encouraged partly by the appearance of H.263 video encoding which was more forgiving of the delay variations inherent in packet networks, and partly by the appearance of the beginnings of QoS support in routers, trials of IP-based videoconferencing were begun, particularly in Wales. In the light of JANET's continuing requirement for real-time video, in late 2000 UKERNA convened a Think Tank to consider QoS for JANET. More specifically, the brief was: to assess the requirement; develop a policy; and suggest a technical approach. As a result of the subsequent report (UKERNA, 2001), UKERNA began what turned out to be quite a long programme of QoS technology trials to establish the viability or otherwise of incorporating QoS support into JANET.

Because of the scaling issues associated with implementing the reservation-based micro-flow model intrinsic in IntServ, the Think Tank had recommended a trial programme based on DiffServ. JANET was not alone in this: indeed, it was merely confirming the direction being taken by every other network with an interest in QoS at the time, including Abilene in the US and the next generation of the European backbone, GÉANT (*see* below). The rather cautious approach taken by the Think Tank reflected the fact that DiffServ technology was only just being deployed and correspondingly the facilities available in different routers were by no means uniform, nor was end-to-end performance as yet well understood. Moreover, it was becoming clear that, in respect of multidomain networks, parts of the model – and therefore the technology – were incomplete. JANET, as has been

12. SuperJANET: Realisation with WDM

Figure 12.6. JANET backbone, geographic view, March 2003.

emphasised, is a multidomain network since the backbone and each of the regional networks and campus networks is under separate, autonomous management, each therefore constituting a single domain. As the trials progressed, it became clear that although the DiffServ technique could be made to work, configuring it for use throughout the multidomain JANET was not a practical proposition. On the other hand, within a single network, to handle a particular application with limited, well-specified behaviour and deployment, it might be of assistance. Within the backbone it was concluded that it would be sufficient, considerably less complex and affordable simply to ensure that there was enough capacity to ensure congestion never – well, hardly ever! – developed.

Although we have focused here primarily on the issue of providing support in a packet network for time-sensitive applications, particularly real-time videoconferencing, some of the techniques are also applicable in other areas of network engineering to provide a range of performance options not necessarily related to time-sensitivity. One of these, the managed bandwidth service, has already been mentioned; indeed, a limited service of this sort was offered using ATM during SuperJANET III. Stimulated partly by the UK e-Science programme, particularly the projected requirements of data dissemination from the CERN LHC, there was also interest in how an IP-based managed bandwidth service (MBS) might be engineered. One technique for engineering an MBS channel over an IP network is to combine DiffServ and MPLS – sometimes referred to as DiffServ-enabled MPLS. Technically, both DiffServ and MPLS make use of packet classification. In DiffServ a packet is 'marked' with the result of the classification by using a field in the IP header. In MPLS, the classification results in assigning a packet an MPLS header with the label for a specific path. There is some room in the header, enough for a few codes which could be associated with a subset of the DiffServ classifications. In this way, packets associated with a particular path could in principle be afforded specific treatment. Some testing of this was undertaken using the SuperJANET4 test network which became available after the backbone was upgraded to 10Gbps. However, from an operational perspective it appeared that the combination of techniques was insufficiently mature for deployment, particularly in a multidomain environment. In the case of DiffServ, classification is intrinsically domain-specific because packet-handling behaviour is peculiar to a domain; it also lacked any automated means of managing capacity allocations across domain boundaries, which meant that policing limits had to be set up manually. MPLS had no means for managing paths with cross a domain boundary: path engineering is again domain-specific so paths terminate at a boundary and classification has to be repeated on ingress to each domain. Both technologies are examples of a collection, all of which have yet to be fully developed for use in the multidomain context: we shall encounter others.

International access and telecomms industry shakeout

Almost in parallel with planning and implementation of SuperJANET4 was the construction of the European next generation backbone interconnect, again co-funded by the EU and the European NRENs. Planning, under the project title GN1, began in November 2000 and resulted in GÉANT, the Gigabit European Academic Network, which began service in December 2001, using 10Gbps links in the main backbone. Within the year, Abilene in the US also upgraded from 2.5 to 10Gbps and the JANET connection to the US was being provided by two 2.5Gbps links. GÉANT was succeeded by GÉANT2 in 2005, following the successful outcome of the planning and implementation project GN2.

However, apart from the relentless capacity upgrades being applied and consumed with monotonous regularity at every level in networks all over the world, there was a new, entirely non-technical issue to deal with in 2002 which, from a JANET perspective, first appeared in the transatlantic international context. Hardly were the new circuits to the US in operation when Teleglobe went into administration in May 2002. There had been news reports for some time of the turmoil in the telecommunications markets, associated to some extent with the rapid change in business model being forced on the industry by the deployment of the long-awaited global optical-fibre infrastructure. No sooner was the Teleglobe situation being addressed than KPNQwest, which provided much of the non-European, non-US global Internet capacity for JANET, also went into administration. It was replaced by Sprint and Level 3 Communications, respectively, with contingency cover provided by Cable & Wireless as a stand-by reserve supplier. Then, in June 2002, came revelations about WorldCom's financial status – a more nerve-wracking situation since now the main backbone of JANET was threatened. At the end of June 2002 WorldCom announced that it had overstated its profits by $3.8 billion.[4] In the event it turned out that the UK arm of WorldCom was sufficiently separate from its US parent that SuperJANET4 remained undisturbed. However, as a result of these particular events emphasising the instability of the telecommunications industry at the time and, perhaps more importantly, the dependency on a single supplier which SuperJANET had developed, UKERNA and JISC now initiated the provision of contingency cover to guard against such events in the future.

UKLight

We've seen already, in implementing SuperJANET4, that although fibre and WDM transmission and switching technology were on the verge of wider deployment, they were not quite mature enough for JANET to invest in them directly.[5] However, it was clear that close involvement with early exploitation of optical networking would be of advantage to the JANET development programme and the scientific community would be in a position to make use of such facilities for parts of its programme. This conclusion was reinforced by international activity in optical networking already under way in 2001, particularly in Canada (CANARIE[6]), The Netherlands (SURFnet[7]) and the US (Abilene and several others), which were planning collaboration in creating an experimental optical communication network. Transatlantic collaboration was focused on two centres: Chicago and Amsterdam. In Chicago, the Science, Technology, And Research Transit Access Point, STAR TAP, provided a national and international peering point for many networks and was already implementing an optical peering point under the generic title of StarLight which would include Abilene (Internet2), National LambdaRail (NLR), CANARIE and a number of others. Others were also planning participation in optical networks, including, in Europe, TERENA and the Czech Republic. Among the international scientific community, particle physics and astronomy were particularly interested: the former because of the need to transport data from the next generation of accelerators, including LHC at CERN, and the latter because of new possibilities for enabling more rapid synthesis and analysis of VLBI observations.

From the beginning of 2002 there was considerable activity, led by Peter Clarke,[8] to make the scientific case to fund UK participation, the activity and

---

4   WorldCom, Inc. subsequently underwent fraud investigation in the US, as well as investigation by the Securities and Exchange Commission. In July 2002 it filed for reorganisation under Chapter 11 of the US Bankruptcy Code, to restructure debts in excess of $30 billion. (*UKERNA News*, No.20, September 2002.)

5   An aspect of this is that JANET remains a community operation, rather than a partnership with industry, as some other NRENs.

6   The Canadian NREN.

7   The Netherlands NREN.

8   Then Head of Particle Physics, UCL; latterly Professor of e-Science, University of Edinburgh and Director of the National e-Science Centre.

subsequently the national and international part of the network becoming known as UKLight. The participants in this were drawn from particle physics, photonic communication, electronic engineering, computer networking and UKERNA. UKLight entailed three components: the first two, international connectivity and national access, would provide for pilot use by demanding users and technology trials for future service development; and a dark fibre[9] segment to enable field trials of research developments in photonic communication. From a JANET perspective, the type of service under trial in UKLight had the potential to be a component of future service. From the perspective of those using the facility, management, procurement and operation was needed. Combining these perspectives made UKERNA the natural agent to take overall responsibility for UKLight. The source of funding presented a rather different sort of problem. In terms of research use, EPSRC (with responsibility for photonics, electronics, computer networking and core e-Science within its portfolio) and PPARC (particle physics and astronomy) might be candidates; whereas JANET provision and development in support of research is squarely within the JISC remit. But the level of expenditure required for UKLight – a few millions over a few years – was in nobody's budget in the sort of timescales needed. Indeed, funding for this sort of facility which crossed the boundary inherent in the UK dual support system between Research Councils and funding councils had encountered similar difficulty (aside from any questions of merit) in 1999, when funds had been sought for a national communications research infrastructure through the Joint Infrastructure Fund, JIF. JIF was a fund created in 1998 in an effort to correct the imbalance which had arisen between direct and indirect support for research. JIF was replaced in 2001 by the Science Research Investment Fund, SRIF, with the same broad intent and funded (as JIF had been) by the Office of Science and Technology, the Wellcome Trust and HEFCE.[10] A part of the SRIF funds had been made available to JISC and, the technical and scientific case having been approved in principle by the JCSR toward the end of 2002, it was to this fund that the bid for UKLight support was directed, where it was approved in February 2003.

The initial installation phase of circuits from London to StarLight in Chicago and NetherLight in Amsterdam, together with some initial access, were completed in April 2004. Access was extended in late 2004 and early 2005 to include Net North West (Manchester), EastNet (Cambridge), YHMAN (Leeds), CANLMAN (Lancaster), Daresbury Laboratory and RAL.[11] These later stages overlapped with the design and procurement of the next phase of the JANET backbone, SuperJANET5, in which experience with UKLight had already proved useful. At the same time, several of the regional networks had been pursuing similar approaches in terms of moving to the use of optical fibre and these initiatives would all combine in SuperJANET5. Meanwhile, what of the dark fibre segment to enable photonic communication research field trials? Eventually, approval for this was granted in 2006, by which time the projects which would use it had moved on since the early days of 2002, or even 2004; and UKLight was about to

---

9   'Dark fibre': term used to denote fibre only, without the optical transceivers needed at either end to enable communication using light – hence the alternative term 'unlit fibre'.

10 See the Second Report of the Select Committee on Science and Technology (http://www.publications.parliament.uk/pa/cm200102/cmselect/cmsctech/507/50702.htm), under 'Research Funding in the UK', accessed at http://www.publications.parliament.uk/pa/cm200102/cmselect/cmsctech/507/50705.htm, March 2010.

11 *See* Appendix B for the names of the regional networks.

become part of SuperJANET5. The next chapter of this part of the story will be taken up in the next section.

As we shall see shortly, SuperJANET5 signalled the achievement of the original goals of SuperJANET of 1990: JANET would at last have the opportunity to invest in a transmission infrastructure, if not necessarily with ownership of fibre then certainly with access to multiple wavelengths. The issues involved in this were wide-ranging, from the technology itself to the taxable position of fibres and wavelengths: ultimately, the analysis was sufficiently complex and the market still sufficiently volatile that it seemed prudent to postpone procurement of SuperJANET5 for a year. Accordingly the contract for SuperJANET4 was extended for a further year, to December 2006.

Meanwhile, in the later years of SuperJANET4, now that processing power had become more dispersed the University of London had concluded that the building in Guilford Street which since 1968 had housed ULCC and generations of machines designed by Seymour Cray[12] and Gene Amdahl[13] was surplus to requirements and should be disposed of. The NOSC, which had been born of the ULCC NOC in the days of Shoestring and JIPS, and had worked closely with UKERNA for 15 years, looking after the operation and participating crucially in the technical development of JANET since its conversion to IP, now merged with UKERNA: network operations moved to new premises in Grays Inn Road in 2007, the new JANET Network Operations Centre. The merger had been on the cards for some years and effectively represented an alignment of mission and organisation. Fortunately, all this was decided in 2006, before the final (fifth) phase of SuperJANET development, so that the next generation of cables could terminate in the new premises in Grays Inn Road (about half a mile from the old premises in Guilford Street). The alternative hardly bears thinking about: for those of a horticultural turn of mind, visualise attempting to transplant a mature thicket of bindweed, ground elder or bracken!

12.2 SuperJANET5: meeting a broader remit 2007

The march of technological and scientific events in the first years of the second millennium seem – in retrospect, if not at the time – to have been particularly rapid. The acquisition, processing and storage ability now available to science demanded that all should be able to access everything from everywhere – fast! Simultaneously, the coordinated 'farm' paradigm for achieving vast processing scale had been confirmed: from meteorology to particle physics, processing was being distributed across processors in the hundreds, thousands, millions and beyond. And above all this was the threat of the 'deluge of data': mere terabytes

---

12 Chief designer of the 6600 and 7600 for CDC before starting his own company making machines in his own name.

13 Chief designer of the 360 and 370 series for IBM before starting his own company making 370-compatible machines in his own name.

were a thing of yesteryear, now it was petabytes and more – though whether that was in a week or a year was a little vague. Still, it was a lot!

Of course, it was not all science, nor art, nor research, nor learning, nor teaching. JANET now needed to support the virtual university, the virtual college, the virtual school: virtual education in the widest sense. Without becoming too philosophical about what constitutes education, the pragmatic observation was that the educational JANET was now required to provide not only access to the formal aspects of education but to the informal, recreational and social aspects also.

The planning and architecture development of SuperJANET5 took a while. Fifteen years before, when SuperJANET began, the imperative was 'simply' to release the capacity inherent in optical fibre and enable the nascent applications of the future. In 2002, a year after SuperJANET4 began service, things were no longer so simple. Education now took for granted a stable communications infrastructure – indeed, was now willing to spend the more substantial sums needed to make it even more reliable and available than it already was: 24 hours a day / 7 days a week / 52 weeks a year, on the move, at home, as well as on campus, i.e., permanent, continuous availability was now demanded. Alongside this, development, upgrades and 'problem resolution' (from code 'bug fixes', through resolving routing problems, to dealing with malicious activity) were all needed with no break in service – and with security of communication assumed, naturally. Apart from emergencies, these maintenance and development activities were all now taking too long because of the paucity of time allocated – two hours on a Tuesday[14] – when the network could be put at risk. And furthermore, what about packet network research, some of which needed access to, even reprogramming of, the network switching elements? What about the provision of non-standard or dedicated communications for the type of research for which a regular (or even enhanced) IP service was not sufficient? And then there was the question of support for entirely new optical communication and transmission research field trials. Had the stage been reached where we could actually construct a JANET which had a chance of satisfying all or most of these conflicting calls upon it? The answer was 'Yes' (with one possible exception) but engineering it was not so simple.

The general idea of what to do had been known for years but had not been practical. If one controls a layer of the communications infrastructure then generally one can implement some set of relatively independent communications facilities for different users or communities of users above that layer. So, for example, over a single ATM network, several IP networks may be implemented. Using time division multiplexing, if one owns a cable network, one can implement a number of circuit networks. And if one owns a network of ducts covering the country, one can lay independent cables to create a number of cable networks. There are some notable basic features of this concept. The lower the layer one controls, the more independent of each other the networks may be and the less they may interfere with each other – two aspects being performance and security. However, the lower the layer, the greater the cost so that, at the lowest layers of ducts or cables, this had never been practical for JANET on a national scale (though it certainly had already been exploited on a campus for several decades, and sometimes regionally in the 1990s). The principles were certainly appreciated in 1989: use

---

14   Tuesday historically stemmed from SRCnet's planned maintenance time, which in turn had its origin in the two hours reserved on that day for system development on the IBM 360/195 at RHEL – the only change over the years being that from evening to morning.

of WDM was contemplated but was not in general deployment; use of TDM was sought but, in the event, proved far too expensive, with SDH only just beginning deployment; and so ATM was pursued, as we have seen, as the only layer at which multiple services might be provided, initially with some promise. But now there was a change: fibre services were being offered by a whole range of suppliers, together with a variety of WDM (including DWDM) services and equipment. The beginning of this had been observed during procurement of SuperJANET4, as has already been noted. By 2002, when strategic discussions were turning to thoughts of SuperJANET beyond 2005, it was clear that investment in wavelengths, perhaps even fibre, would be possible and needed investigation. Note, incidentally, that unless fibre were obtained, one aspect of the support desired above would not be possible: field trials of optical communication systems research (the source of the exception noted above).

Developing strategic requirements

Understanding the requirements and developing the strategy for SuperJANET5 began with a couple of network strategy workshops. The first, at Warwick in December 2001, was quite wide-ranging, covering issues surrounding network use, including legal and economic: the latter included charging as well as the geographically variable cost of broadband access. In addition there was discussion of user requirements such as the needs of the Grid, the UK e-Science programme, and access to centralised information stores. The second workshop, in Glasgow in December 2002, served to introduce the growing set of stakeholders and service providers to each other's activities, as well as laying out some of the technical issues and technological developments which could be expected to influence and even shape the next generation of JANET.

2003 saw the start of gathering and analysing user requirements. This was a much more formal process than hitherto, partly because the full effects of FE participation were now being experienced, partly because schools participation in JANET was now agreed by the Government, and partly because deciding what to procure was going to be quite difficult and complex: with a greatly expanded spectrum of users and requirements, it was going to be important that no strategic needs were missed in deciding on the architecture of the network and how to build it. UKERNA, the JISC and, indeed, most of those who had participated in providing the JANET range of services had up to 25 years experience of HE, maybe 5 to 10 years experience of FE and almost no experience yet of schools – indeed, the amount of experience was somewhat inversely proportional to the sizes of the respective communities. And with a much expanded range of stakeholders, obtaining funding authorisation needed a methodical approach.

The results of the requirements gathering was summed up during 2004 under five generic, strategic headings, some of which were specific to SuperJANET5 while others harked back to earlier, long-standing goals that had now been reinforced with some specific needs.

> **1. Reliability:** increased levels of reliability were needed, particularly in respect of the levels of resilience at regional network to backbone, and institution to regional network connection points.

The main JANET backbone had always been multiply connected. Over the years its resilience had steadily improved, with multiple, automatic fail-over routes, redundantly engineered routing equipment, and uninterruptible power supplies. These improvements had also spread to the regional networks. However, while considerable resources had been expended to provide resilient access to external networks, the same was not true of the interconnection points between the backbone and regional networks or an regional network and its constituent institutions: these represented single points of failure which now needed attention. While such resilience might not be afforded everywhere at once, the architecture for SuperJANET5 must enable such single potential points of failure to be eliminated.

    2. Scalability: upgradable capacity was required at controllable cost.

This could scarcely be thought of as new: for the first decade of its existence, JANET had been increasing its capacity as fast as it could afford. The spur for the original SuperJANET programme had been evading the tyranny of capacity prices which did not reflect the substantial reduction in capacity costs brought about by availability of fibre. JANET link capacity was referred to in kilobits per second in its first decade or so, megabits in the second, and gigabits during the third. Leading up to SuperJANET5, usage of JANET was doubling every nine to twelve months, with no signs of any let-up, with the community continuing to expand, new applications, new styles of use and ever-increasing amounts of data and information. The Government had announced that all schools should be connected to JANET by the end of 2005; beyond schools, adult and community learning centres were connecting to JANET. Access to educational resources like Curriculum Online, the British Pathé archive and the BBC Digital Curriculum materials was becoming increasingly important and only just beginning. Schools were not only beginning to use videoconferencing but also to participate in research using Grid technologies.

    3. Separability: multiple network services were needed to meet the diverse requirements of the research and academic communities.

SuperJANET had been seeking to do this since 1990 when use of TDM had been sought to provide the basis of configurable, separate network services for research and development without risk to production traffic. This was now an increasingly pressing requirement. Several communities were using the network for extensive flows of data, including parts of the research community, notably particle physics and astronomy, and the teaching and learning community; and new flows, such as TV, including campus TV, and distributed orchestra were about to begin. Large, high-speed bulk data flows must not disrupt other, possibly real-time, interactive usage on the network – that this could happen had been amply demonstrated during 'speed record' bulk data transfers, during which other users of parts of the network had on occasion been locked out. Pioneering high-speed transfer techniques over intercontinental distances is important to some communities but if it locks out other users, it is indistinguishable from a denial of service attack to those others. The ability to provide separability or segregation was by now generally accepted as being necessary to support a variety of experimental use, including network research, service development and commissioning, which in each case required an infrastructure sufficiently dedicated that its users could break it without disturbing other JANET community use, either experimental or service. It was also noted that segregation could be a tool for the provision of security between different communities of users. Various techniques at the cell or packet layer were already familiar (*see* Chapter 13) for providing a degree of mutual security; whether this might be required more generally remained open.

> **4. Flexibility:** the network architecture needed the flexibility to respond to requirements for additional network services in a timely manner.

This requirement was somewhat related to the previous one for separable services. Given that such services were open-ended, the network needed to be able to respond to new service requirements on a similar basis – indeed there had already been occasional requests for new, dedicated services, typically to support a particular community, project or experimental facility.

> **5. Visibility:** real-time information about the performance of the network was needed to provide support for a developing breed of applications that adapt to network conditions.

Monitoring of JANET has always been required for a variety of purposes. Operationally, for example, it is needed to provide visibility of basic functioning and to assist in diagnosis when a fault develops. For management purposes, performance records are needed, for example, to enable traffic patterns and growth to be observed, to provide part of the basic information for future planning. And observation and measurement have consistently been of importance in network research for understanding network behaviour. At a somewhat higher layer, performance monitoring and measurement relating to particular applications or services – such as data transfer or videoconferencing – are needed. A particular aspect of future applications being developed was to couple Grid computing systems with real-time performance observation of both the network and potential Grid system components so that distributed assembly of resources could adapt to make best use of temporal or spatial regions of network and resource where capacity is available. Thus, there was now need to provide network visibility not only for internal operational use but also for external use by end-systems and users.

Architecture

A key concept in developing the SuperJANET5 architecture was that of the 'flexible transmission platform'. Effectively this provided an abstract projection of the requirements in terms which described an architecture without reference to specific implementation or technologies: it is shown diagrammatically in Figure 12.7.[15] In parallel with gathering requirements, UKERNA also explored the developing market for fibre and WDM services, including commissioning an independent study, to assess the options available for implementing the transmission platform. This served to confirm that obtaining a fibre-based WDM

---

15 The diagram went through many versions during development of the requirements and architecture of SuperJANET5; that shown here is taken from *SuperJANET5: An Architecture for Diversity*, UKERNA, 2004 (http://webarchive.ja.net/sj5/requirementsanalysis/an-architecture-for-diversity.pdf, accessed June 2010).

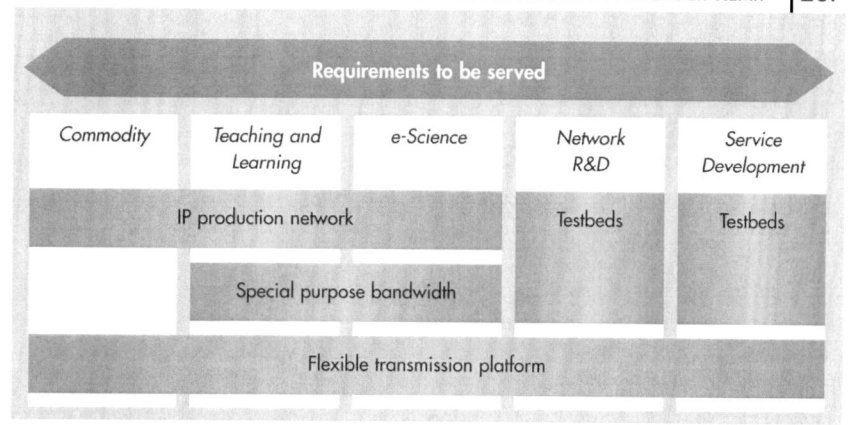

Figure 12.7. SuperJANET5 flexible transmission platform.

infrastructure would certainly be possible. The major issues now revolved around the options available for managing and operating this infrastructure under contractual conditions which provided UKERNA with sufficient control to meet the requirements of the flexible transmission platform. The low-level technical requirements of the transmission platform could now be summed up as:

- Operational visibility to JANET Operations, enabling improved reliability and reduced complexity.

- Ability to increase capacity and reconfigure the transmission platform to respond to developing requirements at predictable cost.

- Ability to configure parallel transmission networks within the platform.

- Capability to adopt new optical transmission technologies in advance of public service offerings.

In exploring the available technologies, the architectural outlines for implementing the transmission platform now began to emerge. The implications of what has already been described were that the platform would consist of a set of (point-to-point) configurable (digital) channels, which could be used to construct a number of digital networks – the major one being the main IP service. In order to implement the necessary configurability, the platform backbone would be split into a 'transmission core' and a set of 'collector arcs' attached to two core connection points (see Figure 12.8). The core could be viewed as a distributed, re-configurable channel switch, capable of providing a large number of digital channels of configurable capacity and transmission format between any two core connection points. Implicit in this was that the core connection points would be multiply connected and therefore the platform core could be considered resilient. The collector arcs would provide channel distribution to the RNs regional networks (and any other aggregated connection points), resilient by virtue of collector arc dual connections. Regional networks would in turn generally be dually connected to a collector arc. An essential aspect of this transmission architecture was this capability of constructing individual networks, each of which could exploit the underlying potential for transmission resilience to a configurable degree.[16]

---

16 The transmission architecture for SuperJANET5 was very much a collective effort which took some time to develop; however, on a personal note, the author would like to acknowledge the contribution

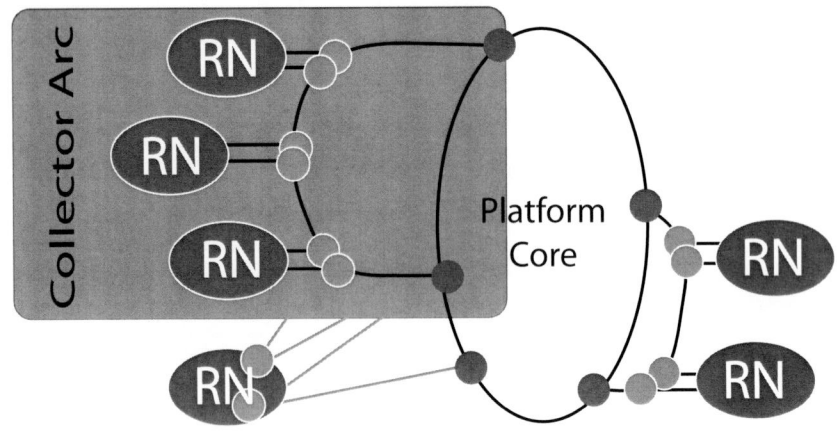

Figure 12.8. SuperJANET5 transmission architecture.

Elaboration and procurement of the backbone transmission platform

The concept of the flexible transmission platform together with the transmission requirements and architecture formed the service implementation model, and this was then used as the basis for elaborating and procuring the specific implementation. This more formalised approach to defining and building SuperJANET5 was not only helpful in developing what was a more complex network than any previous version of JANET, it also fitted in well with the (then) relatively new two-stage public procurement procedure which was used. However, by mid-2004 it had become clear that this was all going to take longer than on previous occasions – and this was certainly not the moment to try to rush things to have the new network in place by the end of 2005, particularly since SuperJANET4 was still operating satisfactorily following the core upgrade to 10Gbps in 2002, and estimated traffic growth suggested that capacity would be sufficient for another year. Consequently, at this point the SuperJANET4 contracts were extended by a further year to the end of 2006.

Armed with the service implementation model, the first stage of the procurement of the transmission platform was launched in October 2004 with the issuing of a public 'prequalification' questionnaire. From the responses to this, a set of potential suppliers would be identified and then, with their help, UKERNA would define a formal operational requirement which would form the basis of an invitation to tender. The remainder of the procurement would then follow the familiar pattern. The introduction of the prequalification stage in effect formalised what had often happened informally in the past in cases of complex technical procurements (including for SuperJANET4), when rounds of pre-tender talks under non-disclosure were held with potential suppliers, primarily to identify the imminent state of complex technological products so that an operational requirement could be constructed that would achieve a balance between the impossible (immature) and run-of-the-mill (passé), to elicit a correspondingly credible short-list of suppliers.

---

made by Duncan Rogerson, who sadly died on 5 March 2009 from the recurrence of a brain tumour. The author recalls with pleasure a number of illuminating discussions on the architecture in front of a whiteboard with Dunc, pens in hands.

In terms of constructing the backbone transmission platform, although the service implementation model had laid the groundwork for the implementation, it had not gone so far as to specify precisely what was wanted from the provider in terms of exactly who owned and operated what: this was open for exploration and negotiation. Two options had been identified but they did not exhaust the possibilities, one of which was for UKERNA to lease a managed fibre infrastructure: it would own, operate and manage the infrastructure, perhaps through a contract with a third party. Another scenario was for the platform to be built to order as a managed facility on a dedicated infrastructure selected and owned by a telecommunications provider. Although the provider would manage and operate the infrastructure, UKERNA would retain overall control. Both these options avoided UKERNA undertaking direct physical provisioning, management and operation of the platform, for which UKERNA did not have the relevant experience.

The concept of the flexible transmission platform was all very fine but, as we have seen, in the years since the original SuperJANET began, the wide-area component of JANET had evolved into a mature two-level hierarchy, with around 20 regional networks participating in inter-institutional end-to-end delivery. As elaboration of the architecture moved to the next stage, so it was possible to exploit the service implementation model to provide a variety of realisations of the network and transmission architecture to suit an individual regional network and its subscriber institutions. The key to this was that it was not necessary that a regional network peer with the backbone at every available layer or service; it could choose which it needed. This also allowed each regional network a degree of flexibility to follow its own timescales for development and procurement. In the simplest case where a regional network did not need dedicated channel services and only required the regular, 'commodity' IP service, it could just peer with the router providing that service. Additional services could be provided later. Just as funds would be needed to re-provision the backbone to enable the new services, so too funds would be needed for the regional networks; and the process of allocating these had already begun by the end of 2004.

The other aspect needing study was routing equipment for the main IP service. SuperJANET was already operating backbone links at 10Gbps and certainly wanted to ensure that the routers it obtained for SuperJANET5 would be capable of incremental upgrade to handle a substantial number of access links at 10Gbps, and a number of backbone links at 40Gbps. JANET had for some years been operating a network backbone essentially as fast as technology would allow; moreover, the extent of the backbone network was comparable with the UK public network. This put the whole operation into the category referred to as 'carrier class' and there had recently been considerable developments in carrier class routers. During 2004 a formal Request for Information was issued following informal discussions with manufacturers, as a result of which routers from Chiaro, Cisco and Juniper were chosen to undergo trials during the latter part of the year. In respect of constructing the service IP network, there was already apparent an inherent advantage in the transmission platform architecture: it would be possible to implement the IP service with considerably fewer routers than had been used in SuperJANET4 – a distinct advantage in view of the cost of carrier class routers, as well as simplifying and reducing the effort needed to maintain and develop the network in future.

The transmission platform procurement now proceeded with the prequalification questionnaire being published in October 2004, which resulted in the selection of six potential suppliers in the first quarter of 2005: Energis, Geo,

240 | 12. SuperJANET: Realisation with WDM

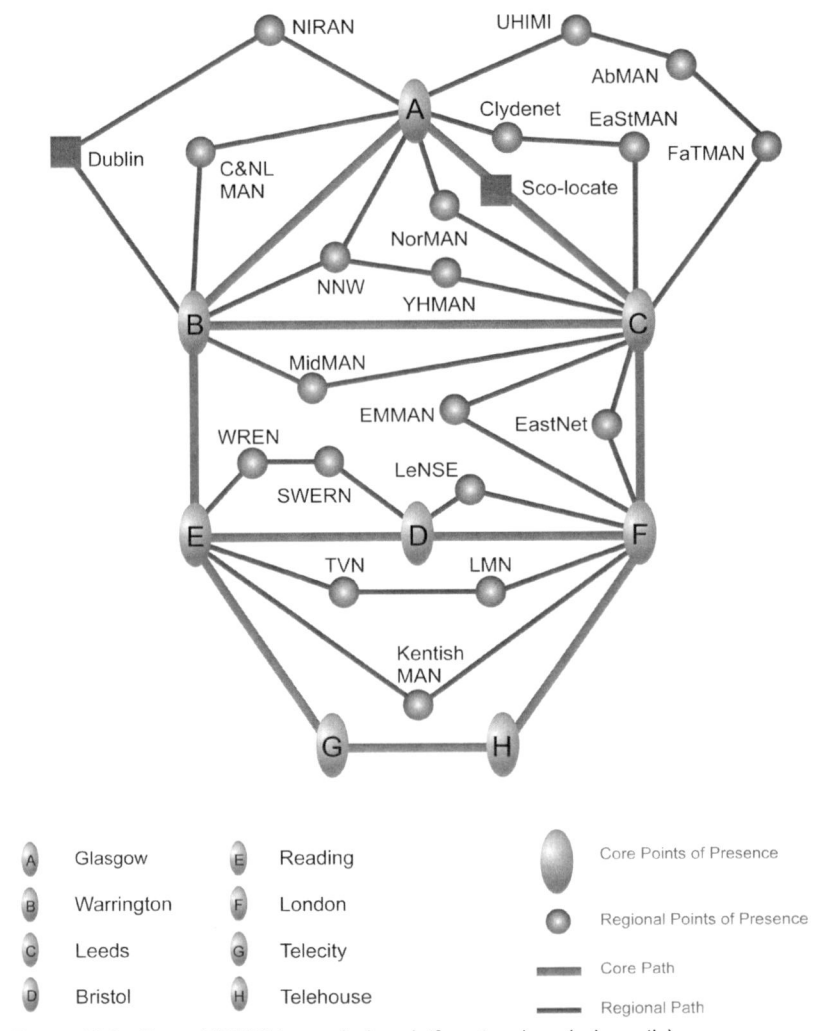

Figure 12.9. SuperJANET5 transmission platform topology (schematic).

A	Glasgow	E	Reading	
B	Warrington	F	London	
C	Leeds	G	Telecity	
D	Bristol	H	Telehouse	

Core Points of Presence
Regional Points of Presence
Core Path
Regional Path

Global Crossing, MCI, ntl:[17] and THUS. Working with these companies, UKERNA then published the SuperJANET5 Operational Requirement in June 2005 as the formal Invitation to Tender, to which, as part of the negotiated procedure for public procurement, only the selected six suppliers could respond. The result of the procurement was a contract placed in December 2005 with MCI, by then known as Verizon Business.[18] Meanwhile the backbone router trials had been completed and this was followed by the procurement in the first quarter of 2006, resulting in the choice of Juniper T640 IP routers to be supplied by Lucent.

---

17 Now Virgin Media at the time of writing.

18 MCI and Verizon Communications finalised a merger at the beginning of 2006. Former MCI business and government customers would now be served by a new unit of Verizon, called Verizon Business, which also incorporated the former Verizon operating unit, Verizon Enterprise Solutions.

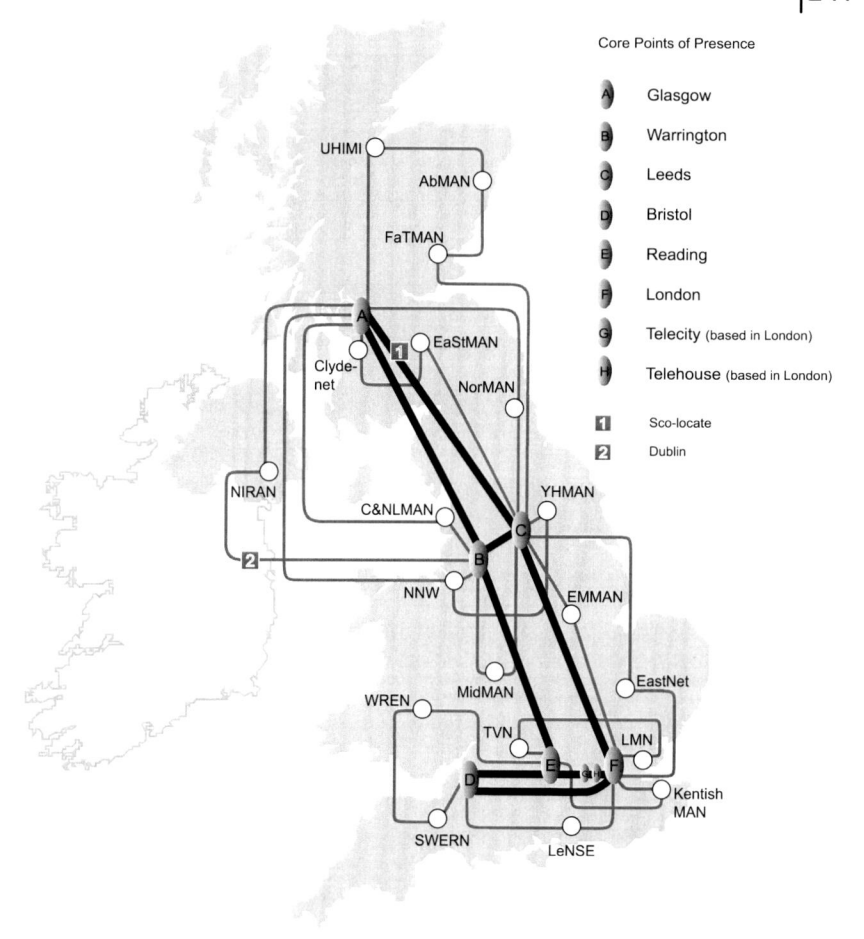

Figure 12.10. SuperJANET5 transmission platform topology (geographic).

Implementation and continuing development

The mode chosen for the provision of the SuperJANET5 transmission platform was the second outlined above: Verizon would own, manage and operate a fibre-based DWDM transmission infrastructure for dedicated use by JANET, with overall control being retained by UKERNA. The transmission core (*see* Figures 12.9 and 12.10 for schematic and geographic views, respectively) consisted of six core locations (designated C-PoPs as for SuperJANET4) at Glasgow, Warrington, Leeds, Bristol, Reading and London, with the addition of Telehouse and Telecity included in the core, thus providing external peering arrangements with direct access to the full capabilities of JANET, including resilience and scaling. Each of the core locations was connected to two or three of the others, providing the basis for resilience for any network constructed over the platform. Distribution to the regions was accomplished by 12 collector arcs, each serving up to three regional networks. Channels, operating at transmission rates of 1, 2.5 or 10Gbps, may be reconfigured or added, typically within days, between any two points on the platform: within a given fibre, the digital channels are provided over WDM wavelengths. Additional channels could be included at the marginal cost of any additional equipment. During 2007 Verizon would conduct trials with UKERNA of channels operating at 40Gbps, in anticipation of these being required in service during 2008.

The underlying transmission platform was sufficiently complex that considerable thought had been given to it, both in conception and then in planning installation and transition to its use – in consequence of which there is remarkably little to be said about bringing SuperJANET5 into operation! Acceptance testing of the Juniper routers took place during the second quarter of 2006. There were seven of these: six for use in the backbone and one for test and development and 'hot' standby. (There was also a router at each of Telehouse and Telecity for interconnection with external networks.) Although there had only been eight backbone routers in SuperJANET4, the total number in the backbone, including the edge routers, had been 30: since there were no backbone edge routers in SuperJANET5, regional networks connecting directly to one of the six C-PoP routers, a major reduction in JANET IP complexity had been achieved, with corresponding reductions in maintenance effort and future development lead time. Moreover, because of the dual-entry resilience, in many cases it would be possible to consider scheduling routine maintenance outside the traditional 'at risk' period. Meanwhile, following the signing of the contract with Verizon in December 2005, preparation and installation of the transmission platform had been taking place: it was completed and handover took place in October 2006. During the last quarter of 2006, the routers were installed and the transition of all the regional networks was made to the new backbone, so that by January 2007 the major part of SuperJANET5 was in service. There remained a number of connections which had never been via one of the regional networks: these were connected to the new backbone during the first months of 2007 and, by the end of March 2007, the final vestiges of SuperJANET4 were switched off and decommissioned. The whole operation had proceeded in a remarkably streamlined way, a tribute to the meticulous planning which had been undertaken over a period of several years.

Even while the service IP network was being constructed over the transmission platform, the first (pilot) service to be implemented over SuperJANET5 was provided directly by the transmission infrastructure, before the end of 2006. It will be recalled that the particle physics community had been preparing to handle the data from LHC at CERN since the late 1990s. An important requirement was to ensure operation and performance of a bulk data channel connecting CERN and RAL. Tests of the distribution infrastructure had been taking place for two years over four 1Gbps circuits using the UKLight and NetherLight facilities. During the last quarter of 2006 a pilot, dedicated wavelength channel was established between RAL and CERN via TVN, SuperJANET5, GÉANT2 (see below) and SWITCH, the Swiss NREN; the channel was configured for digital transmission using 10Gbps Ethernet framing. Others would follow as SuperJANET5 took over from UKLight the provision of dedicated high-performance, special-purpose channels under the new JANET Lightpath service established in 2007.

UKLight, though expensive, nevertheless played an important role in enabling parts of the user community, as well as JANET, to participate internationally in field-trials of the next developments in networking, several years ahead of their adoption for regular use in research and education networking. The remaining goal of UKLight had been the provision of a dark-fibre network to support advanced optical communications field trials. However, as with UKLight itself, the SuperJANET5 transmission infrastructure was based on WDM channels, not dark fibre, and support of this sort was as yet still in the future (*see* Chapter 14).

Development of the transmission core platform began almost as soon as the SuperJANET5 IP service was stable. As had been incorporated into the contract, testing of 40Gbps transmission began during 2007 on the link between the London C-PoP and Telehouse. The initial trial was conducted on a dedicated fibre path, to avoid possible interference with existing DWDM channels. Since the purpose of the development was to be able to support the predicted IP traffic growth on the service network and between JANET and its peer networks, two additional Juniper T640 routers were obtained, together with 40Gbps PoS[19] interfaces for testing. Following the satisfactory outcome of testing using the isolated fibre and standalone routers, the routers were installed at Telehouse and Telecity so as to incorporate these two JANET peering nodes fully as part of the JANET backbone. The 40Gbps interfaces were then incorporated into the routers at the London C-PoP and Telehouse and connected using a new WDM channel in the JANET transmission core. Following confirmation of the stability of the new channel, production traffic began using the new circuit at the end of the summer in 2007. The traffic predictions made at the time of procuring the transmission core were by 2007 being born out; accordingly, following the successful deployment of the first WDM channel at 40Gbps, further upgrades were made to the core during 2008. These upgrades were to the peering links and to provide a 40Gbps closed-loop path connecting London, Bristol, Reading, Warrington and Leeds, effectively promoting the central part of the IP backbone to operation at 40Gbps. And the upgrades to the peering links had now provided JANET with dual connections at 40Gbps from both its entry points at London and Reading to both its peering points at Telehouse and Telecity. This was a resilient arrangement using the highest capacity links to widely separated entry points on the backbone – something JANET had sought hard to achieve over the years.

International

As already implied, JANET was by no means alone in adopting a mode of network provision which would allow for parallel, non-interfering transmission channels and networks to be set up in addition to supporting the regular academic Internet service.

In September 2004 GN2 began, the project managed by DANTE to put in place the successor to GÉANT. GN2, like GÉANT, was co-funded by the EU (under the Framework 6 programme) and the European NRENs, initially for four years. To provide time for development and enable a period of overlap for transition to GÉANT2, the GÉANT contract was extended until June 2005 at which time GÉANT2 came into service, based on DWDM and with the ability to extend across Europe a service similar to the Lightpath which JANET would shortly provide. The GÉANT2 contract was for a further four years; in 2008, the GÉANT programme would be funded for another four years (under the EU Framework 7 programme) and GN3 would begin with the launch of the third generation of GÉANT in December 2009. With the basic transmission platform now stable, the focus for GÉANT would then shift to the provision of higher-level services

---

19 PoS: originally, Packet over SONET; subsequently used to refer to either SONET or SDH.

to support the diverse requirements of the European education and research sectors.

12.3 SuperJANET applications and services

The expression 'Necessity is the mother of invention' certainly applied to the origin of SuperJANET. In 1989, the need for capacity was particularly desperate but the direction taken to alleviate it was strongly coupled to the belief that new services would flourish. Some of those, like real-time video, were self-evident at the time and were believed to be a necessary basis of many other collaborative activities. e-Science (and e-everything-else) had not really been thought of. A few sample applications have been mentioned already, particularly those in the early days of SuperJANET. Here, the stories of some of the developments which were intimately bound up with the development of the network are continued to the time at the end of 2006 when SuperJANET5 finally realised the essence of the original concept. The huge increase in the community over the same period also brought about very significant expansion and evolution of the way in which JANET, the concept, the network and its services was purveyed. More recent services, and services like roaming which have little to do with the developments described in this chapter, are described in Chapter 14.

JANET videoconferencing

In retrospect, it is easy to observe that this service bore the brunt of the effects of the technological development of the underlying network as it continued in search of a multiservice bearer. Research demonstration (be it QoS models, ATM or IP) is not the same thing as service or product deployment; and underlying it is the continuing debate about just how much state (knowledge) the network should have about the service or application it is conveying. The debate is primarily an engineering one: if one knows what applications one is optimising the network for (real-time, data, etc.) and the parameters of the underlying communications (cheap or costly transmission), design is relatively tractable. But if the parameters of the communications are shifting and the applications are unknown or 'everything' (data, real-time, whatever), things are more difficult.

SJVCN, the SuperJANET VideoConferencing Network inaugurated as a studio-based pilot using the pilot ATM network set up during the initial stages of SuperJANET, had been sufficiently successful in a number of teaching and demonstration situations – including carrying the ceremony for SuperJANET's BCS Award in November 1994 – that by 1995 there was pressure to develop an overall strategy and develop a service. There were a number of issues, which might perhaps be summarised as: how to deploy a studio-based service more widely than possible with the pilot; the relationship between studio and desktop videoconferencing; and what of the various technologies and standards?

In the studio-based pilot, one aspect had become obvious: help was needed in both choice of equipment and training in setup and operation. Perhaps the most pressing aspect of this related to echo cancellation. This is a problem as old as telephony and arises from the round-trip delay in telecommunication being long enough to disturb voice human communication in the presence of echo – see

The need for echo cancellation

In audio communication, if there is an echo which is loud enough to be perceptible then there is a range of echo delay in which the speaker hears him- or herself just at the moment of beginning the next utterance, causing the speaker to stop and start utterances almost as if stuttering – a feedback effect somewhat akin to the 'kangarooing' a learner driver may typically experience in a non-automatic car. In old analogue telephone systems, the round-trip echo path had sufficient attenuation that, even when the delay was long enough to be a potential problem (as in intercontinental telephone calls, for example), the echo was sufficiently reduced in volume as not to cause a problem. There is no such attenuation in digital systems. Moreover, packetization (filling up packets with speech samples), buffering to remove jitter, and the processing associated with video compression and decompression all contribute to produce delay, which may be troublesome, even over short distances. (It may be noted that, although encoding and decoding of audio is rapid, moving picture compression and decompression is not, and for video, the audio needs to be synchronised with the pictures.) These all add up to a round-trip delay measured in hundreds of milliseconds, particularly if MCUs are in use. In these circumstances echo cancellation is essential. However, setting up echo cancellers, especially during the 1990s was not particularly easy, partly because the effect of incorrect setup is not perceived by the site responsible: instead, it wrecks the videoconference for everybody else. Their use is altruistic but essential!

panel 'The need for echo cancellation'. The first steps towards a studio service were taken during the life of the original pilot SJVCN when it was recognised that two aspects needed urgent attention: provision of advice and assistance; and expansion of the service to enable wider participation. It will be recalled that only 14 sites had ATM access:[20] by this time, even for these sites, the three MCUs at Cardiff, Edinburgh and UCL were insufficient. There were a number of options for expanding the service, of which perhaps the most obvious was ISDN. Since the initial introduction of H.261 video codecs at the end of the 1980s, commercial videoconferencing using PNO-provided ISDN services was now widely available, including internationally – the latter of particular importance for parts of the scientific community. In spite of the call charges, a number of universities were already using the service; moreover, WelshNet – an initiative which became the prelude to the Welsh Video Network, the nationwide deployment of video-based learning and teaching resources a few years later – was also providing a service based on ISDN6,[21] managed from an MCU in Cardiff, and it was possible to link the two video networks there by linking the two MCUs. However, although Edinburgh operated a booking system for SJVCN, which had been adapted from that developed for LiveNet in London, there was no booking system which covered the combined networks.

During 1996 considerable time was spent in reviewing the options available for expansion of the video network, along with gathering more detailed

---

20 Although UMIST had remote access via a 2Mbps leased line to the MCU at Manchester; and there was some use of access via the analogue Livenet in London using a codec at UCL.

21 ISDN6 is the marketing name for the entry-level ISDN service based on six 64Kbps data channels (known as B-channels) giving a 384Kbps video channel (using H.261 at the time).

requirements for a videoconferencing service. A videoconferencing strategy was drawn up towards the end of the year which the JCN approved the following year, in June 1997. Unsurprisingly, particularly in view of the international requirement, the main recommendation for expansion beyond those with direct access to SJVCN was to augment the service with ISDN. This was not regarded as an exclusive solution, rather just one of several technologies which needed support as part of the overall strategy. Indeed, the ATM-based service was expanded with additional MCUs bought in 1996. Another requirement identified in the strategy, to which we shall return, was the need to link the studio-based service with desktop videoconferencing provided by the IP-based MICE/UCL/LBL tools, also commonly referred to as the MBone tools or MBone video.

In Scotland the regional networks, the Scottish MANs, had utilised JISC MAN funding to exploit ATM for regional networking and for interconnecting the Scottish regions. Following this it was decided to exploit ATM-based video further for teaching. So, in 1996 UKERNA conducted a procurement for what would become the Scottish MANS Videoconferencing Network, the SMVCN. This resulted in equipping the SMVCN with Cellstack[22] video codecs, using Motion-JPEG, a form of encoding which provides very good quality (typically substantially better than the VHS quality of 2Mbps H.261) at the expense of a typical transmission rate of 15Mbps, with peaks at around 30Mbps (though these figures might be reduced if the scene was relatively static, as in meetings or some lectures).

What was now needed was a switching centre capable of achieving interworking amongst these: thus the JANET Videoconferencing Switching Service, JVCSS, came into being. Its brief was to provide interworking amongst: SJVCN (2 Mbps); the public ISDN service using ISDN2 and ISDN6 (128–384Kbps); SMVCN (15–30Mbps); WelshNet (384Kbps); and MBone video (variable rate but typically less than 384Kbps) – shown diagrammatically in Figure 12.11. And to make this practical, it needed the support of a booking service which covered all of these, including the possibility of access by non-JANET organisations which were collaborating via public ISDN. Edinburgh was awarded the contract to operate the service, which began in September 1997. Technically, switching between all these services was not simple, even amongst those using H.261 encoding, because differing rates are involved. M-JPEG, though incorporating some of the same underlying transform encoding principles within each frame, is an entirely different form of encoding, so decompression and recoding (transcoding) is needed. The ability to interwork with IP-based videoconferencing using the MICE suite of tools (which supported a form of H.261) seemed to offer a way for desktop participants to join a studio videoconference. However, even here, the form of H.261 used had the constant bit-rate feature removed (it being of little relevance in a packet network), making it yet another variation. *In extremis*, decoding into the analogue domain and then re-encoding is always possible, whatever the encoding, but with consequent loss of picture quality which may be unacceptable.

To meet the other requirement, that of help and advice, some sort of support centre was needed. The initial task was to survey equipment and publish the results. This was undertaken by the University of Newcastle upon Tyne and UCL in 1997 and resulted in the creation of the Videoconferencing Advisory

---

22 There were two companies in Cambridge making ATM M-JPEG codecs, both spin-offs from the Computer Laboratory; K-Net (which marketed Fore switches in the UK at the time) acquired one and the other, Nemesys, was acquired by Fore.

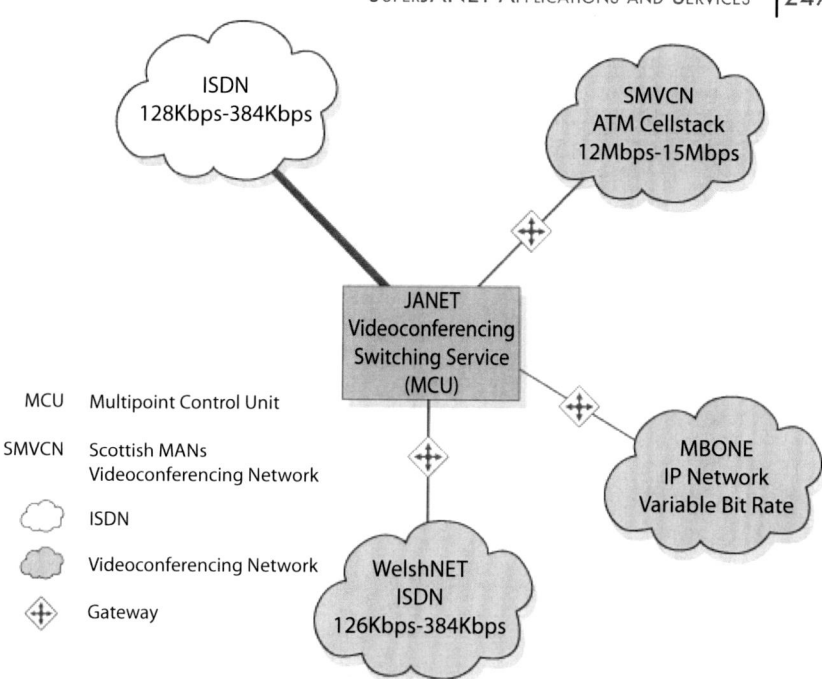

Figure 12.11. The JANET Videoconferencing Switching Service.

Service, VCAS, in 1998, also operated by Newcastle and UCL, with a helpdesk at Newcastle. Of course, product advice was just one aspect; advice on studio environment, interworking with the various services and operation were also needed. Subsequently, as video content delivery grew in importance, VCAS changed its name in 2000 to Video Technology Advisory Service, VTAS, reflecting the more general range of advice then being offered.

In 1997, as it became clear (see Section 10.3) that SuperJANET III would not be able to deliver the ubiquitous video service required, including international connectivity, the decision was taken effectively to replace SJVCN with an ISDN service. The service would now be referred to as the JANET Videoconferencing Service (JVCS), with the JVCSS being subsumed as part of the overall service capability provided by the Videoconferencing Management Centre (VCMC) in Edinburgh. All this, combined with the growing requirement to deliver video content, triggered a further review of the video strategy in 1998.

In the same year, a review commissioned by HEFCW in 1998 highlighted the need for a comprehensive video network in Wales. UKERNA undertook a study of the combined needs of the FE and HE sectors there, which confirmed the need, particularly for distance learning and teaching. The plan was initially to equip around 40 studios and to begin with a service based on a combination of ISDN6 and ISDN2, much as for the former WelshNet. However, the number of studios would soon double, as matching EU funds were obtained late in 2000; and right from the beginning the goal was to evolve to use of IP-based video since, in common with the rest of the UK, ISDN call charges were proving a barrier to use. The service, called the Welsh Video Network (WVN), would be registered with the JVCS VCMC in Edinburgh, which would provide the booking service but would have its own support centres provided by the University of Wales at Aberystwyth and Swansea. The target for initial service to begin was 2001, around the time SuperJANET4 would come into service.

So, as it became clear that SuperJANET III would be relatively short-lived and be replaced by an IP-based network in 2001, so the goal of IP-based videoconferencing gathered force. Moreover, SMVCN was due for decommissioning in 2001. As a result, in 2000 a programme to develop IP videoconferencing was announced, based on the H.323 framework; and the first step was a demonstrator project with participation from England, Scotland and Wales, co-ordinated out of Edinburgh. Since IP videoconferencing would potentially require IP Quality of Service, particularly where shortage of capacity might lead to congestion, in late 2000 the QoS Think Tank was also set up (Section 12.1). All this activity culminated in June 2002 with the opening of the pilot H.323 IP-based videoconferencing network which now covered the whole of the UK. It was declared a full service in March 2003, and by then included transcoding between ISDN and IP videoconferencing. The next couple of years would see a number of developments as use of the service grew. As could be expected, the booking service needed an upgrade in 2004 to cope with the growth, particularly in IP conferencing. And in 2005, conferencing on demand (that is, without prior booking) was introduced.

But probably by far the most significant development was in relation to schools. In the same year as the JVCS IP videoconferencing was launched as a full service, 2003, Charles Clarke, then Secretary of State for Education and Skills announced that schools would use JANET to provide the interconnect amongst the English RBCs (*see* Chapter 11). This effectively provided confirmation – and signalled the eventual provision of funding – for what had been in planning for two years. However, the real significance of this was that now the school sector would begin to have access to the same services as the rest of the education and research community. Although each of the constituent countries of the UK would take somewhat differing approaches – as outlined in Chapter 11 – the overall goals were much the same, and one of the major services which schools began to exploit was videoconferencing. Later the same year, for example, the University of Cambridge began using the service for a schools mathematics project to encourage interest in the subject, engaging school participants from 7 to 18 years of age on topics from actuarial science, through magic and financial planning, to knots.[23] Then in 2004, WVN was extended to cover a group of Welsh-medium schools, the beginning of another thread of development. At the same time an initial one-year project in England was mounted to pilot a country-wide schools videoconferencing service. Following the successful conclusion of this in May 2005, a further one-year pilot was mounted, funded by DfES, C2k, LTS and the Welsh Assembly Government. As a result, JVCS became the videoconferencing service for the whole of education. A side-effect of this has been that the JVCS booking system became the UK National Videoconferencing Directory for Schools: a consequence of this is that the directory contains not only all participating schools, colleges and universities but also video content providers, such as regional and national museums, archives and galleries – to mention just a few.

In September 2005 the largest schools videoconference to that date took place – London Live 2005, organised by the London Grid for Learning, LGfL, on the theme of 'Living in London in 2005'. There were over 120 participating venues. Apart from the schools themselves, participants included the Science Museum, a member of the London Symphony Orchestra talking about the double bass, the National Archives at Kew, the National Maritime Museum, the British Paralympic

---

23   *See UKERNA News*, No.24; and http://motivate.maths.org/content/, accessed March 2010.

Association and the Deputy Mayor of London. Viewers at other sites could watch the proceedings via a live video stream.[24] This was the first large-scale schools videoconferencing event and has since been followed by many more.

e-Science applications

The UK e-Science programme had an influence in several areas of networking. The expectation of increasing amounts of data to be handled was a basic driver which contributed to demands for increased capacity. As the science programme developed, so applications began to exploit the increased capabilities of the network. Facilities enabled to support bulk data transport for particle physics have already been described in Section 12.2. Realisation of advances in communication for astronomy are described below. Quite a different contribution resulted from e-Science adopting an experimental form of videoconferencing for collaboration: Access Grid (see below).

A third area has not so far been mentioned. The software and the corresponding engineering techniques for dynamically assembling computing and data resources to form loosely coupled distributed application systems over a network are generally referred to as 'middleware', from the idea that the techniques and associated software are positioned somewhere between the application and the underlying network. Although work in this area had been in progress for some twenty years before Grid computing, at around that time it took a particular direction which resulted in the development at Argonne National Laboratory of a middleware system called GLOBUS. Similar work in the context of Web technology for constructing derivative distributed applications resulted in a more general approach, a technology subsequently referred to as 'Web services'. Both initially developed independently, though Web services built on much that had gone before in Web technology, distributed systems and software engineering technologies. An ingredient that both had in common, with each other and every other use of the Internet, was the increasingly urgent requirement to incorporate security in a systematic and, as far as possible automated, way – a topic we take up more generally in the next chapter.

Access Grid

Video development was by no means confined to deployment of standards-based services already outlined. Although not the only mode of operation, the commonest way of operating early constant bit-rate videoconferencing systems was to display one venue to all and to choose which was displayed on the basis of voice switching – whomsoever shouts loudest gets on screen! For two-way meetings and lectures, this was generally satisfactory, though the lecturer tends to have to approach this as if on TV because feedback is minimal. For question and answer sessions or multiple-venue discussion it is distinctly limiting: the voice switching typically gets in the way of the rapid interchange of debate – and even if it manages, the rapidly changing view is distracting. Large meetings under this

---

24 *See* the article 'London Live 2005', by Rob Symberlist, in *UKERNA News*, Number 33, December 2005, for further details.

sort of system are best compèred, as on stage: this was how London Live 2005 was organised.

An alternative approach was adopted at Argonne National Laboratory where Access Grid was developed in the latter part of 1990s around the time when Grid computing was also being developed. At Caltech, similar ideas led to the development of the Virtual Rooms Videoconferencing System (VRVS).[25] The paradigm was that of seeking a virtual form of meeting place, where the concept that everybody could see each other and speak to each other at all times was preserved. To make this work, feedback amongst participants would need to be good enough for visual cues to be picked up as in a face-to-face meeting in a single physical location. Both developments took as starting points the MBone videoconferencing tools. VRVS became popular within the particle physics community, while Access Grid was initially promulgated amongst the Grid computing community. Within the UK e-Science programme, Access Grid was adopted as the system of choice for collaborative videoconferencing meetings. From an end-user perspective, amongst those who use either Access Grid or VRVS, the paradigm which results in everybody being able to see everybody else at all times during a meeting is generally regarded as considerably superior to that outlined above for traditional videoconferencing. From a practical service perspective – certainly during the UK e-Science programme – it required expert set-up and management and it made use of network multicast. The latter, ever since its development in 1989 (see Chapter 8), had always been the poor relation in service networking development and deployment, largely because it was used by few and is complex by comparison with point-to-point (unicast) communication. In consequence, it had never been as robust as unicast; furthermore, as gradually emerged, there were inadequacies in its technical definition and implementation for operation over a multidomain network. And, on top of all this, Access Grid did not initially interwork with other videoconferencing systems. Nevertheless, within the e-Science community it was gradually adopted quite widely; and because the system was sufficiently different and new – indeed, experimental – in comparison with the regular service, a dedicated Access Grid Support Centre (AGSC) was set up at Manchester University in 2004. Gradually, interworking with other systems became available and, as we shall see in Chapter 14, the new paradigm for videoconferencing would continue to develop and begin to spread beyond the scientific community.

eVLBI

As network transmission rates moved into the gigabit per second region at the beginning of the millennium, so the possibilities for what came to be called 'eVLBI' began to look plausible. eVLBI refers to VLBI operation in which observation data from telescopes are transferred directly over network connections to the image synthesising processor (correlator). Early demonstration of this came in January 2004 (just as UKLight was being approved), when data from Cambridge were transferred via Jodrell Bank[26] to Manchester and thence to JIVE via SuperJANET4, GÉANT and SURFnet: effectively, UK participation in the European VLBI Network

---

25 European support subsequently provided by Pavol Jozef Safarik University, Kosice, Slovakia.

26 The connection between the University of Cambridge Mullard Radio Astronomy Observatory and Jodrell Bank was via a separate fibre network, a part of another astronomy project MERLIN.

(EVN) had begun and the first EVN image had been produced overnight. Over the next two years, development of high-speed data transfer for eVLBI continued, using regular SuperJANET4 and UKLight communication between Manchester and JIVE. As SuperJANET5 and GÉANT2 came into operation, effective transfer rates were raised from 256Mbps to 1Gbps and beyond. So, in December 2007, a few months after the JANET Lightpath service came into operation, a dedicated lightpath connection was established, with the aid of the corresponding facilities of GÉANT2 and SURFnet, connecting Jodrell Bank and Cambridge to EVN, effectively creating a higher resolution radio telescope with more rapid production of images.

JANET operations and service organisation development

We have seen (in Chapter 4) how in 1984 the original ten[27] JANET NOCs (Network Operations Centres) were mostly[28] inherited from SERCnet, with the NOC at RAL fulfilling the role of the NCC, the Network Control Centre. The NOCs all reported to the Network Executive, which had been created in 1983 as a result of the recommendations of the Network Management Committee in 1982.

The Network Executive initially reported to Mike Wells as Director of Networking but, with the appointment of Peter Linington to become Head of the JNT in 1983, it was decided that an additional senior post in the Network Executive at the same level as the Head of the JNT was unnecessary: the posts were combined and Ian Smith, who continued in charge of the Network Executive, now reported to Peter Linington. In simplistic terms, the JNT would in effect become the development and strategy arm while the Network Executive would evolve from being responsible for operating the network to providing all those parts of the JANET service directly interfacing with both users – the 'customer-facing' part of the organisation in the current argot – and those outside the central organisation responsible for operation of JANET and its component services.

The beginnings of a specific user support role for JANET originated in 1986, when Shirley Wood joined the Network Executive from RAL where she had been a member of the support team for ICF GEC4000 MUM systems (*see* Chapter 2) and SERCnet. It was from this beginning that the Communications & Support Division would evolve over the subsequent two and a half decades. However, it was not until after the creation of UKERNA in 1994, with the incorporation of the provision of user support services into the JANET SLA, that additional staff began to be recruited into user support positions, the JANET Liaison Desk came into being, and it began to be possible to undertake a wider community support role.

On the operations side, the move from X.25 to IP had substantially increased the complexity and number of pieces of equipment which needed service: from eight X.25 PSEs, each looked after by a NOC, to over 30 points of connection for the IP network. Accompanying the network transition, as the pilot JIPS became operational in 1991, we saw the NOC at ULCC become the NOSC (Chapter 6), initially for the IP service and subsequently for JANET as a whole as the X.25 component was phased out in 1997. It was at this point that the need to

---

27 Shortly thereafter reduced to eight again, as PSEs at Bidston and Swindon were removed.

28 The exceptions were those at UMRCC and Queens University, Belfast, created as two extra PSEs were installed in 1983 as part of the process of developing the SERCnet core into the JANET core.

reorganise operational fault reporting brought into being the JANET Operations Desk, the JOD, situated at the NOSC in ULCC. Subsequently, on the user side, Customer Services (formerly with Business Division) came under a more unified arrangement within Support Services, which now became responsible for all external communication of information and training, whether by document, the Web, workshops or training courses.

At the beginning of the 1990s there were many new technologies coming into service in JANET, a situation which would continue throughout the decade. It began with the transition to TCP/IP and the introduction of new technologies onto campus: first FDDI, then ATM and video, followed by switched Ethernet, all accompanied by the evolution of the MANs or regional networks. As these changes began, in 1991 a new series of workshops started: the User Support Workshops. This series would last for ten years – the last in the series being in 2001 – essentially from the beginning of the introduction of IP to the completion of the transition to an IP-based network configured as a set of regional networks interconnected by the JANET backbone. After that, dissemination of information moved away from an all-embracing workshop (apart from Networkshop): the number of technical information and training events increased as the breadth of information to be communicated to an increasingly varied audience militated in favour of a rich set of more focused workshops, presentations and training events.

Over the last decade a pattern has emerged in which overall community support is provided through the Communication & Support Division, together with a number of more specific service units. First-line contact with JANET(UK) is now handled through the JANET Service Desk, operated on the call-centre model. The NOSC merged with UKERNA in 2006, when it became the JANET NOC (all the other NOCs in effect having previously been absorbed into the regional networks); it continued to operate the JOD, the means by which direct operational communication is maintained between the backbone and the regional access networks. The CERT, mentioned briefly at the end of Chapter 9 – of which more in Section 13.3 – is now CSIRT (Computer Security Incident Response Team) and maintains its service through its own set of contacts in JANET's constituent networks, as well as being part of what amounts to a world-wide, semi-hierarchically distributed organisation. Video is another example of a service provided by a number of units, including the management centre at Edinburgh, Access Grid support in Manchester and an advisory team distributed throughout the country: indeed we shall see in Chapter 14 how expertise from the Welsh Video Network assisted the initial extension of videoconferencing to NHS sites in the North East of England.

So, in 2007 SuperJANET achieved its major, original, strategic goals in terms of network infrastructure. It had reached the stage of being able to serve the whole of the education and research community, while yet having the capacity and flexibility not only to entertain novel applications but also to serve requirements and developments incompatible with standard service without disturbance. Development in core infrastructure would continue but now there was freedom to innovate and enable all the many higher-level, end-user applications and services for which there had previously been an insufficiently rich infrastructure. It had taken a while – 18 years from original concept: a seriously slow burn! JANET had come of age: the evolution to a multiservice network had been made and made available to the whole education and research community.

By now the JANET community was generally reckoned to be about 18 million. It covered a number of sectors, had stakeholders which were now numbered in double digits, and was operating not just a basic network infrastructure but a rapidly increasing range of sophisticated services. Unsurprisingly, it had outgrown its premises. In March 2007 UKERNA moved out of the Atlas Centre – which, as the Atlas Laboratory, had been purpose-built and named for the original Atlas Computer[29] – and moved into a purpose-built new building a couple of hundred metres away, in this case named for the modern association of light[30] with high-capacity communication: Lumen House. And shortly thereafter, in June 2007 a new trading name, JANET(UK), was adopted by the JNT Association, reflecting the single 'brand' of JANET as the name associated with the network, the operation, and the UK-wide community of JANET users.

---

29   Developed at the University of Manchester (with Ferranti and Plessey) in the late 1950s and early 1960s.

30   The Latin for 'light' having provided the name of one of the several units relating to radiated light intensity or power.

PART D

MULTISERVICE JANET

INTRODUCTION

JANET was now, in 2007, an established multiservice network. In the conventional sense of being a multimedia network, it had achieved this some years before; but it could now also offer multiple, specialist services based on channels using individual wavelengths. JANET had reached the stage where many new services were being offered throughout research and all stages of education; and these services were now capable of supporting not only all aspects of formal education but increasingly many of the other, more informal – including leisure-oriented – activities which contribute to an all-round education.

JANET had now been a part of the Internet for over 15 years. A less welcome consequence of the Web-induced take-up of networking on a world-wide scale as part of the infrastructure of life in the 21st century had been that the Internet, as well as supporting the notion of the 'information superhighway' (as the promise of increased capacity tended to be dubbed in the 1990s), had also taken on some of the more undesirable aspects of 'the street'. There was anti-social behaviour on the Internet as elsewhere. In consequence, the topic of security in its widest sense had come to the fore and been added to the regulatory framework in which information and communication services now had to operate: some coverage of how this affected JANET is given in Chapter 13.

Finally, in Chapter 14 we begin by looking at a selection of recent developments. Of course, some of these had been developing for years but had only recently entered service: key examples include roaming – by the turn of the century computers had not only become personal, they had also become cordless and peripatetic – and interconnection of JANET with the NHS network, something which had been needed for many years but whose difficulty of realisation arose primarily from issues surrounding confidentiality.

Given the apparently insatiable demand for network capacity – and the prospect of ultra-high definition video needing 1–10Gbp per channel signals, the continuation of that demand – one may speculate that the requirement for an intelligent optical core capability in the network is fast approaching. Recently, the third part of UKLight was realised as the wide-area optical fibre facility JANET Aurora, which has already seen national and international use.

Parts of the final chapter serve to report specific developments in JANET services. Others are more in the nature of review and commentary and, as such, reflect a personal view on aspects of the JANET story so far.

13

REGULATION AND SECURITY

In simplistic terms, most of what is generally termed regulation and security has to do with who is allowed to do what, ensuring the rules are not broken, and defending systems, users and information against misuse, malicious or otherwise. Systems in this context include computer systems for processing and storing information and communications systems for conveying information from one computer to another; and, more pertinently, from one person or organisation to another. Here we shall be concerned primarily with the aspects relating to communication systems.

Computer communication effectively inherits problems from the domains of both computers and communication. Much of the regulatory aspect originates from what was already established for the public telephone system, as was already discussed in Part A in respect of the regulations about providing public communications in the UK. Stored information is a valuable commodity, so attracts those with both benign and malicious intent. Enabling information to be transferred amongst computer systems for benign use also offers an additional opportunity for malicious interference, namely during transfer.

In respect of computer system and network development within the research community, it is illustrative of the changing environment in which both operate to observe that, during the formative development of both, security was not initially a major concern, perhaps because the community was small and people knew each other. Compromising a multi-user computer system initially consisted of little more than causing it to crash and the appeal was primarily the challenge of finding a new way of doing it. On one university system in the USA, where the research focus was not specifically on operating system security, the strategy to combat the problem was simply to implement a 'CRASH' command, thus removing the element of challenge and reducing the problem to manageable proportions. On the other hand, in the early days of systems research at a UK university which was interested in operating system security, there was at one time in the 1960s a standing, open challenge to users to 'crash' the system.

Many of the hosts on the early ARPANET were Tenex systems. A new sort of challenge to the curious was offered by virtue of their being connected: it was possible to set up an endless loop of communication amongst a group of these systems and then exit and leave them to it. Since the intent was primarily as a joke rather than a malicious act, it was not arranged that the systems would try to saturate the communications links, as would be the case in later 'denial-of-service' (DoS) attacks. In an environment where many, even most, of the developers knew each other, little if any of the information on any of the systems was of either personal or financial value: the goal was generally to make things work and security was simply not a serious consideration – and in the basic Internet architecture of 1982, which was effectively defined by the set of less than 20 RFCs published as the Transition Workbook (SRI, 1982), security was nowhere addressed. This situation changed drastically during the 1980s and 1990s as the Internet spread world-wide; with continued exponential growth – maintained during the 1990s by the arrival of the World Wide Web and its widespread adoption into many activities of everyday life – security became a serious issue as penetration and malicious activity generally became increasingly common.

13.1 Regulation

As the term implies, regulation is about the rules governing who is allowed to do what. So, for example, the telephone system was originally to enable people to make calls and talk to each other. Can anyone set up and operate such a service? Historically, the answer to this question has varied over time, as with many public services and utilities. In the UK, originally, the answer to this was 'yes' and, just like the train service and the London Underground, there was a collection of private companies providing the service. Since telephones needed cables all over the place and handling people's private conversations demanded a degree of responsible care, government stepped in to provide rules about both care and installing cables around the country, including organising who could do it by means of licensing. As we have seen, by the time JANET was forming, with the exception of Kingston-upon-Hull, the UK telephone system was operated by a government monopoly and the answer to the question above had become 'no'. Fortunately for JANET, at the end of the 1970s the pendulum had begun to swing back again, so that by the time SuperJANET and the regional networks began tendering in the early 1990s, there were a number of transmission network suppliers, all operating under government licence.

Pursuing the example of the public telephone system, the next issue is who may use it? Here, the answer is basically simple: anyone willing to pay, so authorisation to use the system is by payment, in particular, knowing who is responsible for the payment. Naturally, avoiding paying is stealing and against the rules. So, when touch-tone dialling came in, whistling down the phone to emulate a method of call establishment which avoided the charges was against the rules – though the first of those caught doing it are reputed to have escaped penalties in exchange for revealing the technique! In respect of the basics of such a service, there was initially no need for specific regulation: there was already a whole set of general rules for buying and selling services, of which the telephone was just another example.

Of course, the telephone service has always afforded anti-social elements opportunities to make unwanted telephone calls. What constitutes 'unwanted' varies from person to person and may depend on the caller or the content, or both; however, there are some types of call content which are sufficiently widely regarded as unacceptable as to be against the rules, either altogether or unless requested. Nevertheless, it has to be born in mind that notwithstanding any rules, the issue of what constitutes unacceptable content is typically both contentious and context dependent, making regulation and enforcement habitually difficult. Finally, there is the telephone variation of the confidence trick by which the caller seeks to obtain information, typically personal. Since obtaining things fraudulently is generally against the rules, nothing additional was needed for the telephone variant. All these activities would develop their computer communication analogues, with the added feature that the Internet is world-wide and without even the disincentive of any distance-related charging to end users.

An activity peculiar to communication in general, and the telephone in particular, is that of listening to other people's conversations – 'phone-tapping' – or, more generally, intercepting information. We have already mentioned above that responsibility needs to be exercised in handling private conversations, in particular, ensuring that the correct and only the correct parties are connected to each other; indeed, being allowed to engage in what is referred to as 'third-party' switching is one of the regulated activities for which a telecommunications operator licence is required in the UK, at least in part because the operator in charge of switching is in a very good position to 'listen in'. This is one activity in a whole class relating to mishandling information, all of which are generally regarded as a Bad Thing and therefore strongly against the rules – with the singular exception that governments generally reserve the right to indulge in this activity on the *prima facie* basis that it is needed for criminal investigation and national security.

Another activity more or less peculiar to communication is preventing people using either the communication system itself or a particular service by arranging to flood all or parts of the communication system with traffic which has no other purpose – similar to flooding the highways of the transport system with large, slow-moving lorries. For a network service, the same thing can be arranged by sending so many requests or so much data that the service or the network becomes overloaded and response time becomes so long that the service is effectively useless: an activity termed a 'denial-of-service' (DoS) attack. For the telephone service, the commonest non-malicious occurrence of either of these effects is when there is a disaster somewhere and a large number of calls focus on the relevant area or a particular service as a result of people attempting to enquire after relatives and friends. Creating a similar effect maliciously by the orchestrated use of a large number of systems distributed across the network is known as a 'distributed DoS' (DDoS) attack.

So, for our example of a public communication system, we observe that: service providers need to be licensed; use needs to be authorised; rules are needed to prevent abuse, whether of the system itself or of other users, whether clients or services; and regulation is also needed in respect of content and its handling.

Turning to JANET, how has all this affected the management of the UK education and research network? For most of the period up to JANET coming into operation in 1984, there had only been one (licensed) public operator. During the 1970s, star networks which developed to enable organisations to use computer bureau services were deemed not to be contravening the monopoly position (and needed no licence) because they did not switch data from one user to another. Initially, this was also the case for SRC and regional computer centres. However, by 1980 it was clear that all were technically in contravention of the BT monopoly in this particular respect at least. The situation, and its eventual regularisation in the after-effects of the process of telecommunication deregulation during the 1980s has already been described in Part A (*see* Sections 2.5 and 4.4). In effect, JANET became a Crown-operated community network in the public interest, dedicated to research and education, and didn't need a licence in the conventional sense. Later, in the 1990s, as the regional networks became independently-run networks carrying a mixture of JANET and other traffic, it became important to be able to distinguish between the two since non-JANET traffic was (and continues to be) ineligible to be carried over those parts of the infrastructure which are JANET-specific: principal among these being the JANET backbone.

Regulation of JANET beyond its original creation has been touched upon already: firstly, as JANET joined the Internet (Chapter 6); and again in discussing the governance and operation of UKERNA during and just after its formation (Chapter 9). JANET has always been an 'open' network in the sense of placing as little restriction on its use as possible, consistent with legal requirements, its sources of funding, the needs of its community of users, and avoiding conflict with network licensing regulations. An early statement of the general philosophy was that:

> '... any traffic which originates in one academic organisation and terminates in another will be regarded as acceptable.'[1]

Correspondingly, connection to JANET was expressed as being available for three groups:

> '(i) universities; (ii) Research Council laboratories or institutes; and (iii) holders of Research Council grants in polytechnics or other Institutes of Higher Education.'[2]

Though expressive of the general idea at the time of JANET's creation, both statements needed refinement. The first was at once too all-embracing – 'any traffic' was not intended to include that which was offensive or illegal – and not wide enough: for example, for some years before JANET there had been a number of gateways in operation which allowed exchange of email with non-academic sites. Moreover, in respect of connecting to non-academic organisations, there had been sites connected to SERCnet which were not academic but which were at the time participating in research with academic sites.[3] An additional issue was funding for JANET connections.

In terms of refining the rules for use of JANET, the starting point is of course that use must be legal. So, while use might be 'open', the majority of the undesirable usage described above is unacceptable because it is *prima facie* illegal on any network. To define further who may use JANET and what forms of usage are unacceptable, there was initially just an AUP (Acceptable Use Policy). However, as it became evident that 'use' of JANET and 'connection' to JANET, though intimately related, were not quite the same issues and had different ramifications, dealing with both together became too cumbersome. Recognition of this distinction in the early 1990s led in 1995 to revision of the AUP and the creation of a separate Connection Policy for JANET.

Connection Policy

Historically, the first major anomaly in terms of connection to JANET lay in the treatment of polytechnics. Though having a clear remit in terms of publicly funded education and eligible for Research Council funding, the sector was funded through local authorities and there was no overall funding body to coordinate sector institutional infrastructure provision of the sort exemplified by JANET –

---

1   Wells (1988), p.153

2   Wells (1984), p.59.

3   For example, systems situated at GEC Marconi Research and BT Research Laboratories were connected to SERCnet during 1982-3 (Burren and Cooper, 1989). HP Laboratories in Bristol and Olivetti Research Laboratory in Cambridge were among a number of other examples in the 1980s.

even though there was a polytechnic IT director[4] on the Computer Board, this was as an independent expert, not as a representative of the polytechnics. However, Research Council grant holders within polytechnics were eligible to access Research Council computer facilities and the Research Councils were adamant that all their research institutes and grant holders should be connected. This resulted in the connection of a Research Council grant holder in a polytechnic being affected by a direct line from a nearby JANET switch to the polytechnic's local network. Although this had the merit of providing unified access to local and remote facilities for a grant holder, it resulted in a low-capacity line, because the cost was born by the grant holder's project and furthermore it was shared with the whole institution. And a polytechnic which had no grant holders, though eligible to connect to JANET, had to find its own funds to do so. The effect was to create an anomalous category of 'second-class citizens of JANET': a situation which only began to be properly addressed with the creation of the PCFC in 1988.

During the 1990s the increasing variety of organisations with *bona fide* reasons to connect to JANET prompted an overhaul of the policy. By now (1995), all FE organisations were *prima facie* eligible for JANET connection. However, there were organisations which were not part of HE or FE but whose primary activity was nevertheless associated with education or research, individually or jointly with an HE or FE institution. And it would only be a few years before schools were formally added to the list. Beyond these there were libraries, museums, galleries, parts of the NHS, some professional bodies (such as the British Computer Society, the Institute of Electrical Engineers, the Institute of Physics, etc.) and university spin-off companies. So far (that is, in the version of the Connection Policy current at the time of writing[5]), three categories of organisations beyond the prima facie education and research sector have been defined in an effort to promote an open and equitable basis for judging eligibility.

The Policy also defines four types of connection to JANET. Two of these effectively existed from JANET's inception: 'Primary' and 'Interconnect'. The former refers to the type of connection which, for example, any HE or FE institution may have, providing all the services defined in the JANET SLA. An 'Interconnect' refers to a connection to another network and may include transit services but is not subject to the JANET SLA; connection to a school RBC network, other NRENs (such as Internet2 or HEAnet[6]) and GÉANT provide examples.

Two other forms of connection, 'Sponsored' and 'Proxy', relate to provision of services at arm's length by an organisation (possibly a Regional Network) with a Primary connection. The services provided are typically limited, not subject to the JANET SLA, and generally the responsibility of the primary organisation.

While the Policy specifies the type of connection to which a given category of organisation may be entitled or eligible to receive, the question of cost is not addressed. However, since all connections are effectively provided only for approved purposes, all connected organisations and their users are subject to the AUP.

---

4   Dorothy Nelson.

5   The JANET Connection Policy, dated June 2007: http://www.ja.net/documents/publications/policy/connection-policy.pdf, accessed October 2009.March 2010.

6   Ireland's NREN; *see* Glossary.

Meeting the costs

As we've seen, from its inception the Computer Board established a policy whereby computing services were free at point of use. This was naturally extended to networking in the 1970s when its sole purpose was seen as enabling access to remote services. This was easy enough to achieve because funding communication was just an extension of the top-slicing mechanism used for central provision of computing. Although the JANET community has widened dramatically since the 1970s and early 1980s, the growing set of stakeholders have continued to subscribe to the top-sliced funding model in respect of those for whom the stakeholders are directly responsible: 'free at point-of-use' has remained a general principle for JANET, though there have been periods of debate over the years.

Of course, there have been exceptions to the simple, clean model above. The first appeared before JANET began, when EPSS was replaced by PSS and initiating a PSS call could only be made by providing authorised account details which BT could bill. There was a period of use of individual, institutional and even community accounts which withered as PSS became increasingly irrelevant to the JANET community. The need for IPSS accounts remained longer, into the 1990s, until made irrelevant by the spread of the Internet.

The unfortunate position of polytechnics described above was never intentional – rather, it was because that there was no stakeholder to implement the top-slicing until the PCFC in the late 1980s. In the early 1990s FE colleges also found themselves in this sort of position: until 1997, there was no coherent central funding for FE connections to JANET, with the inevitable effect that only the larger colleges, with larger budgets to absorb the costs, could afford to connect. Apart from these unintentional 'disenfranchisement' situations, each of which nevertheless took a number of years to address, the general principle of 'free at point-of-use for all' has been maintained throughout all the devolutionary developments in funding for the whole of the UK education and research sector.

Principles, however worthy, also come under strain when the cost escalates! During the latter part of the 1990s the continued growth in transatlantic traffic, coupled with deregulation of the international market not quite yet having taken effect, caused a serious increase in the cost of provision of transatlantic capacity. It came at a time when there was still substantial expenditure on SuperJANET, together with more European interconnection capacity still being needed and the FE connection programme having recently begun. In 1998 JANET was facing a budget shortfall and this triggered another round of debate about charging. At a JANET policy level it was not a question of charging individual users; nor was there any desire to see substantial resources consumed by fine-grain usage accounting. However, there was usage monitoring already in place as part of operation and planning. Moreover, the cost associated with the provision of transatlantic capacity was comparable with the shortfall, and both the cost and usage on an institutional basis were fairly easily separated out. As a result, 'transatlantic charging' was introduced as a reasonably simple and cheap way of dealing with a budget deficit. Perhaps it should have been called a 'transatlantic surcharge': it was levied purely to deal with an expensive item and it was considered to be reasonably fair for the majority of the costs to be born by the institutions associated with the heavier use. Any apportioning of costs within an institution was strictly a matter for the institution.

The other area of exception relates to those who are eligible to use JANET but are not funded as part of the established education and research sector. Again, given that JANET has always operated on a not-for-profit basis, the general approach to this has always been to recover the marginal costs. When faced with the polytechnic situation, the Computer Board reacted much as SERC had done in the early 1980s in connecting collaborating companies: the institution paid a one-off charge to cover the direct costs of line installation and network switch interface equipment; it also paid the line rental; but there was no attempt to apportion or recover network operation and maintenance costs. However, even as the community expanded, the network grew in complexity and the type of organisation and the purpose of connection became more diverse. Because the purpose of connecting non-education, non-research organisations has remained primarily the benefit of the education and research community, and essentially excludes use for any other purpose, there has never been any move to recover other than the direct marginal costs.

Acceptable Use Policy (AUP)

As we have seen, from its inception JANET has sought to be as open and unrestrictive a network as possible consistent with the primary goals of the sector, which amount to the combined aims and policies of all the user organisations and their funding stakeholders participating in provision and use of JANET. Within these limits – which are very wide – a user activity, corporately or individually, is essentially unrestricted provided the activity is neither illegal nor upsets other users. The JANET AUP[7] refines the sense of this into a more precise form to provide explicit guidance about what is or is not allowed and to provide the basis for determining if a breach has occurred. All users have responsibility for ensuring use of JANET is in accordance with the AUP; corporate users have the responsibility for enforcing acceptable use by its members, as do networks subject to the JANET SLA.

As originally specified the AUP included matters to do with eligibility, now dealt with under the Connection Policy, and it was these which caused a number of the early revisions. Among the activities explicitly listed as unacceptable are a number of those which deliberately get in the way of other people's work or use of the network. While not intended as an exhaustive list, these together with a number of specifically illegal activities are listed in the AUP to avoid doubt and to draw attention to some of the commoner forms of network abuse.

Given that unacceptable use of JANET may occur, particularly in view of the open nature of the network, there arises the issue of dealing with it – whether directed toward prevention or responding to incidents. The topic forms a part of what is generally referred to as 'security' and applies to a variety of personal and organisational risks, activities and consequential results which, in the context of JANET, form the subject of the next section. However, addressing the security of JANET depends on cooperation throughout the community; moreover, use of JANET also brings with it certain legal responsibilities in respect of security. Since this is an area which has seen considerable change and development since the early 1990s, elaboration of user responsibility has in recent years been made the

---

7   The JANET AUP, Version 10, dated April 2008: http://www.ja.net/documents/publications/policy/aup.pdf, accessed March 2010.

subject of a separate JANET Security Policy.[8] Responsibility to conform to the Security Policy is included in the AUP.

The legal regulations surrounding security have mushroomed in recent years in an effort to combat malicious activity associated with communication and computer systems. The legislation stems from a variety of international contexts – human rights and the European Union – as well as UK law. The remarks made at the start of this chapter, that much telephone misuse was already covered by existing criminal legislation, provide a gross summary of the starting point for prosecutions in respect of computer and network misuse. However, 'progress' in the technology of misuse resulted in the need for increasingly 'creative' interpretation of laws never intended for application to computers or networks. This resulted in 1990 in the first of a range of legislation to deal with the situation; since this was (and remains) a new area with considerable technical ramifications, development of appropriate practical legislation is taking time, particularly as it also takes time to accumulate the body of case law to compliment the legislation. Broadly, the areas covered by the legislation are misuse, content and liability, and law enforcement. In respect of networks, an early attempt to address part of the latter in 1985 had been effectively directed at the duopoly public service of the time, leaving out all private networks. The further deregulation which began in the early 1990s soon resulted, through a combination of circumstances, in most networks being effectively beyond the law. What was needed was legal regulation which balanced personal privacy, stemming from human rights legislation, with criminal investigatory powers. The initial legislation addressing this area came into force in 2000 and, like the data protection legislation begun in the 1980s, has brought a range of new responsibilities to both users and operators of JANET, including particularly in respect of traceability. An introduction to the issues involved and the relevant legislation is provided in Appendix I, 'Computers, communication and the law'.[9]

13.2 Security

The security threats to an information and communication system handling (potentially) valuable information almost all have their non-information technology analogues and include breaking and entering, defacing, stealing, defrauding, trashing and so on. More specific to the communications component are interception, masquerading (typically in an effort to conceal identity while indulging in activity against the rules) and altering the communication stream by changing, deleting or inserting (mis)information. None of this is new or even peculiar to modern technology, to the extent that examples are reflected in fiction: Alexandre Dumas (1884) has the Count of Monte Cristo insert false information into the French optical semaphore network for the purpose of influencing financial markets;[10] and, in the next century, in 1973, the film *The Sting* has an example of

---

8   JANET Security Policy, dated June 2007: http://www.ja.net/documents/publications/policy/security.pdf, accessed March 2010.

9   Kindly contributed by Andrew Cormack, Chief Regulatory Advisor to JANET(UK).

10 The story begins in 1815 just over 20 years after the invention of the Chappe mechanical semaphore or optical telegraph (*see*, for example, Salford CNTR (2008)); the events are related in Chapter 62, 'The Bribe'.

tampering with information, in that case delaying the relaying of horse racing results for the purpose of rigging betting.

The fictional examples focus upon security breaches within the communication system. In the context of JANET the network may act as the unwitting conduit for security breaches of end systems. A compromised end system, apart from any damage which may be caused to its data or system, becomes a potential source of further damage spreading via the network. Although specific risks and defensive action may relate to specific components, in assessing the security threats the whole system of network and connected hosts needs to be considered.

The taxonomy of risks included in the JANET Security Policy, while not claiming to be exhaustive, identifies five categories: breach of confidentiality; loss of integrity; failure of availability; damage to reputation; and legal liability. The most obvious examples of the first relate to intellectual property and personal information. Unauthorised access to the former typically results in loss of value; and in case of the latter, consequences range from loss of privacy to theft of identity. Integrity applies to both data and systems, the latter including the network. Loss of integrity not only means that output from the system cannot be relied upon but that control of the system may have been subverted, implying that affected parts may be used to affect other parts: the metaphor of the spread of an infectious disease is apt. Failure of availability refers to activity which results in (partial or total) loss of network services. The (distributed) DoS attack which maliciously overloads servers or (parts of) the network has been mentioned already. However, there are more subtle forms of loss of service which may result from bad administrative practice: one example being lack of care to ensure that a system is not used to spread unsolicited bulk email (UBE or 'spam'[11]); lack of care in this instance can result in a system or the set of systems belonging to an administration being black-listed with the effect that email service is, in effect, partially lost. In this case, regardless of any view regarding the origin of the UBE, the administration suffering loss of service is not guilty of malice, rather a lack of due care and attention. It is this latter aspect – of not taking reasonable precautions and hence becoming the source of infection which affects others – which can lead to a system administration or network operator being regarded as unreliable or, worse, an easy vehicle for the pursuit of further anti-social or malicious activity. The latter may include activities beyond those already identified; examples include purveyance of nuisance, offensive or illegal material; causing financial or other loss to others; and generally disobeying the rules, such as the JANET AUP and the law. In some of these cases the administration in charge of such a system or network may become liable in law.

Evidently one aspect of both regulation and security relates to protection of resources: controlling who may access processing power, storage, network communications, indeed, network services generally. Naturally, this had been recognised in a general way since the earliest computer services in the 1960s. By 1981 it was clear that there was at last going to be a national academic network and the need to control access to network resources in a UK academic context came into more prominence as part of the general need to manage network services as a whole. There were as yet no standards or non-proprietary products for network management and the JNT initiated discussion and study of what was needed. At the time a cause of particular concern was that while local network capacity was relatively cheap and plentiful, wide-area capacity was precious, being expensive and slow: access control to the latter was required and added to the long list

---

11 *See* Glossary.

of management requirements.[12] Authorisation had been added, albeit in an *ad-hoc* manner, to the Yellow Book transport service to enable usage charging to be implemented in any gateway between an academic network segment and PSS; however, apart from this, control of access to JANET during the 1980s would depend entirely on access control being applied on all JANET-connected hosts (including terminal concentrators), primarily through the use of user identities and associated passwords.

In security terms the 1980s saw the rise of the 'hacker' – one who gains unauthorised entry to a computer system or network.[13] Although JANET did not at that time experience problems to the same degree as the Internet, towards the end of the 1980s it was experiencing regular hacking, which occasionally reached epidemic proportions, usually associated with a new mechanism in circulation for a particular operating system.

Until the late 1980s, one of the commoner automated methods of break-in was to set a computer to login to another using 'brute force', by working through the complete list of possible passwords of increasing length – which duly gave rise to longer and longer passwords as systems became faster. However, automation in pursuit of unauthorised entry finally got going in earnest with the Morris 'worm'[14] on the Internet in 1988. Then, at the start of the 1990s, as the World Wide Web and the Internet spread, so automated methods of anti-social network behaviour – mediated by software generically referred to as 'malware' – began to develop into a major activity, engendering in turn, if not the birth then at least the rapid growth of the computer network security industry.

As all this began in the late 1980s, JANET was initially insulated from it because the malware and associated techniques were developed for use on the Internet which had a world-wide community: most of the security-cracking technology could not pass through the JANET gateways; and even those that in principle could (via email, for example) would not operate with JANET network protocols. Then in 1991 JANET began to join the Internet – and Pandora's box was opened!

It had taken a decade for security cracking to reach the proportions of a menace; within JANET it would take another decade – during the 1990s and into the current millennium – to persuade people to take security seriously. That JANET should be an open network was, and remains, a carefully preserved fundamental principle. Moreover, people were used to there being little or no problem; even when systems were penetrated, it had been more in the nature of mischief and a temporary inconvenience. There might be waste of time but seldom permanent loss or damage. During the 1980s, email lists for those technically responsible for services provided by particular machine ranges, together with those managing and operating networks, had evolved and these or specialist lists would also circulate information about the latest hacking exploits and incidents. However, by the 1990s administrations were relying on networks which meant that various forms of confidential information, including personal, was now regularly being carried by the network and was accessible via the network. There began to be a general realisation that not only was the problem not going to go away, it was going to have to be taken seriously (see panel 'Taking security seriously').

---

12   Article on 'Network Management', *Network News*, No.11, August 1981.

13   *See* Glossary. The corresponding activity on the phone system is referred to as 'phreaking'; the two developed more or less alongside. Fraudulent phreaking is sometimes referred to as 'wire fraud'.

14   Worm: self-propagating malicious program. *See* Glossary.

Taking security seriously

Someone who has experienced burglary tends to be amenable to expending resources to prevent it happening again. Within the JANET community, until some time in the mid-1990s relatively few people had experience of the potentially serious consequences of broken security. Consequently, there was a period of years during which it was hard to obtain resources with which to tackle the problem. Talks were given in an effort to educate, but there's nothing like a demonstration. One of the commonest techniques used in developing or engineering networks, diagnosing problems or teaching is to observe packets (sometimes called 'sniffing'). This is particularly easy in a shared-media LAN. Like any technique, it may be used for good or ill. At one university with such a LAN, where persuasion was still needed, at a seminar organised in the mid-1990s for senior administrative and academic staff, a live demonstration was arranged of intercepting email by sniffing using software freely available on the Internet. Within a minute or two, several emails had been observed: but when one of them turned out to be from the university's Vice Chancellor, debate ceased!

Resources were going to have to be put into dealing with it, from educating users and defending systems to dealing with incidents.

13.3 Protection and enforcement

We have seen that the community began to organise itself to pool information to help recover from incidents of malicious network activity: as we shall see, this proved its worth and evolved into a network of organisations to assist in responding to incidents. But what about prevention? Considering the various forms which attacks may take, dealing with these divides at a technical level into preventing penetration of a system, whether of an individual computer or a whole network and its computers; preventing denial of service; and preserving confidentiality, in particular of information in transit over a network.

Preserving confidentiality of information, whether stored or in transit, is the primary goal of encryption. However, apart from encryption and cryptanalysis, cryptography may also be applied to authentication. The research for much of this area stems from the 1970s: examples include algorithms such as Diffie-Hellman (Diffie and Hellman, 1976) for establishing a shared secret key over an insecure channel; Rivest-Shamir-Adleman (RSA) (Rivest et al., 1978) for public key encryption of text and its authentication by digital signature; and the Data Encryption Standard (DES)[15,16] block-cipher data encryption algorithm. Penetration of systems also has two sides: authorisation of *bona fide* use and preventing access without authorisation. Initial models of network authentication and authorisation

---

15 The result of work at IBM's Thomas J. Watson Research Center; published by NBS in 1976 but see Burr (2001) for review and description.

16 In the case of Diffie-Hellman and RSA public key encryption, there was earlier work in the 1970s on equivalent systems (classified at the time) at GCHQ (UK) by Williamson and Cocks, respectively.

were also developed at this time, an early UK example being in association with the Cambridge model distributed system[17] in the early 1980s.

Preventing denial of service is intrinsically hard because there is no way of distinguishing such traffic until a pattern has developed which is evidently malicious. Early detection of DoS attack is possible by exploiting knowledge of such patterns. Once an attack has been detected in progress, it can be blocked in relevant routers.

Incident response

As attacks on networked computers and associated services and resources turned serious and became increasingly widespread at the end of the 1980s, people began to pool resources on an ever wider basis. Organisations and communities formed teams to respond to emergencies and pool information about their nature and how to deal with them: such a team was often known as a Computer Emergency Response Team or CERT. In the Internet context, the immediate effect of the Morris worm was for DARPA to initiate the setting up of the CERT Coordination Center (CERT/CC[18]) at the federally funded Software Engineering Institute at Carnegie Mellon University (CMU), Pittsburgh.[19] Although the original role of CERT/CC was focused primarily on co-ordination and information to enable more effective response to widespread attack, it soon broadened to include observation of the state of security on the Internet and a research and development programme into techniques for detecting and countering malicious activity.

JANET had begun its prototype IP service in 1991 and in less than two years it was confirmed by JISC in October 1993 as the JANET service of the future. Following rapidly upon this, JANET CERT[20] was created in January 1994, just three months before UKERNA took over from the JNT. The service was immediately incorporated as one of the services to be provided under the UKERNA SLA with JISC. During 1996 there was discussion sponsored by TERENA about setting up a European-wide CERT. DANTE became a member of FIRST,[21] the world-wide forum for such teams; and in the spring of 1997 EuroCERT, a partnership between DANTE and UKERNA, began a pilot European service. This ran for a year and provided European-wide experience of what was needed, which was primarily coordination among the European NREN CERTs on individual incidents and service developments. By 1999, the major NREN CERTs knew each other well enough not to require external incident coordination, so EuroCERT ceased in 1999. A TERENA CSIRTs Task Force was formed to take forward service development ideas: work it still continues.[22]

---

17 The work was based on *capabilities*, (unforgeable) tokens having some of the properties of modern certificates; *see*, for example, Needham and Herbert (1982), Girling (1983) and Cooper and Burren (1989). The ideas had also been applied to internal computer system security (Wilkes and Needham, 1979).

18 CERT and CERT/CC at CMU are registered names rather than acronyms.

19 *See* http://www.cert.org/meet_cert/ (accessed March 2010).

20 JANET CERT was renamed JANET CSIRT in 2007, adopting what had become the preferred acronym title for such teams: Computer Security Incident Response Team.

21 Forum for Incident Response Teams.

22 Security continues to be an area needing support in a wider context than education and research: one example was the formation of the European Network and Information Security Agency, ENISA,

Authentication and authorisation

Underlying several aspects of security are the questions 'Who are you?' and 'By what right do you do this?' – which typically equates to 'Who says you can do this?'; moreover, the latter (authorisation) typically rests at least in part upon the former (authentication) since it is likely that in seeking confirmation of authorisation, confirmation of the identities of both the authorising and authorised parties is required, implicitly or explicitly.

Traditionally, authorisation to use a system has for over 40 years been by an identifier and password combination: the identifier assigned to an authenticated user by a system administration, together with a password defined by the user and recognised by the system. With the coming of many network services available to a given community of users, the notion of having an authorisation that would work for a number of services became attractive. In the UK, within the JANET community of educational and research users, journal access is an intrinsic requirement, traditionally provided by institutional paper journal holdings in a library. By the 1980s, preparation of journals for publication was already electronic and the availability of ubiquitous academic networks now signalled the era of electronic publication: here was a classic situation for federated access management, with many people in many institutions needing access to many journals from a variety of publishers and learned organisations. In the UK the Athens[23] access and identity management system, originally developed at the University of Bath as part of the Computer Board funded NISS[24] project (for NISS internal use), began providing such a service in 1996: using a single sign-on identifier-password combination, a user could access a range of journals from those publishers which had joined the system, taking into account the specific subscriptions which the user's institution had with the providers. From the user perspective, Athens provided a substantial advance over having to remember a list of individual sign-on credentials for each journal – sufficiently so that the JISC funded the service for the UK education community from its inception until 2008.

From a network perspective there were two weaknesses in this model which increasingly needed attention from the late 1990s onwards. The classic weakness in the identifier-password method of user authorisation by a network service is that it is typically sent unencrypted over the network: valid combinations are easy to sniff, for example, on a shared-media network, like the original Ethernet or any wireless LAN not implementing suitable precautions. The other issue is one of scaling, coupled with a growing appreciation of just how complex are the trust relationships involved in both authentication and authorisation. A single person typically has a large number of roles in a great variety of organisations and contexts. Moreover, the organisations are from a far greater spectrum of possibilities than simply a university, college or school: they embrace project collaborations (an instance of what is sometimes termed a 'virtual organisation'), and joint institutions where a person has specific authorisation by virtue of being a member of a parent or member institution: a joint medical school provides such an example. The situation develops a further dimension when it is realised that each component of a Grid computing assembly operating on behalf of a set

---

in 2004; JANET experience is contributed through participation of Andrew Cormack.

23   *See* Glossary.

24   *See* Glossary.

of people needs to be authorised, possibly authenticated, to all or parts of the assembly. And as the UK e-Science programme developed and the adoption of Grid computing began at the end of the 1990s, coupled with schools joining JANET and the provision of a range of educational content becoming important and beginning to expand rapidly, so the need for a general authentication and authorisation infrastructure to serve the whole of the JANET community became pressing. Grid computing and content provision were, of course, not the only activities needing this infrastructure: collaborative use of video and guest use of a host institution's network services infrastructure – to name but a couple more – were also in need of such an infrastructure. However, in order to be able to operate over the untrusted global Internet, Grid middleware had already had to build in its own incipient authorisation facilities and now demanded such infrastructure in order to continue.

What was needed was a scalable system in which trust relations for both authentication and authorisation could be continually established, rearranged and revoked, all in a secure, decentralised fashion over the network. In 2003, Internet2 released the first version of Shibboleth,[25] an open-source system with the goal of supporting identity management and access control for Web-based resources; it is based on use of SAML[26] and is fundamentally a system for handling authentication: authorisation requires additional functionality. In 2004, as part of its Core Middleware programme, JISC funded a three-year project, Shibboleth Development and Support Services (SDSS) at EDINA[27] (based at the University of Edinburgh) to develop a prototype shared access management for UK education and research use. By the end of 2004, based on X.509[28] certificates provided by GlobalSign,[29] SDSS had begun to provide a prototype service for early adopters within the community. At the basis of authentication is a legal framework of procedures for establishing trust. UKERNA had established such a framework with GlobalSign under which UKERNA was a registration authority (RA) trusted (authorised) to authenticate servers at institutions within the JANET community.

The SDSS project and the prototype service it provided were a success in demonstrating the viability of the technology. In 2006 it was decided that SDSS should form the basis of a permanent service and in November 2006 the UK Access Management Federation for Education and Research was launched as a JANET service with the support of JISC and Becta. Overall management was provided by UKERNA – soon to be JANET(UK) – while technical effort and operation continued to be devolved to EDINA SDSS staff in Edinburgh. With the launch of the UK Federation service, the JANET community now had a trusted service which properly protected privacy during authentication; however, services often require additional information for authorisation. In principle this can also be achieved while protecting privacy by suitable use of certificates (as demonstrated in some

---

25 The Shibboleth project is a part of the Internet2 Middleware Initiative. It is led by the Middleware Architecture Committee for Education (MACE). MACE has international representation: European liaison includes Brian Gilmore (Director, Computer Services, University of Edinburgh) and Josh Howlett (JANET (UK), technical lead for the JANET Authentication, Authorisation and Identity programme). (See http://shibboleth.internet2.edu/, accessed March 2010).

26 Security Assertion Markup Language; see Glossary.

27 See Glossary.

28 See Glossary.

29 UKERNA having the previous year, 2003, already contracted GlobalSign to provide server certificates for the JANET community.

model systems of the early 1980s); as yet, however, additional authorisation information is typically obtained by reference to either a user's institution or the user directly, neither of which necessarily protects privacy. While various systems are currently in use in the UK – Grids typically use personal X.509 certificates and network access typically uses IEEE 802.1x and RADIUS[30] – the Federation has provided the basis for authorisation in an increasing number of contexts in the UK, including that of journal access (with the aid of a bridge for journals only using Athens) and, in 2007, the JANET Videoconferencing Booking Service.

Virtual private networks

Virtual LANs were described (Section 10.3) as a by-product of switched LAN technology. The set of stations on the LAN is partitioned into groups, so that stations appear in only one group. A switch then enables communication within but not between groups. The VLAN is virtual in the sense that the physical infrastructure is shared by a number of groups but to any station the network appears only to connect members of its own group. Assuming wire-tapping is prevented and the switch not subverted then communication within a VLAN is private to the stations in that VLAN, without any recourse to encryption. Thus, a VLAN may be characterised as a virtual private network or VPN.

In the case of a VLAN implemented over a LAN using multiple switches, some way needs to be provided to multiplex multiple VLAN communication over the inter-switch links. The method most often adopted is encapsulation: just precede each LAN frame in such a link with a field identifying the VLAN. This technique will be recognised as being very similar to the encapsulation used by MPLS, although with different semantics; and indeed, MPLS can be exploited to help provide more general forms of VPN than a VLAN.

Of course, if a VPN is to be implemented over a network where the nature and security of the constituent links is unknown, mere encapsulation of a frame is unlikely to be adequate in respect of the privacy aspect: encryption and perhaps authentication and authorisation may be needed, all of which may be provided with the aid of further headers or, indeed, in combination with complete encapsulation, given that there is an authentication and authorisation infrastructure in place. It will be recalled that JANET has used encapsulation several times in the past: IP over X.25, XOT and, for a much longer period, provision of multicast over parts of the network having only a unicast service. Each may be regarded as the provision of a virtual network with the aid of encapsulation: the VPN is yet another example.

Although at the time of writing the provision of some kind of JANET VPN service remains an issue for study, localised VPNs have been in use since the mid-1990s. VLAN use within institutions is little different in the JANET community from anywhere else and includes VPNs for administration, which commonly handle personal data, and guests who most likely are regarded with a different degree of trust, even suspicion, from members of the host institution – a view which may, of course, be reciprocated. Those who work away from their home institute – whether at another institute on JANET or situated somewhere on the public Internet – also typically make use of a degenerate form of VPN consisting of a single encrypted link connecting, for example, a laptop to a server at the

---

30 *See Glossary.*

home institution which effectively incorporates the remote device into the home institution's network. That it is more than likely that the first link in any such chain is wireless (Section 8.2) makes it even more desirable that the link should be secured, typically by encryption. However, in establishing any such virtual private link, the issue of authorisation arises, a topic to which we shall return in Chapter 14 in the context of roaming.

Firewalls

We've mentioned recovery from malicious incidents, authorising use of systems and guarding privacy during communication, but what about preventing unauthorised access? The general, abstract scenario is one in which there are two interconnected system regions: one which is trustworthy, secure and having only benign activity; the other untrustworthy, insecure and with malicious activity. The goal is to provide a wall of protection so that malicious activity in the untrustworthy region cannot penetrate into the trustworthy region. The device to achieve this is generally known as a firewall, by analogy with the physical walls in a building to prevent the spread of fire. Computer and network system firewalls may be software, hardware or a combination. One example is the software firewall incorporated into the operating system of a personal computer to prevent penetration from any network to which the computer is connected. Similarly, at an institutional level, to prevent penetration of the private network and all the systems attached to it, a network firewall may be provided at the point of connection to any external network.

As with other network security developments, the Morris worm in 1988 provided the spur to a major increase in firewall development. Early firewalls were provided by adding features to the router already present to connect the two network regions. Evidently the provision of firewall functionality adds to the processing needed to pass a packet from the untrustworthy to the benign region, since additional inspection of the packet may be needed, together with the application of rules to determine whether the packet should be admitted. Since routers could frequently be a bottleneck for an institution's external communication, it was not long before separate network firewall devices were available at the beginning of the 1990s.

Although basic firewall operation depends on packet inspection to filter out packets addressed to or from inadmissible hosts or ports, this can be supplemented by additional knowledge of applications and protocols. In the case of applications, two basic approaches are possible: all are assumed malicious until demonstrated or declared otherwise; or the converse. Unfortunately, since the latter strategy only defends against new forms of penetration after the damage is done, the tendency is to adopt the former approach, with the consequence that since the beginning of the 1990s, JANET and the Internet as a whole have become substantially less open. Some protocols and applications – especially new ones – typically fail because of firewalls: then a case has to be made to reconfigure the firewall.

So, a common strategy in configuring a firewall is only to allow communication on the ports associated with particular, required applications. While this works satisfactorily for some applications, there are those which assign ports dynamically during setup, to be used thereafter as a means of multiplexing or subprocess addressing within the host. Without more detailed knowledge of such applications to recognise and inspect packets exchanged during setup, a firewall may be unable to accommodate such styles of use. Even if a firewall does have such knowledge, conflicts with other aspects of secure communication may arise. For example, in the case of encrypted communication, it may not be possible to inspect any details of connection setup in order to accommodate dynamic port use: it may be necessary, for example, for the firewall to participate in setting up the secure communication.

Within JANET, even before IP was available there were occasional instances of problems caused by a site defending itself against hackers from another site by configuring an X.25 switch or transport service gateway not to recognise calls from a particular site – an unfortunately gross temporary measure which disabled even email between the sites! However, since the end of the 1990s common problems experienced on JANET with the use of firewalls typically included real-time video and Grid computing. In the case of video, where packet delay is important, the performance of early firewalls was inadequate and a site sometimes had to resort to bypassing the firewall for this traffic. In the case of Grid computing, the middleware typically made use of dynamic ports and encountered the problems already mentioned. While the particular issue of dynamic ports can be overcome by the firewall sniffing the initial setup packets or avoiding such use, nevertheless the problem is symptomatic of security having not been intrinsic to the original network architecture, resulting in a variety of approaches to different aspects of the network security problem but, as yet, no globally accepted approach.

Filtering

A network firewall acts as a filter on traffic entering a site; although it may have some minimal knowledge of the properties of some protocols and applications, its basic operation is primarily at the packet level. In the mid-1990s a somewhat different sort of 'defence' or 'protection' issue arose, having not so much to do with allowing only authorised packet or traffic flows but focused on forbidding particular traffic or access to material on the basis of content.

Actually, it had all started with Usenet News. It will be recalled (Chapter 5) that in the 1980s the University of Kent had been providing this service since before JANET. Early on there appeared news groups disseminating material likely to cause debate about propriety, offence, obscenity and perhaps even with potential to cause liability for those responsible for participating in transmission. As a result these groups were excluded from quite early in the 1980s, if only to avoid any potential liability to those in Kent who were providing the service. Subsequently, this became one of the earliest services to be filtered by JANET on the basis of IWF recommendations (*see* below).

Next came spam. This is almost universally regarded as a nuisance. By the mid-1990s an appreciable amount of network capacity was being consumed and recipients had to spend increasing amounts of time discarding the daily dose of spam which in some cases by now was beginning to constitute the major fraction of email received. The first action was to stop any mail server being used to relay other than for the benefit of its own community of users, to prevent its participation in spreading spam on behalf of unknown sources. Services appeared on the Internet which provided lists of sites not behaving responsibly in this respect: a site so black-listed would find that its mail service was increasingly isolated as other mailers would cease to communicate with those on the black list.

Then email filter programs appeared, followed by dedicated filter servers and external filtering services, which used a variety of heuristic algorithms to classify email as to its probability of being spam (frequently combined with inspection for embedded malware). Users, or whole sites, could assign a variety of actions to the classifications, from 'discard immediately', through 'await inspection', to 'accept'. At the time of writing the majority of JANET sites have the machinery to deal with spam in this way on a site-wide basis.

Spam was easy, largely because, with relatively minor differences, there was general agreement on what constituted it, and that it was a Bad Thing. However, the question of access to material which contravenes an AUP was also raised. There are two quite different issues: attempting to decide whether accessing particular material does indeed contravene an AUP; and what to do about it if it is considered to do so. Unlike the case of spam, the first issue is fraught with problems. Apart from opinions varying widely, there is also context: what may contravene privacy, even law, in one context may be acceptable and necessary in another under, for example, police authority. Considering the difficulty of classification, the general reaction was to stay away from an area which too easily slips into ill-defined and unregulated censorship – a situation illustrated on the international stage in recent years by government action in some 40 countries to bar Web access nationwide to some international sites.[31] Instead the emphasis was on increasing traceability so that, in case of some suspected abuse, the basis of administrative action could be furnished. JANET was not the only organisation to avoid such issues: the Microsoft Network decided to close some of its chat services in 2003 because of the risk posed by those with malicious or illegal intent gaining unauthorised access to the vulnerable, particularly children. In this case, a worthwhile service, much used educationally – as well as socially – by children was lost because of the lack at the time of a way to manage the risk acceptably for such a world-wide service.

So, in the case of JANET, what changed? In a word, schools. The issues hadn't really altered but the particular context of schools having joined the JANET community brought such considerations under renewed scrutiny. There is little dissent from the view that some material easily available on the Internet was best not viewed at too young an age. There are also others in the community who may either be vulnerable or find particular material unacceptable. Given the general agreement on the need in principle, UKERNA set about procuring such a service, with the important requirements that the service should be scalable and configurable to the needs of each group, organisation or part of the community it serves; and that control over the filtering rules should be vested in those with responsibility and authority to exercise it in the various particular parts of the community. The JANET Web Filtering Service was inaugurated in 2004, provided by RM,[32] an Oxfordshire-based company with a long association with education and learning, especially in the schools sector.

With any filtering service there arises the question of just who, exactly, is deciding on behalf of others what may be viewed. A leading contributor in this context in the UK is the Internet Watch Foundation (IWF) formed in 1996. JANET(UK) – together with most UK ISPs – is a funding member. IWF works with several government departments, the police and a variety of agencies, particularly those concerned with child protection. The latter, which was the

---

31 *See*, for example, Deibert *et al.* (2008). (Related information may be found via http://www.oii.ox.ac.uk/people/faculty.cfm?id=3, accessed January 2010.)

32 *See* Glossary.

primary context for its formation, remains the most visible aspect of its work, though there are others such as content contravening UK law in respect of Obscene Publication and Incitement to Racial Hatred. The most obvious output of the IWF is its list of URLs and Usenet groups that it recommends should be blocked. Moreover, perhaps the most important and effective part of its service is the 'notice and takedown' service which it provides to all UK websites to let them know if they are hosting material in the various categories mentioned. Many UK ISPs, as well as the JANET Web Filtering Service, use the URL list, as do some search engines. An indication of the effectiveness of the service can be gained from the observation that the amount of undesirable or illegal material hosted on UK websites is negligible. Nevertheless, the whole area treads an ill-defined line between genuine protection and unwarranted censorship of free speech, as witness the history of prosecution under the Obscene Publications Acts: D.H. Lawrence's *Lady Chatterley's Lover*[33] (1960) and the magazine *Oz*[34] (1971) being two celebrated infamous cases.

From the beginning of the 1990s, as we have seen, the rise of the Web signalled the adoption of the Internet into everyday use for every purpose, corporate and individual, public and private. Correspondingly, data and information of every description began to be accessible and communicated around the world by this means. Publicly accessible content was already subject to regulation; now, in addition, information and content accessible by network became an additional focus for malicious activity so that network operation and electronic content became the focus of increased regulation and legislation, accompanied by renewed research in security-related detection and prevention techniques, growth of the security industry, and development of specialist sections of enforcement agencies.

On the legislative front, the first specific legislation in respect of computer abuse was in 1990 (Appendix I), by which time hacking and automated penetration were well-established. Then, as the nature of content came to prominence, so legislation to regulate content began at the end of the 1990s and continues today. Having established the legal rules, the last few years have seen legislation concerned with what might be termed 'enforcement audit trails': maintaining records in case information is needed by enforcement agencies – an area in which there is a fine line between invasion of individual privacy and society's need to investigate potential breaches of its rules.

From a communication network perspective there has been considerable research and development focus in recent decades on models of authentication and authorisation infrastructure, including aspects such as how to express trust relationships and how trust is established; and in the detection of malicious activity,

---

33 Written in 1928, it was not published in the UK until 1960, when Penguin was promptly prosecuted under the recently relaxed Act (1959), and duly cleared.

34 The editors of the magazine were defended by John Mortimer (creator of *Rumpole of the Bailey*), cleared of the main charge of obscenity at the initial trial and cleared of the remaining charges on appeal, after it was found that the jury had been misdirected in several instances. The issue of *Oz* which occasioned the trial was notable in this context for having been created by minors in the first place.

the study of (traffic) patterns has been the basis of a number of techniques.[35] There has also been effort in increasing awareness that, of course, security is not merely a matter relating to the technology of computers and networks. While such is necessary, it has to be combined with physical security of critical locations where sensitive equipment is located, and good practice by users and service providers alike.

In recent years JANET has exploited substantially new authentication and authorisation technology – the UK Access Management Federation is one of the largest to date – in order to provide a widespread, uniform basis for sound security in use of JANET services, at the same time avoiding excessive growth in the amount of authorisation information to be remembered. However, there are two threads which permeate security: trust and the myth of absolute guarantees. Even were every packet traversing a network required to carry 'unforgeable' authorisation and be 'unbreakably' encrypted, on penalty of being discarded, this only improves but does not guarantee authenticity because there is no such thing as 'unforgeable' or 'unbreakable', only probabilities. And the trust issue extends to all those using, providing and regulating the JANET services. The notion that every packet be secured in the way just mentioned may, in the analogy of the information highway or street, be likened to every citizen being required to carry an ID card at all times or face summary penalties at the first crossroad – no longer an 'open' network in the sense striven for historically by JANET. Evidently a range of security may be needed. Trust is equally slippery. How, for example, is it established? Reputation is one social mechanism by means of which a degree of trust and self-regulation may be achieved; this has been emulated by numerous Internet trading services, eBay and Amazon's third-party market place being examples. But what of trust in regulators, for which the aphorism *Quis custodiet ipsos custodes?* – 'Who is to guard the guards themselves?'[36] – applies? Some of these issues are at the heart of the Internet neutrality debate regarding its control and regulation – a continuing debate of international significance, but beyond the scope of this history.

---

35 *See*, for example, the book by Conti (2008).

36 Juvenal, *Satires*, vi. 347.

14

WHAT NOW?

The advent of SuperJANET5, encompassing the incorporation of the UKLight prototype wavelength service as part of JANET, signalled a turning point. The vision of SuperJANET had now been realised for JANET: the straitjacket of capacity and performance had been removed to the extent that JANET would now focus increasingly on new applications and content delivery, while at the same time there was now a diversity of transmission infrastructure available to the academic and research community.

Which is not to say that technological challenges have vanished from the scene. Ubiquity and virtualisation have begun but exploitation of either is still at an early stage. Societal issues of trust, security, protection and censorship were touched upon in the previous chapter: the topic of 'who controls the Internet?' has been much debated in the last decade and a half. Each is likely to influence the development and exploitation of technology and have an impact on the evolution of JANET.

14.1 RECENT DEVELOPMENTS

Much recent development has been a combination of extension of JANET services to a wider and roving community, while also exploiting the opportunities offered by the beginnings of optical networking. And underlying several of these has been the continuing necessity for incorporating authentication, authorisation and security.

Yet there has still been the opportunity for simple, new services of widespread and obvious utility: JANET txt was just such. By 2006[1] there was already substantial institutional use of SMS[2] by the JANET community as a way of keeping in touch with students, whether in support of teaching, administration or associated marketing. While already popular, there was evidently continuing demand and, with the aid of aggregation, opportunities for improved functionality and substantial decrease in price (compared with the average over a substantial variation then existing). The JANET service began in May 2007, provided by PageOne, and saw rapid take-up in both HE and FE sectors; by September of the same year, discussions were already under way with Becta about extending the service to schools. The system supported group messaging on a scale large enough to include the total membership of an institution, international coverage, and email, voice or pager delivery. Later in 2007 PageOne made available a Web-

---

1 Incidentally, the year in which the popular Twitter site (http://twitter.com/) was created, a World Wide Web-based short message system also accessible using SMS.
2 Short Message Service.

based on programming interface[3] as a basis for integrating JANET txt with other systems such as virtual learning environments (VLEs), management information systems and library systems. The first such integration became available in 2008 when support for JANET txt in the open source Moodle VLE[4] was released.

Roaming

The arrival of the laptop together with the provision of wireless access across campuses had by 2000 begun to encourage internal roaming by staff and students: first just within the confines of the main campus, then further afield to satellite campuses, halls of residence and off-campus student accommodation. The advent of wireless had hugely facilitated unauthorised attempts to connect and thereby emphasised an aspect which had always existed whereby anyone with a cable could plug into any vacant socket on the more accessible parts of a campus network. What was needed was a securely authorised means of extending access to *bona fide* users of an enterprise network who are not permanently connected and may wish to connect from anywhere in the enterprise infrastructure, including remote parts requiring traversing a 'foreign' network. By 2002 this problem had begun to be tackled, in particular at the University of Bristol where Josh Howlett had developed an in-house solution[5] which was already in use at some other universities under Gnu Public Licence terms. Of course, the ability to roam securely anywhere within an enterprise naturally stimulated ambition to roam further within academia.

The general nature of the problem and its solution can be sketched as follows. A person is known to a home institution or network which can vouch that the person is who she or he claims to be by reason of the institution having verified supporting legal documentation in the process of admitting or employing the individual. For future authentication and authorisation, the home institution issues the person with an identifier and secret token. The person now visits another institution as a guest with a laptop and wishes to use whatever network facilities of the host institution are available to guests. Having achieved physical connectivity, the only access made available to the visitor is to authorisation facilities. Since the visitor is unknown to the host, but is assumed to be from a home institution trusted by the host institution, the host then arranges for the visitor to by authenticated by the visitor's home institution. Given that the result of this vetting is positive, that is, the home institution is trusted by the host and the home authenticates the visitor, the host then grants guest access to its facilities – the nature of such guest access being locally defined by the host, but typically including access to JANET, possibly also to local complimentary facilities.

The idea of forming a group to pool knowledge on the subject of campus wireless deployment had been mooted with UKERNA as early as 1999; however, it was not until 2002 that formation of a Wireless Advisory Group got under

---

3 Based on SOAP, Simple Object Access Protocol, one of several message-based Web interface specifications used to provide an application programming interface over which further Web services may be constructed.

4 *See* http://moodle.org/ (accessed March 2010).

5 Kit Powell & Josh Howlett, private communication, December 2002.

way. It was finally established in 2003 with James Sankar[6] as chair, with roaming (termed Location Independent Networking at the time) as one of its remits. By now the need to roam widely within the community was becoming increasingly necessary to underpin everyday academic and research functioning. An outline architecture to accomplish this using a two-tier deployment of RADIUS[7] servers was published in June 2004,[8] followed by a call to participate in a national trial. The latter was established in January 2005 with 36 participating organisations, each with a RADIUS server, together with two national servers to provide federation of the organisational server tier. After a year of pilot experience, the JANET Roaming service went live in spring 2006 – and enabled international roaming via eduroam. Apart from AARNet in Australia, the roaming service was initially available mostly in European countries; by 2009, international roaming included most of Europe, much of the Asia Pacific rim, and the Americas.

The JANET Roaming service is primarily an exercise in secure authentication and authorisation, something which the UK Access Management Federation infrastructure seeks to provide on a UK-wide basis. Work on roaming in the UK predated Shibboleth. Considering that roaming now has an international dimension, it remains to be seen whether (JANET) roaming may eventually benefit from exploiting the Federation infrastructure.

Aurora

The outline plans for UKLight (Chapter 12), which received initial JCSR approval in principle in 2002, had always included the provision of a dark-fibre segment to enable research field trials of all-optical network systems. The International Review of UK Research in Computer Science[9] published in 2001 had been critical of the level of investment in infrastructure provision. The UKLight programme, particularly the dark-fibre segment, had been stimulated in partial response to this. Aurora was the name given to this dark-fibre segment of UKLight at the time when in 2006 it finally received approval to proceed on the basis of a two-year initial period of operation to provide a national facility funded by HEFCE through JISC. However, within the confines of a finite budget, until completion of provisioning of the national and international segments of the UKLight prototype wavelength service (which became JANET Lightpath in 2007), the budget available for the dark-fibre facility was uncertain and hence its scale unknown.

During early discussions from 2002 onwards a number of universities had participated in the early planning; by 2004, tentatively, five (Aston, Cambridge, Essex, Southampton and UCL) had been identified as potential hosts for the dark-fibre facility. In the event, during the procurement process in 2006 it became clear that only three nodes could be afforded and the subset consisting of Cambridge, Essex and UCL were chosen. NTL (subsequently Virgin Media) was selected as the supplier in December 2006 and construction began in 2007. As well as

---

6   Then Network Access Project Manager, UKERNA and co-chair TERENA Mobility Task Force; since 2005, Director, Applications and Services, AARNet, Australia.

7   *See* Glossary.

8   Location Independent Networking (LIN), James Sankar, *UKERNA News* No.27, pp.10–11.

9   Sponsored by the BCS, EPSRC, IEE and The Royal Society; available at http://www.ukcrc.org.uk/grand-challenge/about.cfm (accessed June 2010).

the provision of intermediate co-location access points in the fibre runs, where additional equipment could be sited, it was important that the fibre should be compliant with recent specifications to enable the facility to support advanced experimental transmission equipment. Although it had initially been planned that experimental use of the facility would begin in October 2007, timed to coincide with the start of the academic year, provisioning took longer than expected and the facility was completed in December 2007 – incorporating 350 km of dedicated fibre, making it one of the largest facilities of its kind in Europe.

Exploitation of the new facility began early in 2008 through HIPNet (Heterogeneous IP Networks), a joint university and industry project funded by EPSRC and DTI[10,11] to explore an integrated multilayer architecture for support of future wired and wireless networks. The core layer of the architecture incorporates transparent optical switching capable of handling multiple protocols, bit-rates and modulation formats. Subsequently, later in 2008 it was announced that funds had been approved to extend the facility to Aston and Southampton – achieving the original five-node network – and to continue operation into 2011. In 2009 the Aurora infrastructure was used in an international collaborative[12] demonstration of optical TDM,[13] in which three 43Gbps channels, each using a separate wavelength, were time-multiplexed onto a single wavelength channel of 129Gbps without conversion to or from the electronic domain.

Video-based services

The last decade has seen IP-based videoconferencing established throughout the community – both geographically and at all levels of education – as a routine part of the infrastructure for teaching, learning, collaboration and administration. As with any other successful service initially deployed as an operator service requiring booking – like the transatlantic telephone service of half a century ago – the natural development of this was to remove the necessity for booking and to reduce the need for operator support, thus increasing the accessibility of the service and managing its scaling while controlling costs.

A pilot Conferencing on Demand (CoD) service was launched in September 2004 and immediately proved popular. Based on the same H.323 technology as the booked service but without the requirement to pre-book conference bridge resources, users could essentially set up their own videoconference at the time of use, but might occasionally be disappointed because bridge resources were already engaged. The pilot became a service – Instant Videoconferencing, IVC – under the JVCS umbrella in April 2005, with the addition of a quality assessment tool. It will be recalled that from the outset of the JANET Videoconferencing Service it had been found essential to provide *in situ* quality assurance testing of participants' audio-video equipment setup and environment. With the growth

---

10 Partners: Aston, Cambridge, Essex, Leeds and Swansea universities, together with Ericsson, Freescale Semiconductor and Emerson.

11 *See*, for example, http://gow.epsrc.ac.uk/ViewGrant.aspx?GrantRef=EP/E002382/1 (accessed March 2010); the project was completed in September 2009.

12 Partners: the Universities of Essex and Southampton, UK; University College Cork, Ireland; the University of Karlsruhe, Germany; the Athens Information Technology Centre, Greece; and Nokia Siemens Networks, Portugal.

13 Time Division Multiplexing: *see* Glossary.

and development of the service, it was now time to provide an automated quality assessment system and this was also launched in preliminary form in 2005. Initially the system for booking resources for an IVC session was separate from the managed JVCS booking system. However, subsequent developments integrated the two so that it became possible to access the IVC service as an option available through the Web-based JVCS booking service.

Meanwhile, international growth in use of both Access Grid and VRVS had been continuing and, particularly because both have the same paradigm based on the concept of shared use of virtual rooms (VRVS) or venues (Access Grid), there was increasing pressure for interworking between the two, as well as for H.323 support to be provided in VRVS. (Access Grid already had it.) Both systems exploit multicast; but whereas Access Grid uses network-level multicast (generally provided by routers), VRVS uses application-level multicast provided by reflector nodes which use unicast to distribute both to users and other reflectors.[14] Developments at Caltech had enhanced VRVS reflector node software to provide both H.323 transcoding and Access Grid gateway facilities. In 2006, JANET launched a pilot VRVS service and Access Grid gateway. However, both systems continue to be developed: in 2007, VRVS began to be replaced in the HEP community by Caltech's next-generation system, EVO (Enabling Virtual Organizations). At a technical level the latter also supports SIP, reflecting a growing trend away from H.323 for session management. Originated by the international scientific community, use of this mode of videoconferencing is spreading to a wider community, as for example, use of Access Grid in the UK for seminars in the social sciences and humanities.

In line with the growing focus on multimedia content delivery over JANET enabled by the achievement of a delivery infrastructure of sufficient capacity and serviceability, an increasing range of such material began to be made available over JANET, led by such ventures as the Pathé News archive and the British Library Sound Archive. By 2004 UKERNA had begun exploring with the TV industry the provision of TV-style content over JANET and in 2006 trial provision of some of the Freeview TV channels began, provided by Inuk[15] to a distribution point on the JANET backbone for onward multicast delivery over JANET. This was effectively an extension of technical trials which had been conducted for some time with public TV broadcasters, including the BBC. In the same period, use of IPTV for dissemination of student-originated TV also became more widespread. By 2008, what was generally referred to as IPTV might include delivery of real-time programmes, whether public service or community oriented, or on-demand services ranging from near-real-time to long-term archive, and material in all these categories was increasingly being made accessible over JANET. Of course, this was very much a reflection of what had been happening in the wider context of the TV and related industries, where the balance between what was delivered in real-time over the traditional broadcast radio channels and the Internet, as well as what might be made available on demand, was in a state of flux. Alongside these developments in the nature of content delivery over JANET, there is also interest in newer, higher definition TV formats, including digital cinema (2-4k pixels of horizontal resolution) and the proposed Super Hi-Vision or Ultra High Definition

---

14 Essentially the same architecture as CUSeeMe originally developed at Cornell University for the Macintosh in 1992.

15 Inuk Ltd: UK-based company specialising in VoIP and (via its 'Freewire' service) Freeview channels over closed broadband IP networks. Acquired by Move Networks, Inc., based in Utah, USA, in April 2009.

(8k pixels of horizontal resolution). Of particular relevance in promoting, developing and exploiting the capabilities of these formats in a collaborative research context is CineGrid,[16] a not-for-profit subscription organisation of which JANET(UK) is a network member. While demonstration on an experimental research basis has already taken place, wider access is as yet in the realm of future services, since an individual stream requires end-to-end transmission capacity in the region of one to tens of gigabits per second.

JANET NHS gateway

The field of medicine and healthcare has always encompassed the hospital and the university. Research may be carried out by those based at either or both; study and training for a career in the field may be embarked upon via a teaching hospital or university medical school and typically requires periods of study in both locations. As JANET became part of the accepted academic infrastructure, so full access to it from National Health Service locations became increasingly necessary for university students and staff who might be on placement for clinical training or be working from both locations. Equally, access to NHS network resources became increasingly important from departments in universities. As these requirements expanded in the early 1990s, there was discussion of the position of JANET in respect of the NHS, particularly since the latter was another publicly funded sector. This led to a review in 1994 by the ACN and the JISC of JANET's connection policy in respect of NHS organisations, which essentially re-affirmed the position of the latter as being the same as any other non-HE organisation working with the HE sector, with the exception that university and Research Council staff located in an NHS organisation would not incur any additional charge.

While this clarified the position under which JANET connection could be made, the issue of any interconnection of JANET and the NHS network was not addressed at this time: the two remained physically separated – though, of course, both had access to the wider Internet. While the latter meant that from either side of the divide the public facilities on the 'other' side could be accessed, this was of little or no help for staff temporarily located on the 'wrong' side who needed full access. In consequence duplicate workstations, PCs and connections to JANET and the NHS network (NHSnet at that time) appeared in both HE and NHS organisations, a situation which became increasingly cumbersome. Towards the end of the 1990s, at around the time when the e-Science programme was in gestation and the extension of JANET to the FE sector got under way in earnest, informal discussion began on how this situation might be alleviated, led by Roland Rosner[17] who had taken a leading role in the creation of JANET over 20 years before. By 2001, SuperJANET4 had just entered service and within the next few years a new NHS network (N3) would begin implementation. The timing coincided with consideration of the first round of project proposals to MRC as part of the UK e-Science Programme, which had raised afresh the issues of how to share resources and medical data effectively between HE and the NHS, this time in the context of the application of the Grid computing paradigm for

---

16 See http://www.cinegrid.org/ and http://www.terena.org/activities/media/ws2/slides/Krsek-WIE.pdf (both accessed March 2010).

17 Professor Roland A Rosner, then Director of Education and Information Support at UCL.

which ubiquitous information networking infrastructure was a *sine qua non*. In November 2001, a key meeting of those with policy and funding responsibility for information technology provision in both sectors took place and the NHS-HE Forum was created under the chairmanship of Roland Rosner, a position which he held until his retirement in 2008. The NHS-HE Forum has no funding of its own but has served to champion information technology initiatives of mutual benefit to the NHS and HE: chief amongst these being the issue of network interconnection between the two sectors.

JANET and the NHS network are separate networks. While this can be regarded as a natural consequence of being funded and managed by different agencies of government for different purposes, there are evidently two more fundamental reasons. Although networking is essential to both, the dependence of HE on JANET nowhere near approaches the mission criticality of communication within the NHS; this difference dictates a substantially different approach to the engineering and provision of each. The other aspect is security. Individual, personal, patient medical information is intrinsically private and the corresponding need to handle such information responsibly long predates any technological aids. While any network needs a degree of security to protect itself from wreckers, JANET has an ethos which encompasses the greatest degree of openness reasonably achievable; by comparison, the NHS network needs to place preserving patient confidentiality at the highest position consistent with its mission.

In addition to the need to provide access for HE and NHS staff and students across this divide, there are also areas of content procurement and management which could with benefit be shared. As a basis for the provision of mutual access to shared resources, as well as cross-network access to the resources of the home organisations of staff and students, in 2002 Andrew Cormack[18] proposed several model architectures for interconnecting the two infrastructures so as to preserve security requirements (Cormack, 2002). In particular, the role of a gateway between JANET and the NHS network was described. Now there was a need to elaborate NHS procedures relating to handling of sensitive data, particularly patient personal data, to embrace the opportunities offered by operation in this broader context.

In 2004 BT was awarded the contract to manage the provision of the next generation NHS network, N3, a substantial enterprise which would see the upgrading of the network over the next few years. With this under way, the time was ripe to pursue implementation of an interconnect, an initiative which acquired the title of the NHS-HE Connectivity Project. To this end, in December 2004 Malcolm Teague was appointed to the post of NHS-HE Co-ordinator, jointly funded by the NHS and JISC, based at UKERNA, where he would work full-time on the Connectivity Project from February 2005. A working party was formed in September 2005 under the auspices of the NHS-HE Forum to specify the N3-JANET gateway and, by the end of 2006, there was a preliminary gateway specification and a proposal for a pilot implementation, funded by UKERNA / JANET(UK), to enable early adopters to gain experience on which a full-scale interconnect service could be based in the future. There were initially two trial communities of users: one situated in the North East[19] led by the University of

---

18   Then Chief Security Advisor at UKERNA.

19   CETL4HealthNE: *see* Glossary.

Newcastle and the other in the South West[20] led by the University of Bristol. The initial gateway was situated in Telecity where both JANET and BT had points of presence and it began operation in October 2007. The strategy adopted to begin with was to open up the interconnect on a service-by-service basis as the trial progressed: in order to comply with the regulations agreed for initial use of the interconnect, applications would be initiated from the NHS side of the interconnect and no patient identifiable data (i.e., only anonymised data) would be transferred over the interconnect. The initial services chosen for trial were Web access to enable access to VLE support and terminal service for access to chosen client applications.

The first year of operation of the pilot interconnect saw considerable success, including building up sufficient confidence in the regulatory procedures for maintaining security of patient data that by mid-2008 the gateway was carrying patient identifiable data. By the end of 2008 JANET(UK) had announced funding for a further year of operation of the pilot interconnect; Edinburgh University was using the gateway (to make VLE and library services available to students on placement); and JVCS operation had been achieved across the gateway from the South Tees NHS Trust[21] – an enterprise led by Geoff Constable of the now well-established Welsh Video Network. This was a notable achievement because the real-time traffic has to cross several networks (JANET, N3 and the local NHS Trust network) with different managements and intervening firewalls while maintaining a high level of security and providing acceptable video quality. And during 2009, plans for the provision of a full service gateway, with dual-entry connect in London and Manchester, were under development and JVCS use was achieved for the whole of CETL4HealthNE, enabling full videoconferencing participation among the consortium members.

We have seen how the JANET community has continued to grow over the decades, as new sectors have joined in use of JANET: by 2008 the estimated size had reached 18 million users. Accompanying this have been the continual increases in the amount of traffic generated by new styles of use in support of research, together with exploitation, particularly in the school sector, of video communication of increasing quality. This combined growth from both numbers of users and ever hungrier applications has been responsible over recent years for a doubling in the amount of traffic carried by the JANET backbone every 18 months. By 2008, many of the backbone links had already been upgraded to 40Gbps. In line with estimates made during the planning of the current backbone, testing of the next generation of transmission facilities anticipated to be needed during the life of the current backbone has already begun: successful initial testing of transmission at 100Gbps was completed in April 2009 over a 103km section of fibre between the JANET points of presence in London and Reading, using WDM to carry the test channel together with two others, carrying 40Gbps and 10Gbps, respectively.

---

20 The Universities of Bristol and the West of England, together with the NHS Clinical Academies in Avon, Gloucestershire, Somerset and Wiltshire. (*See* http://www.nhs-he.org.uk/NHS - HE forum Nov 2007 UoB.pdf, accessed March 2010.)

21 A participant in CETL4HealthNE.

14.2 Retrospective

As we have seen, academic computer networking in the UK arose out of the necessity posed by computers being too expensive for every university to have one – or at least, not a 'top-of-the-line' big, fast one: sharing was essential. By removing the need either to travel to the computer or suffer 24-hour turn-around based on job submission and retrieval by courier, what early networking did was to remove the need to be situated close to a computer to use it, coupled with providing access to a variety of different computers from a single terminal. These developments went hand in hand with the deployment of time-sharing multi-access computer systems, starting in the late 1960s and continuing throughout the 1970s. Alongside interactive terminal access and RJE, since files needed to be transferred between systems, file transfer appeared; all of these were effectively aids to use of the computer for what was perceived as its fundamental purpose of calculation and data manipulation. Perhaps the only widespread new application which began to appear at this time was email. But there were also the visionaries and dreamers who, encouraged in part by the example of the multimedia event in 1967,[22] imagined networks that in future might do far more: as a result of which, during the 1970s, research into packetised continuous media and distributed systems began.

JANET experience

The initial model of networking sketched above was realised in the UK during the 1970s and early 1980s; by the time JANET was officially born in 1984, such networks were alive, well and in operation in all the regional networks and SERCnet.

Of course, by 1984 the goal posts were in rapid motion. Personal computers and workstations had appeared: the notion that processing was so expensive that a computer must be kept busy at all times was breaking down. It might be true for the most expensive machines which were still needed for the largest tasks but no longer for the computer in the laboratory or, increasingly, on individual desks: computers could now await people.

Initially this had little effect on the concept of the services offered by a wide area network of national extent such as JANET. But it did introduce onto campuses a collection of systems and networks which did not support the JANET protocols, many of which would never support the Coloured Books or X.25. More importantly, everyone now began to use computers, the majority for purposes other than computation. It also introduced people to what could be achieved using networks a thousand times faster, even if only locally. While the technical direction in which JANET would begin to evolve was still not clear, the community built up a wealth of experience on how to fund, manage and operate a network for the education and research sector.

It will be noted that the start of JANET coincided with the beginning of the Internet and right from the beginning there was pressure to be able to continue network communication with the US in support of research. This had already been provided by UCL for over 10 years. It took a few more years, almost until the end of the 1980s, for penetration of campuses by Internet protocols to build

---

22   *See* Chapter 1.

up as a result of workstation and Ethernet LAN deployment. During this period, JANET had had time to forge a national identity and gain experience of operating a national network, while continuing to have links via the UCL gateway with the US and the Internet which were superior to others at the time. The alternative, of JANET having to address the issues immediately, would have been difficult both politically and technically.

By the end of the 1980s, public networks, private networks and community networks, like JANET, were all reaching after 'universal communication' – communication systems that handled all communication amongst people and computers, for work and play, whether at the place of work or at home. 'Convergence' was here, which could be taken to mean that 'everything was going digital' and that computers, computer networking, telecommunications, TV and radio would all begin to share technologies – though there were differing degrees of enthusiasm and belief, and a wide range of perceptions about timescales.

In the 1990s JANET became an early adopter of these concepts as a service network – even if the goal posts were still shuffling about a bit. With the experience of having built its own network in the past, JANET set out to do it again, though this time all the parts would need to be ready made: with a community now totally dependent on the network and 10 to 100 times the size it had been less than a decade before – and set to embrace all of education in the foreseeable future – the 'DIY' era was well and truly past.

This time it was not just a matter of making the basic network begin to function: to realise the potential of the concept, whole new end-user application services would need to be developed to the stage where they were routinely usable – and not just by those who habitually used and programmed computers, as had been the case a decade before. This would take longer and cost a lot more: only as the 21st century dawned would all this begin to become reality.

Quests

During the course of JANET's development so far there have been a few recurring threads, or quests, of which perhaps the most apparent are performance, functionality, universality and organisation.

For much of its first 15 years, JANET found itself very much in the position articulated by the Red Queen (Carroll, 1871):

> 'Now, *here,* you see, it takes all the running *you* can do, to stay in the same place. If you want to get somewhere else, you must run at least twice as fast as that!'

The trouble was that when JANET began, it had already been needed for a decade. When it finally appeared, the wide-area interconnect was based on SERCnet which used 9.6Kbps lines at best: its community was a tiny subset of JANET's and it couldn't have afforded more anyway. On top of this, part of the political and funding argument for JANET had been based on 'rationalisation' of lines. Admittedly there was room for tidying up the topology, removing duplication of lines and reducing overall line rental charges but the immediate effect was to overload the incipient backbone. This, coupled with the steady growth of the community – just the number of sites roughly quadrupled in the first decade – meant that even when SuperJANET first began service, SMDS was in many cases only relieving pressure: restoring and maintaining performance rather than offering a level of performance which would enable new, hungrier applications. Only with SuperJANET4 was a level of capacity reached which began to enable universal access to quality video, as well as providing the foundations for the next generation of scientific research infrastructure.

The international scene was even more parlous in the first decade of JANET, in part owing to the cost of (international) line capacity. In Europe, there were a plethora of networks and initiatives which sprung up, providing a hotch-potch of facilities and patchy connectivity. Not until TEN-34 did coherent connectivity appear, along with the promise of adequate capacity. Equally, although Peter Kirstein had achieved a link to support research use to the USA a decade before JANET, funding for this had been difficult – ranging from an attempt by the Treasury to charge VAT on academic experimental research equipment;[23] through the SRC requesting Professor Kirstein to desist from network research on what would become TCP/IP;[24] to use of the link for JANET email being on a courtesy basis for several years prior to April 1989 because funding of the link was provided to UCL by the MoD for research. Only in 1988 did the funding and operation of the link achieve a JANET service basis (via the Computer Board, SERC and NSF) – though it was only effected in 1989 owing to delays associated with TAT-8. Thereafter, as already recounted, there have been continual upgrades: as of 2009, the quest for increased performance has continued, partly to match the relentless development of technology, partly to keep pace with community growth, but also fuelled at intervals by the appearance of new applications and services.

The concept that JANET is to serve everybody in its community is self-evidently central. Sharing infrastructure is generally cheaper; and for some functionality, such as communication between people, whether it be email, telephony, television or videoconferencing, it is of the essence that everybody is able to reach everybody else. But for many purposes, individuals or sub-communities often 'prefer' to have their 'own' facilities: that way they can be tailored more closely to requirements and are less susceptible to interference from others. Early interactive time-sharing multi-access systems of the 1960s sought to give each user the illusion of having a dedicated system: the physical realisation of this only became possible with the personal workstations of the 1980s through miniaturisation and mass production. An important feature of having dedicated systems is that users cannot starve each other of resources: a continual thread of debate, sometimes bone of contention, in the story of JANET.

In the late 1970s and early 1980s there were a variety of perspectives regarding networking. Researchers funded by SRC made considerable use of DEC systems and DECnet, in part for compatibility within the national and international community, including NASA and the SPAN network. Users of IBM systems found EARN and BITNET useful. Abandoning these network infrastructures was not regarded as an advantage, particularly as first SRCnet and then JANET were shared and did not necessarily provide directly equivalent functionality. The determining factor each time was ultimately cost: for example, SRC (and subsequently SERC) funding did not stretch to subsidising more than one network – either its own or, later, JANET.

Parts of the computer science community were also critical of the strategy adopted for the development of JANET, in particular, the creation and

---

23 The UCL ARPA TIP in 1973 – Peter Kirstein, private communication.

24 Private communication.

implementation of the Coloured Books. International research collaboration in computer science was primarily oriented towards the USA; the main platform consisted of Unix, ARPANET and the incipient TCP/IP; and a clone of ARPANET might have suited some better. The strategic reasoning behind the choices made at the time have been described in Part A. While it is pure speculation what might have happened if the Computer Board had decided to base JANET on ARPANET, it should be remembered that in 1984, although IP routers and third-party support for TCP/IP in various operating systems were emerging in the US, there was at the time neither TCP/IP support for British mainframes and minis, nor any national network in the USA for the academic or research community beyond computer science departments and those with projects funded by DARPA.[25]

That the needs of research and service should reveal differences of opinion and approach was nothing new: tensions between computer science and computer service requirements, approaches and imperatives were quite widespread and well known. They were also to be expected since the driving aspirations were – and remain – different; and they only really eased, at least insofar as computer research and service were concerned, as the cost of the technology reduced enough to enable separate infrastructure for each. But the issue of wide-area research network infrastructure remained: wide area network infrastructure is costly. Thus, while the early 1990s saw, with the arrival of IP in JANET, a network platform suitable for both general use and some aspects of wide-area network systems research and development, as well as a basis for enabling adoption of new network applications being developed worldwide, the service-oriented imperative of JANET did not allow for experimentation with the transmission and switching infrastructure that JANET itself also had to depend on. From the beginning, SuperJANET sought an infrastructure that would enable a degree of sharing between research and service networking, while simultaneously providing sufficient separation to safeguard the service: this was one of the prime motivations from the outset, even though it proved impractical to the point of being, if not forgotten, at least suspended for more than a decade.

The general idea, originated around 1990, is that JANET cannot allow experimentation with the lowest level of infrastructure it controls, else the service is jeopardised; experimentation with layers above may be countenanced with care. In 1989-1990, as we saw, WDM was not a practical possibility. However, SDH/SONET was just appearing and, at least in concept, seemed to offer the possibility of controlling TDM channels sufficiently flexibly to enable them to be configured for service, research and any other special purposes which might appear. In the event, not only was it too early in terms of the state of development in 1990s, it was also too expensive and in consequence JANET only had access to channels, not SDH switching or multiplexing facilities, so could not provision end-to-end sub-multiplex channels. Nevertheless, using part of the original SuperJANET infrastructure JANET did pursue its established practice of being an early adopter of emerging international standard technology. However, although early trials were conducted with ATM switching and multiplexing – ATM being the technology ratified in 1988 by the CCITT for use in future broadband ISDN services – events were to take a different turn during the 1990s: the rise of the World Wide Web, coupled with deregulation of the telecommunications industry and the long-awaited appearance of routers with gigabit performance, caused

---

25 The author was working in the USA at the time and was present in 1984 at one of the early strategic meetings at NSF when the deployment of an IP backbone – to become NSFnet – was under consideration.

the IETF TCP/IP protocol stack to achieve world-wide *de facto* standard status, to the exclusion of both ITU-T B-ISDN and ISO standards. The effect was felt in SuperJANET III: realisation of the policing and packet transmission scheduling technology necessary to achieve the full potential of ATM QoS in broadband ISDN never came to fruition and this, together with gigabit Ethernet and gigabit routers, spelled the end of ATM and any possibility of providing an infrastructure sharable between service and research, even at the ATM level. Moreover, the effect of the small ATM cell size was also manifested by the end of the 1990s in lack of ATM switch performance.

JANET now found itself again in the position of the Red Queen, this time not just in respect of capacity (throughout the network: backbone and regions) but also in respect of having committed to providing a multiservice network and needing either IP QoS or substantial additional capacity to reduce the probability of congestion sufficiently to enable acceptable continuous media services, particularly video. The leap in capacity which came with SuperJANET4 went some way to relieving this situation, at least in the backbone. Support for configurable special channels to enable either packet network research or special applications – whether VLBI radio astronomy or the massive data transport needed in future by particle physics – was achieved with the incorporation of WDM into JANET in 2007, with the advent of SuperJANET5. The new services released as these developments in the basic infrastructure were deployed have already been described.

Organisational evolution, at national and regional level, has been another continual feature of JANET. Driven initially by cost to share facilities, regional networks first began to appear in the late 1960s, together with some of semi-national extent, all centred around regional and national facilities funded by the Computer Board as a result of the Flowers Report. SRC also needed to provide national access to its expensive facilities. Once the need for a universal network infrastructure to share facilities throughout education and research was established, evolution of an organisational structure to manage and operate it began in parallel with the construction and subsequent development of the network itself. Less articulated has been the continual interplay between organisation and network architecture. Although the original regional networks initially vanished in 1984, the same regional centres continued as the JANET switch locations, providing operational cover locally as well as acting as a focus for advice, support and representation. However, in its first decade or so, growth and development of JANET was essentially organised and managed centrally, with the wide area network considered as a single 'flat' network, though as local networks evolved, it was recognised that this WAN was primarily an interconnect for campuses rather than individual computer systems. Then as the network and the community expanded during the 1990s, so the regional concept was reintroduced quite explicitly into both the organisational structure and the network architecture, in many ways as a means for managing the scaling of both as well as accommodating differing regional priorities.

Initially, the architectural split resulted in an implementation in which regional edge routers connected to the backbone: an arrangement which provided clear demarcation between the region and the backbone but which also resulted in two extra routers in any inter-regional path – a feature which militated in favour of neither cost nor performance. This was the architecture of SuperJANET III; while there were a number of arguments in its favour at the time, perhaps especially in view of the amount of technological change then occurring, it was eventually phased out in favour of a more streamlined architecture in SuperJANET5.

However, there were two other effects which the regionalisation of the 1990s would have and which would gradually become more apparent. The first of these was technical and arose from the way in which IP network technology developed at the time. Multicast development began around 1990 while DiffServ QoS support and MPLS date from the mid-1990s. Each of these requires some state to be introduced into the network;[26] and in each case the initial model of how this state might be managed was implicitly restricted to a single network domain (AS). JANET duly encountered the consequences of this in respect of both multicast and DiffServ.

The other effect of regionalisation was slower to emerge though its seeds were always there. Although a regional network was bound to deliver the JANET service as defined in an SLA or Service Level Description, its management was independent of any other region, the backbone or the campuses. In particular the means of delivery, its funding and the timing of upgrades are all independent of other JANET constituents. By the time came for introduction of WDM-based services, planning, maintaining and operating a coherent end-to-end delivery architecture had become increasingly difficult – and this at a time when JANET had achieved an infrastructure which would enable a greater focus on realising the long-promised delivery of multimedia content. This is not the only aspect of difficulty in relation to the organisation and management of JANET which has surfaced in recent years. Structural inefficiencies – among which may be counted the duplication in operational effort inherent in a regionalised organisational structure – can be observed. Moreover, with an end-user constituency numbered around 18 million, spread over an institutional constituency in the tens of thousands, there is now a sense of only vestigial connection between the centre and the periphery.[27] Coupled with a time of economic recession, these concerns have led at the time of writing to a review of the current overall JANET organisational and management structure.

Timescales: slow burns and flash-overs

In 2009 JANET celebrated reaching its quarter century, though the idea of JANET, a network for education and research, stretches back half as much again. Is this a long time or impressively rapid? It might be five or more generations of some digital technology but in human terms it is perhaps a single 'educational' generation: from the concept to the reality of a network pervading the whole of education, serving 'digital natives' from first-year primary onwards! However, subjectively, at the time, there have been some things which seemed to proceed at a snail's pace while others happened almost overnight.

The first example is the creation of JANET. If 1973, the year of the Computer Board delegation to visit ARPANET and the first Wells Report, is taken as the beginning of the creation of JANET then it took 11 years to come to fruition in 1984. SuperJANET, from its original concept in 1989 to final realisation at the beginning of 2007, took 18 years. And the creation of the current regional

---

26 Multicast requires routers to have knowledge of branching points in the delivery tree; DiffServ requires (edge) routers to perform admission control and policing at entry to a network to prevent overload by a particular class of traffic (maliciously or otherwise); and MPLS requires the establishment of LSPs.

27 Science fiction enthusiasts may be reminded of aspects of Isaac Asimov's *Foundation* Trilogy, originally written in the late 1940s and early 1950s!

networks, which began in 1994 and was completed[28] with the establishment of NIRAN in 2004, took 10 years. If these are all examples of the slow-burn, there have been other developments which might equally be described as examples of a flash-over. Perhaps the clearest was the transition of JANET to use of IP protocols which was in the nature of an avalanche to experience, although it actually took three years from the start of Shoestring in 1990 to the official acknowledgment in 1993 that IP was the strategic protocol for the future. More recently, perhaps particularly in the five years since 2004, the take-up of content delivery services has proceeded apace. Here the significance is on the transition from a particular type of application being technically possible to the appearance of an expanding range of supported services being on offer.

The rapid, flash-over type of event, as epitomised by the IP transition and, much more recently, the growth of both social networking[29] and content delivery, tends to be characterised by pent-up demand having built up which is then released with comparatively little effort, little in the way of technical development, modest cost and relatively little organisational impact. If we compare the slow-burn events, particularly JANET and SuperJANET, we observe that four of these five characteristics are at the other end of the scale: substantial effort, considerable technical development and large cost over a substantial period. Organisational impact is rather different: in the case of JANET itself, a new organisation had to be created; in the case of SuperJANET, the time coincided with community expansion and the need to create more substantial foundations for the future service. In each case it is common experience that where even one of these characteristics is substantial, the endeavour takes longer. As it happened, in the cases of JANET and SuperJANET there were (different) additional obstacles. At the time of the first Wells Report the concept of a network – more especially what it might be for – was so new that few had any experience and it was consequently viewed in some quarters rather as a way of saving money at the cost of loss of service differentiation to meet specific requirements, loss of service autonomy and potentially loss of jobs, instead of as a means to new opportunities. Moreover, since these 'new opportunities' were scarcely capable of articulation at the time it was only too easy to overlook or even belittle them. Persuading people of the value of the opportunities took time. SuperJANET had a different problem: in essence, technological promise was some way ahead of delivery.

Religious wars and other debates

In Britain most people, most of the time, drive on the left-hand side of the road. World-wide, most screw threads are right-handed. Mostly, the usefulness of such conventions is self-evident – the former perhaps more so than the latter in the afore-mentioned examples. In deciding such conventions, debate of a Lilliputian character as in 'endianarity' may arise, and only occasionally is there a 'good', 'correct' or 'right' answer to such questions *ab initio*. From a historical perspective, it may be of interest to understand how the choice came about but to attempt anything more may be akin to tilting at windmills.

---

28   At this point all delivery was via a regional network; evolution and rearrangement of regional networks would continue.
29   Examples include Twitter and Facebook; the purpose is social interaction.

The evolution of JANET has occasionally encountered such choices. Were big-endian names 'wrong'? With the clarity of vision afforded by hindsight, it's pretty obvious that a good deal of trouble and inconvenience over a period of fifteen years could have been avoided if the opposite convention had been adopted in the first place – and to be fair, as already noted, there were some who argued for this at the time. But naming is a particularly rich minefield, and not just in respect of endianarity. 'UK' versus 'GB' is settled in JANET (for now!), but with domain names embarking on home language forms, involving not only variations on the theme of order but also the wonderful world-wide variety of scripts and ideograms, JANET's travails in the past regarding long- and short-form names, domain order and country-level designation may come to seem small by comparison.

Of course, there are questions of more substance which seem to kindle unreasonable fervour, even entrenched positions. Standards are such an area. Few would deny that standards are important, especially in a subject such as networking where communication can only be achieved by agreement on protocols. But when it comes to which standard, how widely it should be adopted, how broadly interpreted and whether it might even be international, the aforementioned screw from Archimedes onwards is testimony to the magnitude of the task and its unending character: the variety of 'standard' screws is seemingly endless, and growing! In the context of JANET: 'Should we have copied the original ARPANET?'; 'Was X.25 and Coloured Books the right decision?'; 'Was X.400 a waste of time?'; and 'What about ATM?' From a business or political stance, standards bring potential leverage; but the choice of standard is something of a gamble. ISO and CCITT network standards effectively received government backing around the world in the 1970s and 1980s, perhaps partly in attempts both to minimise the risks associated with the choice by 'going with the crowd' and to provide a level playing field of maximum dimensions for national industries: unfortunately, standards landscapes are littered with failure. So what can be said about IP? After all, it wasn't confirmed even as the ARPANET standard until 1982. In retrospect, three observations may be made: the Internet architecture could naturally encompass LANs; basic support came free with Unix; and from about the mid-1980s onwards new network applications were increasingly developed primarily for the TCP/IP platform, the crowning example being the World Wide Web.

Costs and charging

As has been alluded to several times, a persistent pragmatic principle associated with JANET has been to avoid charging at the point of use. Originally, in the 1960s and 1970s, communication was merely an overhead associated with access to computing which was itself not charged for at point of use; so the notion of charging for part of the means of access never really arose. Then, part of the argument for JANET was that ultimately shared provision of all-to-all, rationalised communication was in the nature of an efficiency saving; as such, even when it acquired its own budget, charging for this means of access overhead while the computing resource was itself not charged for would have seemed exceptional.

This is not to suggest that charging, whether at individual or institutional point of use, has not been debated: it has, quite regularly over the years, and beginning in relation to mainframes well before JANET ever came on the scene.

The topic is one which causes heated debate on occasion because it is an area where education, politics and economics intersect: a meeting point for widely differing views. Pragmatically, to date, in the context of JANET there has, in practice, never been much appetite for individual charging; network access was to be seen as transparent as possible: any individual charging would act as a disincentive, effectively providing a barrier to use. Moreover, as mentioned earlier there is a substantial cost to implementing charging on any sort of usage basis, a cost which has been seen as a nugatory diversion of resources away from the network. Usage charging, whether based on time or volume, was perceived not only as providing a general disincentive but also specifically targeting those newer, hungrier applications, like video, which were generally supposed to be encouraged. It will, incidentally, be recalled that the volume-based tariff imposed by BT on PSS usage was also ultimately one of the major nails in its coffin in respect of any possibility of its use as a basis for JANET. A decade later, towards the end of the 1980s, as the JANET infrastructure was coming to be viewed as being as necessary to education and research as buildings or utilities, centralised top-sliced funding was already regarded as cost effective and appropriate to coherent national provision, effectively confirming for JANET this model of funding, which had originally been instituted by the Computer Board for national provision of high performance computing.

Of course, there have been exceptions, the most notable being the funding of the transatlantic link towards the end of the 1990s when, in 1998, charging for its use was introduced.[30] It was perceived that transatlantic capacity was becoming an expensive item which was coupled with continued growth. It coincided with a shortfall in the overall JISC budget. In the event, a form of charging was brought in, although the cost of international capacity not only did not continue to rise, it actually fell. There are other areas where an element of charging or budgetary subscription has been introduced, such as in the case of non-education bodies not eligible to a full JANET connection as of right, in order to recover directly incurred connection costs and make some contribution to operational costs. Notwithstanding these areas, the overall direction has been to provide a service in which innovative use is encouraged with as few barriers as possible.

In terms of the cost of transmission capacity, it has always been important to JANET to obtain this at a price as near as possible to the cost of its provision. When JANET began, transmission capacity was only available from telephone companies and the tariffs were close to the price of multiple telephone lines, which made them very expensive – a feature not peculiar to the UK. Indeed, since computer traffic is intrinsically bursty by comparison with traditional constant bit-rate telephone traffic, this had been part of the origin of packet networking in the first place. Further, in the debate between packets and circuits, the economic facet of this debate, encapsulated by Larry Roberts (1978) as: 'if lines are cheap, use circuit switching; if computing is cheap, use packet switching' still holds, and it is increasingly feasible for enterprises to obtain not merely TDM or WDM circuits but also fibre and to build their own networks, often a combination of circuits and packets. Since there are other aspects to the packets versus circuits debate, what this amounts to is that it is now possible to exploit both to engineer a variety of user services. In the case of JANET, historically it has been possible on occasion to use its unique position as an educational resource to obtain transmission capacity at favourable rates. The original impetus in 1989 behind SuperJANET stemmed

---

30   It will be recalled that PSS and IPSS were charged for at point of use; however, this was regarded rather as an unfortunate exigency, not as a mechanism to be adopted for JANET.

from a need to break away from standard tariffs and to achieve greater control over costs. Initially the break with standard tariffs was achieved in some measure in the 1990s by virtue of what amounted to educational deals; and from 2007 onwards, with the help of competition in the fibre communication services market, JANET has been able to achieve greater control over costs through purchasing wavelengths on which to base the network.

However, in respect of obtaining international capacity there have been other changes since 2003, following the shakeout in the communications market when several providers went bankrupt. Until that time JANET had purchased its own circuits and implemented its own peering arrangements with other networks, including with Internet2. Since then, JANET has ceased performing its own peering, instead buying capacity to other networks on the market and just buying more capacity as needed (and afforded). And at about the same time, the mode of funding transatlantic capacity was fundamentally altered. Historically, the larger universities have been the larger consumers of capacity. In 1998, the charging which was introduced was on the basis of overall intitutional use of the transatlantic link, charged a year in arrears. In a compromise between top-sliced and usage-based funding, a levy system was introduced in 2003 by which a university contributed at a rate proportional to its total revenue – as a gross indicator of size and activity – the sector levy rate being determined effectively by the cost of providing the capacity needed to support the overall level of peering traffic. This has the dual merits of being relatively easy to measure and providing a university with a predictable figure for which to budget. It is also consistent with avoiding disincentives at the point of use and maintaining information and communication services in the sector as 'enablers'.

No doubt the issue of charging will continue to be debated from time to time. There have always been those who regarded usage charging as a means of balancing supply and demand, a tenet at the basis of the market economy and commerce. But JANET is not supposed to be a commercial service: part of its remit includes encouraging innovative use and support of curiosity-driven research, and sometimes that is greedy and uneconomic by commercial judgement; but the returns can be unexpected. The Internet and the Web didn't come out of commerce, including the telecommunications industry: they came primarily from the government-sponsored research sector, much of it academic.

Technologies in the wings

In relating the story of JANET, we have encountered a considerable number of technologies which, at least for a period, have played a role in the provision of the JANET service. In most cases that role has either already ended or it is a current component of JANET. However, there are several important IP technologies of relevance to JANET but whose eventual significance remains unclear – they are, as it were, in a state of limbo. The future will doubtless reveal their full significance but in the meanwhile, at least for JANET, the jury could be said still to be out.

Perhaps the foremost of these is the revised version of IP known as IPv6 – for which the period of limbo is most likely almost over! In this case the issue is one for the Internet as a whole and impinges on JANET by virtue of its membership in that community. Revision of IP was stimulated by the observation in the early 1990s that extrapolation of Internet growth indicated that all the IP addresses might soon be used up; indeed, address rationing was already in force. Moreover, routing tables were becoming very large and there was danger of exhausting the number of networks allowed even sooner. While the timescales suggested for exhaustion varied according to the type of exhaustion and the specific assumptions made, total exhaustion was typically expected in 10 or, almost certainly, less than 20 years; and, while never as imminent as the more pessimistic reports suggested at the time, it nevertheless seemed prudent to plan well in advance: and, as of the beginning of 2010, exhaustion is predicted in a couple of years or so.[31] Discussions had begun around 1991 and these culminated in work beginning in 1994 on a revised version which, apart from supporting a much larger address space, would also take the opportunity to streamline IP on the basis of over a decade of experience in service. Recommendations for what would become IPv6 were in circulation in 1995 following which, in 1996, support was introduced in Linux and '6bone' was launched. The latter was an overlay test network, constructed on the same model as the Mbone by use of tunnelling over IPv4 to enable testing of IPv6. UKERNA joined 6bone in 1997 and by the end of 1998, RFC 2460 had been ratified by the IETF as the IPv6 standard. In 1999 JANET-related testing of IPv6 and its support moved to the next stage with the allocation to JANET of a block of IPv6 addresses,[32] enabling it to join the incipient worldwide IPv6 Internet. A variety of institutions and projects participated in early deployment development projects, notably Lancaster University, Southampton University and UCL in both Bermuda 2 (2000–2002, a JISC-funded UK-Internet2 collaboration) and 6NET (2002–2005, a Cisco-led EU project including Japanese participation, with around 15 NREN organisations including DANTE, and 20 individual participant organisations representing universities and industry).

By 2003 support for IPv6 was sufficiently stable for deployment in JANET. In updating IP,[33] the opportunity was taken both to incorporate features which had had to be added on to IPv4 as separate extras as well as removing some which had become moribund. In addition to extending the address field (from 32 to 128 bits), a number of extensions to addressing semantics affecting routing were incorporated including multicast, scope and classless routing (*see* below). Features for mobility and security are integrated, rather than being add-on extras as in IPv4. And there are fields available for QoS support, whether of the IntServ or DiffServ variety. The control side of IP, ICMP (Internet Control and Monitoring Protocol), being part of the IP specification, was also updated and the opportunity was taken to incorporate IGMP (Internet Group Management Protocol), the control side for multicast. While this is not a complete list, it illustrates the nature of the updates, a general aspect of which is that the only features added were those already known through experience to be required. While laudable in preventing total redesign and limiting changes to the tried and tested, this contained some of the seeds of delay because although it offered a number of technical advantages at the network level, it did not enable any new services for the user, thus removing a powerful lever to overcoming the inevitable barriers to adoption: cost and inertia. Two other factors have also slowed the rate of address consumption: the introduction into IPv4 of classless inter-domain routing (CIDR) and network address translation (NAT). The first allowed the division of the address field between network and host to occur anywhere (rather than being

---

31 *See*, for example, http://www.potaroo.net/tools/ipv4/ (accessed March 2010).

32 Technically, a /48 block.

33 For more details of the early history of IPv6, see for example the book by Huitema (1996) – but note that the book predates the IETF IPv6 standard, RFC 2460.

limited to one of the three octet boundaries of the 32-bit field), which helped to limit wasteful address allocation; and NAT enabled the use of private addresses, of local scope, by any organisation. As well as enabling address re-use, the latter destroys the transparency of IP by effectively hiding a subnet's addresses from public view – something which people suddenly found beneficial in a world of hackers!

All of the foregoing have effectively enabled IPv4 to last somewhat longer than initially expected. However, as of 2010 exhaustion does now seem imminent so that perhaps a rise in IPv6 penetration above one percent (Gunderson, 2008) may soon be expected. The change-over will continue to be a long term affair, during which both will continue to be supported in parallel – whether there will ever be a powerful enough incentive to turn off IPv4 support and, if so, when, remains to be seen. During 2003–4, dual support[34] was deployed in the JANET backbone. With similarly mature support deployed in many other networks by then, the need for 6bone had almost disappeared and it formally ended in 2006. Following a further period of stable dual support, IPv6 (unicast) was added to the JANET backbone SLA in 2007 and to the overall SLA (including regional networks) in 2008.

Quality of Service, in the specific sense of determining packet transmission scheduling for the purpose of achieving transport for multiple services with differing requirements for delay and capacity share, is another area in which the jury is out: and within JANET it is not currently, generally implemented. As already noted (Section 8.2), the state explosion anticipated if IntServ micro-flows were to pervade the Internet was considered unmanageable, in consequence of which the alternative DiffServ model was developed. The micro-flow state is needed to enable suitable scheduling and policing of the micro-flow following its initial admission to the network. In DiffServ, aggregate flows are introduced instead of micro-flows and, because there are only a few of these, the state explosion is avoided – but there is still a flow, and all the arguments about the need for reservation and state hold as for micro-flows, especially in respect of admission and policing to prevent malicious disruption or inadvertent over-commitment on some part of the path across a multi-domain network. During the past decade there has been work on how to address this, encompassing end-to-end signalling and inter-domain resource commitment negotiation, but there is no standardised model for such a mechanism: it is left to the management of a particular network to configure policing limits manually, possibly in conjunction with other manual or proprietary mechanisms for configuration of what sources may use any better-than-best effort behaviour.

Within JANET, as the demise of ATM became apparent towards the end of SuperJANET III and its replacement in SuperJANET4 by IP, a programme was initiated in 2000 to investigate the need for IP QoS. Videoconferencing was the application which dominated: conversational audio might technically be more demanding in terms of delay but videoconferencing also required so much more capacity that, in conjunction with data transmission, occasional audio use would be so insignificant a load that provided a link was not overloaded, audio performance was likely to be adequate. A QoS Think-Tank was convened towards the end of 2000 which reported in 2001, essentially advocating the need for QoS support and recommending DiffServ.[35] Following this, an assessment

---

34   Using dual-stack support in routers, DNS servers, etc.

35   *See* UKERNA (2001).

programme was initiated in 2002 to trial QoS engineering in JANET, a multi-domain, multi-vendor network. The programme lasted until late in 2007. The outcome was that although the low-level mechanisms of DiffServ implemented in a variety of manufacturers' products could generally be configured to produce a given desired effect, this was very much a manual process with relatively little in the way of management facilities and no automated state management facilities available of the sort suggested above. For JANET it was concluded that general implementation of current QoS support facilities was not appropriate: the backbone would continue to be provisioned sufficiently adequately as to avoid more than occasional congestion.[36] The backbone would not implement DiffServ, that is, it would ignore any marking of packets and would neither classify, mark nor remark traffic, and would forward all traffic transparently. In addition, comprehensive advice[37] was issued for those cases where the 'infinite bandwidth' argument was manifestly void and some use of DiffServ was desirable. Use of DiffServ among networks interconnected via the JANET backbone was therefore possible using bilateral agreements and would work as if the backbone were not there.

MPLS was described briefly towards the end of Section 10.3. As SuperJANET4 was developed, there was discussion about whether it might be needed for traffic engineering, the general conclusion being that it would be both possible and much simpler to ensure that at least in the backbone there was sufficient capacity available to avoid any such need. Because MPLS is a tunnelling technology it can be used to construct virtual links or networks. It is also possible to combine MPLS and DiffServ by adding a limited range of marking to the MPLS label field, thus endowing MPLS paths or networks with a degree of QoS. Since MPLS separates traffic, any virtual path achieves an initial level of security or privacy, which can be strengthened by adding encryption. All this makes MPLS a candidate for providing a managed bandwidth service, similar to what had been provided using ATM, or a VPN service. VPNs are in use in JANET, the commonest perhaps being to enable single-link virtual extension of an institution network off-site, working from home being a typical example. However, this requires no additional service from JANET: it is purely a matter for the institution and the end user client, and certainly does not require MPLS. At the time of writing, there seems no definite requirement for a JANET VPN service – and also, no requirement for MPLS on either VPN or traffic engineering grounds.

The issue of managed bandwidth provision developed in a different direction during the lifetime of SuperJANET4. The majority of the demand for managed bandwidth services came from a combination of 'big science' (of which particle physics and VLBI astronomy are examples) and experimental networking. Providing facilities for either via some sort of managed bandwidth on the IP network service was doomed to be unsatisfactory by being disruptive to other users, inadequate to the purpose or, in some cases, impossible. As the possibilities presented by UKLight opened up, so attention moved to focus on what would become the JANET Lightpath service. As soon as the basic arrangements for a manually configured service had been demonstrated and incorporated as part of the JANET service, so interest in how to provide a more automated service, perhaps an on-demand service, arose.

---

36 A level of provisioning sometimes – misleadingly – referred to as 'over-provisioning'.

37 *See* Edwards and Mackay (2008).

As mentioned in Section 10.3, the generalised version of MPLS signalling and control (effectively the MPLS 'control plane'), GMPLS, offered a possible way towards on-demand lightpath provision, and exploration of this became a thread within the JANET Optical Development Programme. As it turns out, MPLS may yet have more to offer future JANET services. Following the removal at the end of the 1990s of the distance limit for Ethernet, the Ethernet frame began to appear in use over the wide area, initially in the 'metro' area, subsequently in long-haul applications: in 2000, 1Gbps came into use in the wide area and then subsequently 10Gbps, following its standardisation in 2002. An issue for use of Ethernet transmission in the wide area is the lack of line monitoring and control, which had been part of the stimulus for SONET/SDH development in the 1980s. As a result, one of the wide-area optical standards for 10Gbps Ethernet makes use of SDH framing. The emergence of Ethernet framing for use in wide-area transmission from around 2000 onwards could be said to mark the beginnings of what would come to be termed 'carrier Ethernet': the engineering development of Ethernet for use in the long-distance, wide-area carrier market.

Ethernet over SDH can exploit SDH management and resiliency features to provide a robust wide-area service, sometimes referred to as 'carrier grade'. However, SDH is complicated and support for it is expensive. WDM equipment also has facilities for management and control, as well as possibilities for reconfiguration in case of failure, so that even without use of SDH, Ethernet over WDM may be able to rely on the WDM capabilities to provide management and resiliency. However, as with the evolution of PDH to SDH in the 1990s, the introduction of Ethernet transmission as a carrier service is a gradual process, leading to islands of (new or old) technology. Commonly, tunnelling, the carrying of one format on another, is used to extend one service over another. The means to carry PDH over SDH, thus enabling the old service to be continued where necessary over the new, was defined from the outset and continues. There was also limited definition of how to carry part of the SDH hierarchy over PDH, to enable early extension of (part of) the new service over the old. Use of tunnelling to extend carrier Ethernet over a variety of technologies has also appeared.

So far, what has been described relates only to transmission. To provide either configurable Ethernet links or to construct an Ethernet WAN, switching is necessary, and it is in this role that MPLS has reappeared: in this application, the MPLS header is prepended to an Ethernet frame, enabling virtual ether-framed channels to be label switched over a variety of underlying technologies. The possibility of using GMPLS signalling to extend control for an Ethernet channel across this variety of underlying technologies also presents itself. Of course, it is also possible to use Ethernet switch technology to provide a wide-area carrier switch, known in this context as a Provider Backbone Bridge.[38] It will be observed that these developments introduce several distinct layers at which switching or routing may be taking place, presenting the possibility of unanticipated effects through their interaction – especially in the case where a further layer of routing may come into play if these layer 2 technologies are used to create IP networks connected to the Internet.[39] Exploration of these areas has recently been added to the JANET Development Programme; and label-switching of Ethernet frames is already in use over JANET Lightpath.

---

38 *See* Glossary.

39 Similar issues were experienced in the mid-1980s when, for example, satellite Ethernet bridges were used to provide long-haul layer 2 links between sites already connected to the early NSFnet.

14.3 Last words

During its first 25 years, JANET has participated in network evolution from provision of relatively limited computer access, serving a specialised community, to the provision of multiple types of service and its permeation of the whole of society. Throughout, it has remained at the forefront in the provision of services, providing the support necessary to research and the opportunities for ever-wider experience in pursuit of education. Whither next remains to be seen: the mockery of prediction in relation to technology is amply demonstrated by such asseverations as 'heavier-than-air flying machines are impossible'[40] and the view in circulation in some quarters c.1980 that 'the limits of magnetic disk storage have been reached'.[41] Perhaps there is safer ground in the mundane observation that Moore's Law of 1965 (roughly, that the number of transistors on a chip doubles every two years) and related exponential growth in a number of other indices relating to processing and communications are expected to continue for some years yet – which is just as well considering JANET's own exponential growth 'law' that the capacity needed in the backbone doubles every 18 months.

If the march of technology has been successful in keeping up with network growth, by whatever measure, there are other aspects which have proved more resistant. Routing, for example, is at the heart of networking; yet, routing algorithm properties such as convergence (Will the algorithm result in a consistent set of routing tables being established in every router? How long will it take? Is the configuration optimum?) and stability (Will the algorithm find a stable configuration or will it continually 'flap' between several possibilities? If perturbed by temporary changes in the physical network configuration, will the algorithm rediscover the previous optimum configuration when the original physical network configuration is restored?) are by no means fully understood. Might metarouting[42] and routing algebra offer a systematic approach to specifying and understanding routing behaviour? Routing behaviour is already sufficiently adaptable to be able to maintain routing in the presence of continual change, whether as a result of failure or growth, in the physical network. How far may automated adaptation be needed in other areas? In the context of wide-area Ethernet switching, a variation of PBB known as PBB-TE[43] has been introduced because traditional switched Ethernet routing can produce unpredictable results; is quite complex in the presence of multiple VLANs; and was not designed for large-scale wide-area use. PBB-TE does not include the automated routing features of typical LAN switches; instead, operation and management takes an approach similar in some respects to SDH/SONET.

One of the more venerable and hardy concepts throughout the engineering development of computers and communication has been virtualisation. In terms of practical realisation, it can certainly be traced back to 1959 with the first project to realise virtual memory in the Ferranti Atlas at Manchester University

---

40 Lord Kelvin, c.1895.

41 At the time, shortly before the advent of Winchester technology, 100 megabytes of high-speed disk storage was provided by a hefty arrangement of some half-dozen platters, heavy enough to require care in lifting, housed in a substantial cabinet comparable to a top-loading domestic washing machine – to be compared in 2009 with magnetic hard-disk storage of 100 to 1000 gigabytes occupying the space of a small coin.

42 *See*, for example, Griffin and Sobrinho (2005).

43 PBB Traffic Engineering; *see* Glossary.

in 1962.[44] Shortly thereafter (1964), at its Cambridge Scientific Center, IBM began development of CP-40, the first virtual machine system and precursor of VM/370 (1972). The concept has continued development and is now incorporated in many products at both application (e.g., the run-time system for the Java language) and system level (e.g., Xen[45]). Recently, in 2007 JANET Training announced use of virtual machine facilities to provide technical training in network configuration, including security.[46]

The concept was also adopted in the early 1970s in networking, initially as the idea of the virtual circuit, at transport and network layers in work leading to TCP and X.25, and has continued strong ever since, example technologies being frame relay, ATM and MPLS. As we've seen, the construction of virtual networks such as the VLAN and VPN is also common, often using encapsulation (tunnelling) to convey packets across an underlying network between routing or switching nodes of the virtual network. These switching nodes may be real, separate hardware (as in the MBone) or effectively virtual, as a particular function provided in a shared system. Virtual networks engineered using tunnelling are also sometimes referred to as overlay networks, in acknowledgement of their being constructed over an existing network as basis. Virtual network techniques have been exploited by JANET a number of times, typically as a tool for transition, a major example being the transition to IP (Chapter 6). Pre-service deployment of multicast and IPv6 (MBone and 6bone) provide further examples, and JANET's use of the technique looks set to continue with exploration of carrier Ethernet and label-switching of Ethernet frames over JANET Lightpath.

During the last decade, virtualisation of services has been increasingly exploited. Virtual Web servers have been familiar for well over a decade. Increasingly, virtual services composed of wide-area distributed systems mediated by high-performance network infrastructure are appearing; and some Grid computing systems exhibit aspects of virtualisation. Not only do such systems enable the logical separation of service provision from the underlying hardware platform: by virtue of access being networked, service provision also becomes location independent, offering the potential for engineering optimisation in new ways. A particular example of the latter is optimisation of energy consumption through relocation of storage and processing to a situation with a 'greener' footprint: a topic recently begun investigation in JANET. Indeed, a study by Logica in 2009 has suggested that a dedicated data centre for higher education has considerable potential.[47] Perhaps those who viewed the inception of a national academic network in 1973 as resulting in greater, perhaps massive, centralisation will yet be proved right – at least virtually!

The separation afforded by virtualisation techniques not only affords an engineering separation from physical hardware but also between instances of systems, providing at the very least a helpful basis for implementing secure systems – providing unauthorised direct observation of hardware is prevented or guarded against. Perhaps virtualisation will become an increasingly prevalent component in the provision of secure environments for virtual organisations and

---

44   *See* http://www.cs.manchester.ac.uk/aboutus/history/, accessed March 2010.

45   *See*, for example, Barham *et al*, (2003).

46   *See* Cook (2008), which also summarises many of the uses of virtualisation, including system development and testing, which was one of the original stimuli for CP-40 virtual machine development.

47   *See* Perry (2009).

communities sharing resources specific to each over the network.[48] However, it has to be remembered that all virtualisation incurs overheads, most often evident in timing and delay performance, which may be significant where interaction in real time is important.

As we have seen, security and the related functions of protection and regulation are set to be of increasing importance in the continued evolution of JANET. There is a fine line to tread with respect to privacy issues. Surveillance cameras are multiplying rapidly in public places, seen as an affordable means of improving public physical security, but not everyone is in favour: the spectre of 'Big Brother'[49] is ever present and to some there is the threat of invasion of privacy, especially in the absence of clear, credible guarantees about how such data may be used and by whom. Similar, even stronger concerns have been raised by recent suggestions that personal Internet access might be monitored, with any alleged illegal activity resulting in Internet disconnection. When location data was first exploited in research in the late 1980s and early 1990s, it was recognised that use of such data posed invasion of privacy issues.[50] As of 2009, explorations relating to the regulatory and legal aspects pertaining to privacy reported as part of JANET's Location Awareness trials[51] suggest that the area remains a sensitive one; part of the issue being that of trust, already mentioned in the previous chapter.

One of JANET's pivotal successes has been in providing an open network, a network which may be used by any member of the JANET community for any reasonable purpose related to education or research. We have observed the developments which have occurred to ensure that users are indeed authorised to use the JANET services, and also the developments taking place in the context of widening participation to the NHS to enable mutual use of appropriate services, while yet protecting privacy relating to medical information. Equally, we have seen how optional, community-defined levels of protection in the form of filtering have been introduced in relation to content for the potentially vulnerable, primarily those in school. In each case there has been a minimalist approach: how little needs to be added, which interferes least? Perhaps the least transparent development has been the introduction of firewalls in the 1990s. It could be argued that, if JANET represents the street, this only represented a closing of the premises' front door – a natural updating to the 1990s of the habits of an earlier era when front doors were often kept open. Others might argue that the firewall, as a network layer artifice – even artifact – is one of the more obtrusive signs of the Internet's heritage from its design in a community at a time when security was not a consideration.

But all of JANET's services are based on an underlying network 'designed to support what has yet to be imagined'. Engineering to maintain this vision is much harder than engineering to a specific brief: over the years, JANET has been successful in this, but it has not always been straightforward. At its inception, we have seen that the open concept encountered difficulties, in part because,

---

48   In an experimental context, Planetlab may be viewed as just such a facility in some respects; *see* http://www.planet-lab.org/history, accessed March 2010.

49   *See* Orwell (1949).

50   *See* Want *et al.* (1992); the abstract notes that 'Location systems raise concerns about the privacy of an individual ...', concerns that exist today for any user of a mobile phone.

51   *See*, for example, Cope and Southard (2009). Other reports from the JANET Location Awareness trials are available at http://www.ja.net/development/network-access/location-awareness/investigations-la.html, accessed March 2010.

as pointed out by Mike Wells[52], large commercial enterprises such as banks and airlines were developing networks to serve one or two specific applications for closed communities, accessed only by specifically trained staff. Such networks could be closely tailored to meet detailed targets and to show a return on any investment, similar to engineering an optimised product or service. By contrast JANET was proposed to serve an open community, which would support hundreds of different applications, accessible to users who were likely, even encouraged, to try anything, typically without any particular training. Such a network could not have detailed targets and it would be very difficult to show any specific return on the investment in a traditional commercial sense. As a result, the initial Wells report encountered repeated opposition from the commercial perspective represented on the Computer Board, which no doubt contributed to its initial rejection.

By the time of SuperJANET we have seen that the climate of opinion had changed: the JANET concept was no longer new, and the idea was welcomed that in an education and research context one should build an open platform suitable for supporting applications, many of which were ill-specified, some hardly imagined, was welcomed. And yet in more recent times JANET has encountered arguments for not stepping beyond the conventional. During debate about the funding of the dark fibre segment of UKLight, later to become JANET Aurora, the argument was made in one quarter that 'it could all be done in the laboratory', an assertion reminiscent of another sometimes encountered: that simulation could take the place of experiment. Of course both have their place: equally, there is no substitute for experimental field trial, which rarely fails to reveal the unsuspected. With the combination of a set of flexible transmission platforms represented by wavelengths supporting the IP and the Lightpath services, and the Aurora darkfibre platform, JANET is now able to offer a rapidly growing range of services to a community now expanding even beyond that initially imagined.

A significant aspect of JANET is that it is shared not just by the whole of education but by research and the whole of education. In the years leading up to JANET, there was a unity of purpose which was fostered at the organisational and funding level by close cooperation between the Computer Board and the Research Councils, coupled with a very high degree of cooperation at the technical level amongst all those designing the protocols and implementing the network. There have been occasions since when the research imperative has caused consideration as to whether a shared facility is necessarily in the best interests of a particular area of research. However, the cost of dedicated facilities has so far always militated in favour of sharing. But the advantages of sharing are more to be found through the enabling of a wider community to share in facilities developed for the most advanced applications in particular sectors of the community. Indeed, the extent of JANET sharing its services has broken new bounds in recent years. For example, the need for content, most often for schools and particularly from the arts, has engendered substantial growth in participation by art galleries and museums. In March 2005, the third in that academic year's series of Gifford Lectures[53] was shared by a worldwide audience including participants in Ramallah, Palestinian West Bank, and Ullapool, Scotland, courtesy JVCS staff at Edinburgh University. Then,

---

52 Mike Wells, interview with Ben Jeapes, 2004.

53 'Illegal but Legitimate: a Dubious Doctrine for the Times' by Noam Chomsky, delivered at The University of Edinburgh, 22 March 2005. (*See* http://websiterepository.ed.ac.uk/explore/video/chomsky.html, accessed March 2010; also available at http://www.youtube.com/watch?v=xEvIDiVheys, accessed March 2010.)

in 2008, two British schools situated outside the British Isles also experienced the capabilities of JANET. The first was Prince Andrew School,[54] St Helena, situated in the South Atlantic, connected to the Internet via satellite. The school joined JVCS in order to help sustain its range of courses in the face of dwindling population, including teachers, and achieved a successful initial trial.[55] And before Christmas 2008, JVCS assisted pupils of the MoD school at Fallingbostel, Germany, and their relatives to talk with family members serving with the military in Basrah, Iraq.[56] 2008 also saw JANET presented with its e-Government Shared Services Award 2007 on 23 January, in recognition of its achievements in sharing throughout the educational sector the advanced services inspired by the requirements of research and higher education.[57] If the BCS Award in 1994 could be said to recognise technical achievement in pursuit of the JANET vision, perhaps the Shared Services Award might be said to recognise achievement in its delivery.

---

54 *See* http://www.princeandrew.edu.sh/, accessed January 2010.

55 *See JANET News*, No.6, 10, December 2008 (http://www.ja.net/documents/publications/news/news-6.pdf, accessed January 2010).

56 *Ibid.*

57 *See JANET News*, No.3, 3, March 2008 (http://www.ja.net/documents/publications/news/news-3.pdf, accessed January 2010); *see also* http://www.civilservice.gov.uk/news/2008/January/Awards-Tech.aspx, accessed January 2010. *Note:* JANET(UK) shared the award with Scottish Government: eProcurement Scotland Service.

Appendix A

Computer Board Network Working Parties

The membership of the two Network Working Parties convened by the Computer Board, chaired by Professor Mike Wells.

Table A.1. Members of the Wells Working Parties

Working Party	Name	Affiliation
Convened: June 1973 Reported: October 1973	Prof M. Wells	University of Leeds
	Prof G. Black	UMIST
	B.R. Taylor	DTI
	W. Walkinshaw	SRC, RHEL
	P.E. Williams	ERCC
	F.P. Verdon	CB Secretariat
Convened: June 1974 Reported: October 1975	Prof M. Wells	University of Leeds
	Dr J.L. Alty	University of Liverpool
	Dr E.B. Spratt	Kent University
	P.E. Williams	ERCC
	B. Zacharov	SRC, Daresbury Laboratory
	F.P. Verdon	CB Secretariat

Appendix B

Regional Networks

The origins of the regional networks stem from the funding by the Computer Board of larger computers at particular universities where a proportion of the service was for use by other universities, typically within some geographical regional grouping, sometimes nationally. The policy began with the inception of the Computer Board in 1966 and, by the start of the 1970s, the regional networks were forming in earnest. Their development continued as independent networks throughout the 1970s. From the mid-1970s until the mid-1980s, influenced or constrained to some extent by computer replacement cycles, the regional networks gradually made the transition to use of X.25 and the process of coalescing into the incipient JANET then took place.

During the 1970s there were quite a number of small networks which evolved to meet local or regional demands; some of these were early campus networks while others were among the first examples of regional networks (*see* Chapter 2). Listed in Table B1 below are some of those in the latter category.

Table B.1. Pre-JANET regional networks

Acronym	Title
CYGNET	Yorkshire Regional Network
GANNET[1]	North West Universities Network
METRONET	London Metropolitan Network
MidNET	Midlands Network
NUNET	Northumbrian Universities Network
RCOnet	Regional Computing Organisation Network
SWUCN	South West Universities Computer Network

1   The acronym derives from the ICL project which developed the technology on which the network was originally based: *see* Chapter 2.

Table B2 lists those regional networks in operation in 2009. The term MAN – Metropolitan Area Network – demands some explanation in this context. It derives originally from the IEEE 802.6 efforts begun in the mid-1980s to standardise shared (cable) media networking technology potentially suitable for coverage of an urban or metropolitan area. The terminology was adopted within JANET in the first half of the 1990s for many of the regional networks as the more explicitly hierarchical structure of JANET evolved, both architecturally and organisationally, into a more formal three-level hierarchy of backbone (spine) network, regional access network and campus network. In the event, none of the regional networks adopted 802.6, though at campus level there was use of FDDI – a competing technology standardised by ANSI rather than IEEE (*see* Chapter 5).

Table B.2. 2008 regional networks

Acronym	Title (operator)
AbMAN	Aberdeen MAN
CANLMAN / C&NLMAN	Cumbria and North Lancashire MAN
ClydeNET	Clyde Regional Network
EaStMAN	Edinburgh and Stirling MAN
EastNet	East of England Network (JANET)
EMMAN	East Midlands MAN (EMMAN Ltd)
FaTMAN	Fife and Tayside MAN
Kentish MAN	(Kent MAN Ltd)
LeNSE	Learning Network South East
LMN	London Metropolitan Network
MidMAN	(West Midlands Networking Company Ltd)
NIRAN	Northern Ireland Regional Area Network[2]
NNW	Net North West
NorMAN	North East MAN
NWMAN	North Wales MAN
SWERN	South West of England Regional Network
SWMAN	South Wales MAN
TVN	Thames Valley Network (JANET)
UHIMI	The UHIMI (University of the Highlands and Islands Millennium Institute) Network
YHMAN	Yorkshire and Humberside MAN

---

2   Also sometimes Northern Ireland Regional Academic Network.

Appendix C

Heads of JANET(UK) and Its Predecessors

The organisations concerned with the formation and operation of JANET were successively the Network Unit, the Joint Network Team and the JNT Association, which operated first under the trading name of UKERNA (United Kingdom Education and Research Network Association) and then of JANET(UK). Heads and chief executives of these organisations, together with the chairs of the Board of the JNT Association, are listed here.

Table C.1. Heads and CEOs of JANET

Organisation	Position	Name	Dates
Network Unit	Head	Mervyn B. Williams	1976–1978
Joint Network Team	Head	Roland A. Rosner	1979–1983
	Acting Head	Barrie Charles	1983
	Head	Peter F. Linington	1983–1986
	Head	Bob Cooper	1987–1991
	Head	Willie Black	1991–1994
The JNT Association			
UKERNA	CEO	David Hartley	1994–1997
	CEO	Geoff McMullen	1997–2001
	CEO	Robin Arak	2001–2004
	Acting CEO	Bob Day	2004–2005
	CEO	Tim Marshall	2005–2007
JANET(UK)	CEO	Tim Marshall	2007–
Board of The JNT Association	Chair	Ernest Morris	1994–1997
	Chair	Tom Husband	1997–2001
	Chair	Geoff McMullen	2002
	Chair	Geoff Peters	2002–2008
	Chair	Roger McClure	2008–

During the transition period as JANET came into operation, taking over SERCnet and integrating the regional networks, Mike Wells was part-time Director of Networking for the fixed period 1983 to 1988. Bob Cooper was then appointed to the redefined position of Director of Networking as Head of the JNT and Network Executive. (The projected positions of Head of the JNT and Head of the Network Executive had previously been merged in 1983 when Peter Linington was appointed; the Network Executive became a Section of the JNT in 1990.) Bob Cooper retained the title of Director of Networking when he relinquished the position of Head of the JNT to Willie Black in 1991 until he retired in 1997; the position of Director of Networking, which had outlived its original purpose of overseeing the establishment of JANET, was then abolished.

Appendix D

On Layering

In the early 1980s, international standardisation of network protocols had become not only very active but also political because there was a rapidly expanding market and governments wished to use them in part to support national industries, as well as in placing contracts using public money. In an effort to help create standards with the widest possible scope – and in hope of this inducing products and support at competitive prices – the UK had made adoption of network standards obligatory for equipment supplied to the university sector in 1979. At the same time, many of those with early experience of networking were encouraged to participate in the international standards process.

The process of standardisation, especially in an international context, is one of seeking acceptable compromise in committee, apparently through technical debate but in the presence of powerful undercurrents of vested interests. Naturally, this gives rise to heat at times. A long-standing traditional antidote to such proceedings has been lampoonery.[1] Reproduced below is one such that enjoyed circulation in the UK technical community around the time of early layering debates, on the twin topics of how many layers there might be and just what was supposed to be in each.

The layering's all gone wrong

CP(83)/01

It is customary after the birth of a new child to have the infant christened. The exact form the ceremony should take when the child consists of 40 people is not stipulated in any available book on etiquette, but it is certain that some form of hymn should be sung in order to thank the gods for allowing the creation of the new life. Again, conventional books on the subject do not include sections titled "hymns to be sung at the birth of a new committee" and so we must compose our own. The great composer SC6 WG2 has spent much time on this problem for us and in a flash of inspiration whilst in The Hague, and aided by his faithful companion Liquid Refreshment, has produced the following masterpiece.[2]

> Mine eyes have seen the glory of the architectural view;
> If you'd been to all the meetings you would understand it too;
> We didn't quite design it; we just watched it while it grew,
> But the layering's all gone wrong.
> We must try to make some standards;   (×3)
> But the layering's all gone wrong.

---

1   See http://www.poppyfields.net/filks/ (accessed March 2010) for computer-related examples.
2   To the tune of *John Brown's Body*.

We've guidelines and conventions which will tell us what to write,
To build a shining document and keep our concepts bright,
But we can't agree the concepts if we argue all the night,
'cause the layering's all gone wrong.
We must try ...

Jean-Pierre has got his protocol; at least it's cut and dried;
We've shifted all the problems to the implementor's side;
We're sure that they can solve them, though we haven't really tried,
'cause the layering's all gone wrong.
We must try ...

The connectionless addendum is the really modern view,
The data's out of order, even when you get it through;
But two more layers of protocol will make it good as new,
'cause the layering's all gone wrong.
We must try ...

We fought on expedited; was it in or was it out?
'Till at last we found a compromise we'd all agree about;
On weekdays we will have it, and at weekends throw it out,
'cause the layering's all gone wrong.
We must try ...

The time has come to speak of things you really ought to know;
They did it all in ECMA 10 or 20 years ago;
Our progress in comparison is ludicrously slow,
But we got the layers wrong all the same.
We must try ...

Each layer will need a service to say what it should do;
The service drafts are many, but the protocols are few.
We've not yet got a structure but an architectural glue,
'cause the layering's all gone wrong.
We must try ...

We've made a lot of progress at our meeting at The Hague,
You may feel the resolutions have all come out rather vague,
But the really tricky questions we avoided like the plague,
'cause the layering's all gone wrong.
We must try ...

The first known public performance of this anthem is believed to have been in a bar under the Rathaus in Munich.

See you all at the christening. Merry Xmas,

Appendix E

The Coloured Books

The protocols in general use over JANET and its immediate predecessor networks in the UK were developed by working groups from within both the academic and wider community, with support from the British Post Office, later BT, and the Department of Trade and Industry. As a set, they were generally referred to as the 'Coloured Books' or the 'Rainbow Books', from the set of distinctive colours adopted for the covers of the defining documents. The first five below could be considered the 'core set' and were in regular use over JANET from its inception until the early 1990s. The next group was concerned with use of local area network technologies in JANET. The penultimate group was developed to augment the original core set but were less widely deployed. The last in the list, the 'White book', was different in nature: it proposed the technical transition strategy for phasing out the UK national protocols in favour of ISO OSI protocols. In the event, this never happened: instead, JANET made the transition to Internet protocols during the first half of the 1990s. For more details, *see* Chapters 3, 5 and 6.

Table E.1: The JANET Rainbow Books / Coloured Books protocols.

Colour	Acronym	Title, etc.
Blue	NIFTP	*A Network Independent File Transfer Protocol* High Level Protocol Group of PSS User Forum, HLP/CP(77), 1977. *Revised edition by:* File Transfer Protocol Implementors Group, DCPU, FTP-B(80), February 1981.
Red	JTMP	*A Network Independent Job Transfer and Manipulation Protocol* The JTP Working Party of the DCPU, DCPU/JTMP(81), September 1981. (Draft version published in 1980.)

Yellow	NITS / YBTS	*A Network Independent Transport Service* Study Group 3 of The Post Office PSS User Forum, SG3/CP(80)2, 1980-2-16, February 1980. *Rev. 2nd ed:* Addendum, July 1982. (Also referred to as 'Yellow Book Transport Service'.)
Green	XXX /TS29	*Character Terminal Protocols on PSS:* *A recommendation on the use of X3, X28 and X29* Study Group 3 of British Telecom PSS User Forum, SG3/CP(81)/6, Rev.1, February 1981. (Original version published October 1979.)
Grey	Mail	*JNT Mail Protocol* C.J. Bennett (ed), on behalf of the JNT Dept of Computer Science, UCL, January 1982. Revised version March 1984, S. Kille (ed).
Orange	CR82	*Cambridge Ring 82 Interface Specifications* W.P. Sharpe and A.R. Cash, on behalf of SERC/JNT, Rutherford Appleton Laboratory, September 1982. *Cambridge Ring 82 Protocol Specifications* J. Larmouth (ed), on behalf of the JNT University of Salford, November 1982. *Based on:*
	TSBSP	*Transport Service Byte Stream Protocol* I.N. Dallas, Computing Laboratory, University of Kent Report 1, August 1981. *Based on:* research notes of the Systems Research Group, Computer Laboratory, University of Cambridge by M.A. Johnson, N.J. Ody, R.D.H. Walker.
Pink	CSMA/CD	*Implementation Details for Protocols on CSMA/CD* *LANs* A Report prepared by the JNT Ethernet Advisory Group JNT/CSMA/CD LAN (85), August 1985.
Peach	OSI CR	*A Functional Standard for the use of OSI Protocols* *over Slotted Rings* 1989.

Fawn	SSMP	*The Simple Screen Management Protocol* Prepared for the JNT by the Computing Laboratory of the University of Newcastle upon Tyne, J.A. Hunter, 1985.
	ATS	*Networking over Asynchronous Lines* ('Asynchronous Transport Service') Transport Service Implementors Group of the British Telecom New Networks Technical Forum, A.M. Chambers (ed), CP(83)/12, February 1983.
White	OSI	*Transition to OSI Standards: Final Report of the Academic Community OSI Transition Group,* P.F. Linington (ed), on behalf of the JNT, July 1987.

One other protocol is deserving of mention, for completeness: Interactive Terminal Protocol (ITP). This was one of the earliest protocols defined by the High Level Protocol group of the Post Office EPSS User Forum in 1975 (reference, HLP/CP(75)2). It was used in the mid-1970s by several of the original regional networks (including RCOnet, NUNET, SWUCN) and SRCnet. It was abandoned later in the 1970s because it was seen to have no future as a potential standard, and was effectively overtaken by TS29/X.29. It was not one of the JANET protocols.

The following is a combined list of those who contributed to the definition (and often implementation) of the JANET protocols. The affiliations are taken from the original documents and are those extant at the time the definitions were published; in a number of cases, contributors to more than one protocol changed institution, in which case all affiliations from the various definition documents have been included. The expansion of the acronym titles for institutions are included in the Glossary. The acronyms used in the Protocol(s) column are those given in Table E.1.

Table E.2. Contributors to the JANET Coloured Books protocols.

Name	Affiliation(s)	Protocol(s)
R.B. Abdarabbani	Glasgow University	Mail
D. Ackerman	ICL	NITS
C.J. Adams	Rutherford Appleton Laboratory	CR82
J.P. Aspden	University of Newcastle	SSMP
E. Aylmer-Kelly	York University	Mail
P.T. Barry	University of Glasgow Computing Service	NIFTP, OSI
K.A. Bartlett	ITSU/Alvey	OSI
M.J. Bayliss	formerly of Kent University	Mail
T. Benjamin	Royal Signals and Radar Establishment	ATS
C.J. Bennet	University College London	Mail, NIFTP

Name	Affiliation(s)	Protocol(s)
S. Binns	University of Kent	CR82, OSI CR
W. Black	Oxford University, for NPNCG	OSI, OSI CR
R.G. Blake	Essex University	CSMA/CD, XXX/TS29
R. Braden	University College London	Mail
R. Bradshaw	UWIST	Mail
T. Bradshaw	ICL	Mail
J.B. Brenner	ICL	NITS
P.E. Bryant	Rutherford Appleton Laboratory	ATS, Mail, OSI
A. Buttle	University of Exeter	CSMA/CD
A.R. Cash	RAL/Swindon Silicon	CR82, OSI CR
A.M. Chambers	University of Bristol Computer Centre	ATS
A.S. Chandler	Computer Aided Design Centre, Cambridge	NIFTP
B.J. Charles	Joint Network Team	XXX/TS29
C.J. Cheney	University of Cambridge	SSMP
T.B.G. Clark	University of Warwick	Mail
L. Clyne	Joint Network Team	CSMA/CD, OSI
R. Cole	University College London	Mail
G. Colouris	Queen Mary College	Mail
B.D. Cooper	AERE, Harwell	NIFTP
C.S. Cooper	RAL	Mail, JTMP, NIFTP
R. Cooper	JNT	OSI
J.A.I. Craigie	Joint Network Team	Mail, SSMP, OSI
P. Cross	Plessey	NITS
M.M. Curtis	Rutherford Laboratory	NIFTP
I.N. Dallas	University of Kent	CR82, OSI CR
A.K. Dand	SWURCC	NIFTP, OSI, XXX/TS29
D.R.H. Davies	Post Office	NITS
N.J. Davies	AUCC/University of Bristol	Mail, OSI, OSI CR
H.M. Dewar	University of Edinburgh	SSMP
W.G. Dixon	Churchill College, Cambridge	SSMP
A. Drabble	ULCC	JTMP
A. Dransfield	LNT	OSI
D.A. Duce	Rutherford Appleton Laboratory	Mail
C.D. Evans	Kodak Ltd	XXX/TS29
K.M. Farvis	ERCC	Mail
E. Fergus	AERE Harwell	Mail
K. Fermor	Rutherford Appleton Laboratory	CSMA/CD

Name	Affiliation(s)	Protocol(s)
L.J. French	University of Cambridge	SSMP
J.R. Gascoigne	International Computers Ltd	XXX/TS29
C. Gerrard	University of Newcastle	SSMP
R.J. Gillman	International Computers Ltd	XXX/TS29
R. Gillman	JNT	OSI CR
B.A.C. Gilmore	University of Edinburgh/ERCC	SSMP, OSI
P.M. Girard	Rutherford Laboratory	NIFTP, NITS
P.J.S. Gladstone	formerly of RAL	Mail
D. Gray	Salford	OSI
N. Gunadhi	Rutherford Appleton Laboratory	CSMA/CD
C. Gunner	Polytechnic of Central London	CR82
M.J.T. Guy	Cambridge University	ATS, JTMP, Mail, NIFTP, NITS, OSI, XXX/TS29
J. Hall	SWURCC	XXX/TS29
N.R. Hammond	Post Office	XXX/TS29
A. Hartwig	University of York	Mail
D.V. Harvey	International Computers Ltd	XXX/TS29
V. Hathway	National Physical Laboratory/ITSU	NITS, OSI, XXX/TS29
N. Hawkes	Queen Mary College, London	Mail
A. Haworth	University of Reading	SSMP
K.S. Heard	AERE Harwell/JNT	CR82, NIFTP, NITS
P.L. Higginson	University College London	NIFTP, NITS, OSI, XXX/TS29
A.J. Hinchley	London University Network Team	ATS, CR82
R. Hughes	University of Leeds	OSI CR
J.A. Hunter	University of Newcastle/NUMAC	Mail, SSMP
D.J. Jackson	ULCC	CSMA/CD, OSI
R.B. John	ICL Dataskil	JTMP
M.A. Johnson	University of Cambridge	CR82
M.G. Johnson	Westfield College	JTMP
A.W. Jones	SWURCC	NIFTP
P. Kemp	University of Reading	SSMP
C. Kennington	University College London	CR82, Mail
S.E. Kille	University College London	Mail, OSI
P. Kirk	Marconi Research Centre	CR82
H.C. Kirkman	University of London Computer Centre	JTMP
P.T. Kirstein	University College London	Mail
K.G. Knightson	Post Office	NITS, XXX/TS29

Name	Affiliation(s)	Protocol(s)
T. Knowles	GEC Computers Ltd	NITS
P.S. Kummer	Daresbury Laboratory	CSMA/CD, JTMP, OSI
J. Larmouth	University of Salford	CR82, JTMP, OSI
T.D. Lee	University of Durham	SSMP
W.D. Lees	Logica Ltd	CR82
K. Lewis	University of Oxford	CR82
J.M. Line	Cambridge University	Mail
P.F. Linington	University of Cambridge / DCPU / JNT / UKC	JTMP, NIFTP, NITS, OSI, XXX/TS29
J.A. Linn	Aberdeen University	Mail
D. Lord	Post Office/BT	XXX/TS29
R. Lowndes	Camtec Electronics Ltd	CR82
R. MacKenzie	Glasgow University	Mail
C.J. Makemson	CAP (Reading) Ltd	ATS
J. Malcolm	GEC Computers Limited	SSMP
N. Matthew	Camtec Electronics Ltd	OSI CR
M.A. McConachie	University of Nottingham	NIFTP
D.E. Miller	formerly of Hatfield Polytechnic	Mail
J.N. Mills	RAL	Mail
B. Mitchell	University of St. Andrews	OSI CR
P.E. Moran	University of Edinburgh	CSMA/CD, SSMP
G. Morrow	British Telecom Research Labs	CR82
R.T. Moulton	University College London	Mail
D.S. Munro	University of Aberdeen	CR82
J.S. Nightingale	National Physical Laboratory	NIFTP
M.J. Norton	Post Office	NITS
N.J. Ody	University of Cambridge	CR82
E. Oskiewicz	Toltec Computer Ltd	CR82
F. Panzieri	University of Newcastle	CR82
S. Perkins	ICL	Mail
C.J. ('Kit') Powell	AUCC	Mail
R. Powell	University of Leeds	OSI CR
J.S. Powers	ULCC	OSI
D. Rayner	National Physical Laboratory	NIFTP
S. Reynolds	Camtec Electronics Ltd	OSI CR
P. Riley	University of Kent	OSI CR
M. Roberts	Toltec Computer Ltd	CR82
T. Robinson	High Level Hardware	OSI CR
R.A. Rosner	Joint Network Team/ULCC	Mail, OSI

Name	Affiliation(s)	Protocol(s)
K.D. Ruttle	University of York	Mail
J. Salter	ICL Dataskil Ltd	XXX/TS29
M. Sands	Post Office	XXX/TS29
A.J. Scolley	University of Stirling	SSMP
M. Scott	GEC Computers	Mail
M. Sendorek	Post Office	NITS
J.D. Service	formerly of York University	Mail
J.A. Seymour	LNT	OSI
M. Shand	Kingston Polytechnic	CR82
W.P. Sharpe	RAL	CR82
A. ('Sandy') Shaw	ERCC	Mail, SSMP
H.T. Smith	Nottingham University	Mail
I.L. Smith	Network Executive	OSI
D. Steedmam	GEC Computers Ltd	NITS
D. Steinitz	London Network Team	Mail
R. Tasker	DL	OSI
N.R. Topham	Computer Aided Design Centre, Cambridge	NIFTP
J.B. Tucker	ITSU	OSI
P.F. Tye	ULCC	OSI
C. Wadsworth	Rutherford Appleton Laboratory	Mail
D.F. Walker	University of Edinburgh	SSMP
R.D.H. Walker	University of Cambridge	CR82
T. Wallace	Prime Computers Ltd	NITS
D. Wanless	University of Keele	CR82
D.E.P. Watkins	University of Birmingham	NIFTP
R. Watson	Cranfield Institute of Technology	OSI
P.R. White	SWURCC	ATS, NITS
S.R. Wilbur	University College London	CR82
T. Wiersma	UWIST	Mail
A.H. Williams	Avon Universities Computer Centre	JTMP
R.P.J. Winsborrow	AERE Harwell	XXX/TS29
P. Wright	Marconi Research Ltd	CR82

Appendix F

Networkshops

The titles of hosting institutions are those current in 2008; in one or two cases they differ from those current at the time of the relevant Networkshop (e.g., Swansea, 1981).

Table F.1. Networkshops

Number	Date	Venue
1	Sep 1977	University of Glasgow
2	Apr 1978	University of Liverpool
3	Sep 1978	University of Cambridge
4	Apr 1979	University of York
5	Sep 1979	University of Kent
6	Apr 1980	Durham University
7	Sep 1980	University of Oxford
8	Apr 1981	Swansea University
9	Sep 1981	Heriot-Watt University
10	Apr 1982	University of Nottingham
11	Apr 1983	Royal Holloway, University of London
12	Apr 1984	University of Bath
13	Apr 1985	University of Sheffield
14	Apr 1986	Lancaster University
15	Apr 1987	University of Edinburgh
16	Mar 1988	University of Reading
17	Apr 1989	University of Warwick
18	Mar 1990	Newcastle University
19	Mar 1991	University of Aberdeen
20	Mar 1992	University of Leeds
21	Mar 1993	University of Birmingham
22	Mar 1994	University of Plymouth
23	Mar 1995	University of Leicester
24	Mar 1996	University of Sussex
25	Mar 1997	Queen's University Belfast
26	Mar-Apr 1998	University of Aberdeen
27	Mar 1999	University of Warwick
28	Mar 2000	Heriot-Watt University
29	Mar 2001	University of Exeter

30	Mar 2002	University of Nottingham
31	Apr 2003	University of York
32	Apr 2004	Keele University
33	Mar 2005	University of Manchester
34	Apr 2006	University of Hertfordshire
35	Apr 2007	University of Exeter
36	Apr 2008	University of Strathclyde
37	Mar-Apr 2009	University of Cambridge
38	Mar-Apr 2010	University of Manchester

Appendix G

Digital Transmission Hierarchies

Listed in Table G.1 are the rates defined for the original digital multiplexing hierarchy, the plesiochronous[1] digital hierarchy (PDH), meaning that the signals on individual links are nearly 'in time' or synchronised. There are various versions of the hierarchy: the international one standardised by the CCITT and that used in North America, Japan and Korea, standardised under the auspices of ANSI.

Table G.1. Plesiochronous Digital Hierarchy (PDH)

International (CCITT)			USA (ANSI)		
Title	Rate (bps)	Abbrev (bps)	Title	Rate (bps)	Abbrev (bps)
(channel)	64K		(channel)	64K[1]	
E1	2.048M	2M	T1 (DS1)	1.544M	1.5M
E2	8.448M	8M	T2 (DS2)	6.312M	6M
E3	34.368M	34M	T3 (DS3)	44.736M	45M
E4	139.264M	140M	(DS4)	274.176M	274M
E5	565.148M	565M	(DS5)	400.352M	400M

---

1   One bit in every byte was often used for other purposes, which gave an effective payload rate of 56Kbps.

As indicated above, in PDH each link has its own timing. When multiplexing several PDH links, extra 'padding' bits have to be inserted or dropped sometimes to accommodate timing differences between links. As the position of the padding is unpredictable, only complete demultiplexing is possible, even if only a single channel is required.

In synchronous[2] multiplexing systems, the links are all synchronised and generally no slippage can occur, and partial demultiplexing is possible: the position of the bytes for an individual circuit is known and fixed. The original set of standards for doing this were proposed by Bellcore in 1985 to ANSI in the USA under the title of the Synchronous Optical Network (SONET). SONET was followed in 1986 by draft international versions from the then CCITT, entitled the Synchronous Digital Hierarchy (SDH). After some compromises to accommodate

---

1   Pseudo Greek, from plesios (πλησίος), 'near' and chronos (χρόνος), 'time'.
2   Greek, from sun- (syn-) (συν-), 'with' and chronos (χρόνος), 'time'.

Table G.2. SDH / SONET rates

SDH level & format	SONET carrier & format	Payload rate (bps)	Line rate (bps)	Abbrev (bps)
(STM-0)	OC-1 / STS-1	50.112M	51.84M	52M
STM-1	OC-3 / STS-3	150.336M	155.52M	155M
STM-4	OC-12 / STS-12	601.344M	622.08M	622M
(STM-8)	OC-24 / STS-24	1.202688G	1.24416G	1.2G
STM-16	OC-48 / STS-48	2.405376G	2.48832G	2.4/2.5G
STM-64	OC-192 / STS-192	9.621504G	9.95328G	10G
STM-256	OC-768 / STS-768	38.486016G	39.81312G	40G
[STM-1024]	OC-1024 / STS-1024	153.944064G	159.25224G	160G

the existing international PDH rates, the standards were agreed in 1987.[3] Table G.2 lists the SONET / SDH rates. STM-0 is occasionally used to refer to the lowest SONET rate but it is not an SDH rate: similarly for STM-8, though even the SONET OC-24 is rarely implemented. STM-1024 has not been standardised and, as of the time of writing, is not now expected to be.

---

[3] The two standards are not quite the same, differing in some of the control information included in the time frame.

Appendix H

School Networks

JANET provides a backbone network interconnect service for schools networks in the UK. Schools are generally the responsibility of UK local authorities. However, school ICT infrastructure for education and learning is devolved so that provision is organised individually in England, Northern Ireland, Scotland and Wales.

In England, the majority (140 out of 150) of the local authorities have formed into 10 Regional Broadband Consortia (RBCs). A few authorities have opted out of this arrangement. Both are listed below. JANET forms the interconnect for both RBCs and the 'opted-out' authorities.

The Northern Ireland schools network is part of the C2k project. Connection to the JANET backbone is in Belfast, as for NIRAN.

The 32 local authority networks in Scotland are linked directly to JANET, together with administrative and content provision sites to form the network infrastructure for Glow.

The 22 local authorities in Wales are interconnected via the PSBA Network, the shared public sector communications infrastructure (initially developed from LLNW as the basis). The educational resources for schools are provided by NGfL Cymru over this infrastructure.

The JANET backbone, in addition to providing the interconnection of LEAs in Scotland and RBCs in England, also serves to interconnect the school networks of the four countries in the UK and provide Internet access.

Table H.1. School regional and UK country networks

UK countries	Organisation	Acronym	Detail
England	Regional Broadband Consortia (RBCs)	CLEO	Cumbria Lancashire Education Online
		EMBC	East Midlands Broadband Consortium
		E2BN	East of England Broadband Network
		LGfL	London Grid for Learning
		NG	Northern Grid
		NWLG	North West Learning Grid
		SEGfL	South East Grid for Learning
		SWGfL	South West Grid for Learning
		WMnet	West Midlands Regional Broadband Consortium
		YHGfL	Yorkshire & Humberside Grid for Learning

	'opted-out' LEAs	Barnsley, Bradford, Derwentside (serving Durham, Newcastle, Northumberland and Sunderland), Oxfordshire, Tameside, Wigan, York.
Northern Ireland	C2k	Classroom 2000; Network infrastructure contracted out; interconnected with other UK schools and the Internet via JANET backbone.
Scotland	Glow	Schools are linked via local authorities to each other and administrative and content sites via JANET. The initial project, Spark, developed the initial network infrastructure, SSDN, which came into operation in 2003. SSDN was officially renamed Glow in 2006, which signalled the start of development of the educational application overlay, which began deployment in 2008. The interconnect infrastructure which connects individual schools to the regional authority points of presence provided by JANET is, as of 2010, known as Interconnect 2.0 and undergoing upgrade; both Glow and Interconnect 2.0 are continuing deployment.
Wales	NGfL Cymru / PSBA Network	Content and educational delivery applications via NGfL Cymru; communications network infrastructure provided by the Public Sector Broadband Aggregation Network for Wales.

Appendix 1

Computers, Communication and the Law

Andrew Cormack
JANET(UK)
October 2009

Until 1990 there was no UK legislation dealing specifically with misuse of computers and networks. Where prosecutions were required these generally relied on criminal offences such as theft, forgery or criminal damage, even though these sometimes required creative interpretation to make them fit the circumstances of computer 'break-ins'. This appendix provides a brief overview of legal developments in this field over the last couple of decades, with particular reference to the legislation and an indication of the ways in which it may affect the JANET community, both users and those providing the services.

Misuse of computers and networks

In 1990 Parliament recognised the need for new legislation and passed the *Computer Misuse Act 1990*, which creates crimes of unauthorised access to computers and unauthorised modification of data on computers. None of the terms 'computer', 'access' and 'authorisation' are defined by the Act, which has allowed courts to apply the law to a rapidly developing technology without too much strain. However the lack of definitions also leads to uncertainty: for example, lacking case-law, it is not clear at what point in the TCP handshake 'access' is obtained. Unlike some other countries, UK law does not require the owner of the computer to have implemented any particular measures to prevent access.

A number of reviews of the *Computer Misuse Act* commented that the Act did not seem to cover denial of service attacks, where the aim is not to access or modify data on a computer but to prevent authorised users from doing so. This was addressed by an amendment contained in the *Police and Justice Act 2006* which replaces the offence of unauthorised modification with an offence of unauthorised interference. This amendment was brought into force on 1 October 2007 in Scotland and 1 October 2008 for the rest of the UK.

Two forms of unauthorised access to networks have been criminalised. The *Regulation of Investigatory Powers Act 2000* created a criminal offence of unlawful interception of a network. Anyone harmed by unlawful interception may also be able to sue those responsible to recover damages. The *Communications Act 2003* contains an offence of dishonestly obtaining a communications service with intent to avoid payment, which has been used to prosecute or issue cautions to a number of people who had been using other people's domestic wireless networks. As with the *Computer Misuse Act* there is no requirement that the network implement any particular protective measures.

CONTENT AND LIABILITY

Many types of statement, when written and published, can give rise to legal liability for the author under laws including copyright, defamation, obscenity, racism and terrorism. The law has historically recognised two types of intermediary who may be involved in the publication of these statements: the publisher, who generally carries the same risk of liability as the author, and the carrier, who is generally freed from any liability for what he carries. Newspaper publishers and postal services represent these two types.

Surprisingly late in the development of networks it was recognised that Internet transmission may involve other types of intermediary for whom neither existing model of liability is appropriate. The first UK law to mention Internet intermediaries appears to be the *Defamation Act 1996*. The *Electronic Commerce (EC Directive) Regulations 2002* (implementing European Directive 2000/31/EC) expanded this to a common regime applying to all types of liability then existing and containing two new types of intermediary:

- **caching** services automatically take copies of transmitted material for the purpose of making future requests for the same material more efficient; they are excluded from liability provided that they respect industry standards that allow the originator of the material to prevent caching and that material is removed from the cache promptly when the cache is informed that the original has been removed or access restricted

- **hosting** services store (and may publish) material for others without examining it; they are excluded from liability until they are informed (or otherwise discover) that a specific item is unlawful. If they do not then act 'expeditiously' to remove or block access to the material then they are treated as if they had published it.

The existing carrier (known in the Regulations as 'mere conduit') and publisher types of liability continue to exist. A report by the Law Commission in 2002 criticised this system as encouraging hosting services to remove any material for which a complaint, however ill-founded, was received. Where universities and colleges provide hosting services there are likely to be balancing statutory requirements to support freedom of speech, but these do not apply to commercial hosting services.

Since 2002 new Acts creating liability have generally included, or been amended to include, similar provisions on Internet intermediaries. For unlawful terrorist publications, the *Terrorism Act 2006* modified the general rule for hosting intermediaries by specifying the form of the notice and how it must be served; it also set a period of two working days within which the hosting service must act to avoid being considered to have approved the publication.

Many fewer classes of material are illegal to knowingly make or possess. These create particular difficulties for investigators since the act of preserving or inspecting evidence on a computer may itself constitute a serious criminal offence. For indecent images of children (illegal to make or possess under the amended *Protection of Children Act 1978*) a 'good reason' defence was included in the *Sexual Offences Act 2003*; the same approach was adopted when possession of extreme adult pornography was criminalised by the *Criminal Justice and Immigration Act 2008*. These defences are very narrowly defined and investigators must ensure they meet the terms set out in a Memorandum of Understanding between the Crown Prosecution Service and Association of Chief Police Officers. Scotland has different Acts on these types of material but it is likely that the same good practice would apply to investigators there as well.

LAW ENFORCEMENT POWERS

It has long been recognised that law enforcement agencies may need access to information about communications networks in order to investigate crimes that may be planned or committed using them. Since such access may affect the privacy of users of networks, society also demands that it be regulated. An early attempt to provide regulated powers was the *Interception of Communications Act 1985*, which permitted the interception of traffic on networks designated as public by the Secretary of State – which at the time included the national duopoly services of BT and Mercury, and the local service of Kingston Communications. Other networks, and in particular private networks inside organisations, were left unregulated. By the mid-1990s a number of cases in the European Court of Human Rights had found that interception not controlled by the law breached the *European Convention on Human Rights*. Since the *Human Rights Act 1998* made that Convention formally part of UK law, the *Regulation of Investigatory Powers Act 2000* was passed immediately before the *Human Rights Act* came into force to regulate interception on all networks and thereby make this area of UK law compliant with the Convention. The new Act provides for interception warrants to be issued by the Home Secretary. Interception may also be authorised by network providers for a limited range of purposes: misuse of this power is subject to criminal sanctions on public networks and civil sanctions on private ones.

Interception gives access to the content of messages and is relatively rarely required. Much more common, for both the police and network operators, is the need to identify a subscriber, know when they have logged in to a service, or know when and to whom they have used it to communicate. This information about the use of communications systems is generally referred to as communications data. For the network provider this information will generally constitute personal data about the subscriber and be subject to the *Data Protection Acts* of 1984 and 1998. There is therefore a general presumption that this information will not be disclosed to third parties. A number of Acts provide ways by which this presumption can be reversed. The *Police and Criminal Evidence Act 1984* allows a court to order any person to disclose to the police evidence that is required by the court and that the person holds. Both Data Protection Acts contain sections (s.28 of the 1984 Act and s.29 of the 1998 Act) that allow a person to disclose information to the police if the person is persuaded that it is necessary and proportionate to do so for the prevention, detection or investigation of crime. Since this route places a burden on the person, and gives the police no guarantee that the information will be provided, an order under s.22 of the *Regulation of Investigatory Powers Act 2000* is now recommended where the information sought is communications data. Under this provision the police make the assessment of necessity and proportionality and the recipient of the order does not have to decide whether or not to comply.

Powers to demand disclosure of information are of little value if information is not kept by network operators, and the move to unmetered Internet services has decreased the amount of data that providers need to collect to operate their business. The *Anti-Terrorism, Crime and Security Act 2001* contained the first legislative step to protect the amount of data available to the police by permitting telephone and Internet Service Providers to voluntarily retain logs collected for business for one to two years for the purpose of preventing and investigating serious crime or terrorism. A European Directive (2006/24/EC) made similar retention schemes mandatory for all providers of public networks in all member states. This was implemented in the UK for voice networks in 2007 and for data networks in 2009 by the *Data Retention (EC Directive) Regulations 2009*. Since these regulations only apply to voice and e-mail services operated by UK ISPs the Government has since launched a consultation, as part of its Interception Modernisation Programme, on how ISPs might also retain data about use of third-party communications services such as webmail.

Glossary

6bone   IPv6 overlay on IPv4. Introduced in 1996 to enable testing of IPv6 over the Internet. JANET joined in 1997. Bermuda 2 (2000–2002) and 6NET (2002–2005) both used it. 6bone support in JANET was withdrawn in 2006, after native support for IPv6 had been introduced in 2003. (See Chapter 14)

A

AAL   ATM (*q.v.*) Adaptation Layer: protocol layer defined for use over ATM cell layer to provide (packet) transport appropriate to an application. There are a variety of AAL protocols defined for data and real-time continuous media transport.

AARNet   Australian Academic and Research Network; created in 1989; operated by AARNet Pty Ltd (APL).

Abilene   Initial 'next generation' US academic and research Internet backbone developed under the Internet2 project which became operational in 1998.

Access Grid   Multiple window videoconferencing system, based on paradigm of shared (virtual) meeting space ('venue'); developed at ANL (*q.v.*) in the late 1990s; exploits network-level multicast; originally derivative from LBL/UCL/MICE system. Adopted for use in the UK e-Science Programme. (*See also* VRVS and EVO.)

ACM   The Association for Computing Machinery. US-based scientific, educational and professional international organisation for computer science.

ACN   Former (JISC) Advisory Committee on Networking; effectively a renaming in 1994 (with the formation of UKERNA and following the reconstitution of ISC as JISC) of the Network Advisory Committee, which was formed in 1983, in preparation for the start of JANET, and reported to the Computer Board. Subsequently referred to as the JCN (*q.v.*).

ACTS   Advanced Communications Technology and Services; one of the development and research programmes managed by DG-III, a directorate of the EU Framework IV programme.

AERE   Atomic Energy Research Establishment. The former major site at Harwell (next door to RAL, both sites now part of the Harwell Science and Innovation Campus) contributed to early network development in the UK, including JANET.

AFRC   Former Agricultural and Food Research Council; in 1981, the remit of the ARC was expanded and it was renamed to reflect this; a constituent of RCUK. *See* BBSRSC.

Glossary

AGSC Access Grid Support Centre; set up in 2003 at the University of Manchester to provide specialist support for the developing Access Grid videoconferencing system.

AHRC Arts and Humanities Research Council, formed in 2005 from the former AHRB (Arts and Humanities Research Board) and responsible for funding academic research in these sectors; a constituent of RCUK.

ANL Argonne National Laboratory; particle physics accelerator laboratory which instigated much of the early Grid development, including GLOBUS middleware and the so-called Access Grid application based on the LBL/UCL/MICE MBone videoconferencing tools of the 1990s.

ANSI American National Standards Institute; standardised FDDI and SONET.

ARC Former Agricultural Research Council, formed in 1931. *See* AFRC.

ARP Address Resolution Protocol; broadcast protocol used to achieve address translation from IP address to Layer 2 MAC address for shared media LANs. An ARP service to mimic this was defined by the IETF for use with ATM, SMDS and other non-broadcast, multi-access (NBMA) networks.

ARPA Advanced Research Projects Agency; former agency of the US Department of Defense; responsible for creation of the ARPANET.

ARPANET Experimental packet-switched network in the USA created by ARPA, beginning in 1969, which evolved into the Internet in the early 1980s.

AS Autonomous System; potentially of wide applicability, here used in the commonly-accepted network context to mean a networking domain managed and operated by an independent organisation. The Internet is composed of many ASs, each having agreements with others on the services they provide to each other. ('Domain' is also sometimes used in this context though it also has a distinguished, specialised meaning with respect to naming.)

ASCII American Standard Code for Information Interchange: widely used character set definition and encoding; very similar to International Alphabet no.5, IA5, and ISO-646, though all have minor options and variations.

ASRB Former Astronomy, Space and Radio Board of SRC. Responsible for the funding of science research in these disciplines during the 1970s and 1980s.

Athens UK access and identity management system. Development began at the University of Bath in 1994 (as part of NISS (*q.v.*), originally for NISS internal use); service started in 1996, initially in the UK HE sector for journal access. It was managed by Eduserv (a spin-off from the University of Bath) from 2000 to 2008. The Athens service was funded by JISC from its inception until 2008. Superseded by Shibboleth (*q.v.*). (The system was initially named for the Greek goddess of knowledge, Athena, but changed to Athens because the former was already registered.)

Atlas Laboratory Former SRC laboratory providing national computing facilities based first on a Ferranti Atlas and later an ICL 1906A. Merged with Rutherford Laboratory in 1975.

ATM Asynchronous Transfer Mode; fixed-size packet multiplexing and switching technology chosen by the CCITT in the mid-1980s as the basis of broadband ISDN.

ATM Forum Not-for-profit international industry association of manufacturers and suppliers of computer and communications equipment formed in 1991 to develop and promulgate ATM implementation specifications based on the CCITT standards. Merged with MPLS Frame Relay Alliance in 2004; subsequently, there have been mergers with the IP/MPLS Forum and the Broadband Forum.

ATS Asynchronous Transport Service; *see* Chapter 3 and Appendix E.

AUCC Avon Universities Computer Centre: consortium of the Universities of Bath and Bristol, formed in 1978 to pool funding and share in joint computer provision by purchase of the first Honeywell Multics system in Europe, installed in 1979.

AUP Acceptable Use Policy.

B

B-ISDN Broadband ISDN (*q.v.*).

BBN Historically, sometimes referred to as BB&N: originally Bolt, Beranek and Newman of Cambridge, Massachusetts, USA, known at the time of writing as BBN Technologies; the company began as acoustic engineering consultants but moved into computer-related work in the 1960s. BBN designed and built the ARPANET IMPs (*q.v.*).

BBSRC Biotechnology and Biological Sciences Research Council, formed in 1994 as part of the Research Council reorganisation, responsible for funding academic research in these sectors. It took over the AFRC remit together with the biological science and technology parts of the SERC programme. A constituent of RCUK.

BCS The British Computer Society.

Becta British Educational Communications Technology Agency. Responsible to government for co-ordinating ICT provision in schools.

Bellcore Full name: Bell Communications Research; typically referred to by this official contraction.

BEN Backbone Edge Node: term used in SuperJANET III for routers operated by the NOSC on behalf of UKERNA, which were situated at the edge of the JANET backbone network and acted primarily as the points of attachment for the regional networks (termed MANs at the time).

BGP Border Gateway Protocol: one of a number of routing protocols, in this case intended for routing between different autonomous networks; e.g., different NRENs or, in the case of JANET, between regional networks and the JANET backbone.

big-endian Originally, satirical title appearing in Jonathan Swift's *Gulliver's Travels* to describe the inhabitants of the island of Blefuscu (an island 800 yards off the coast of Lilliput) who opened their eggs at the larger end; similarly, *mutatis mutandis*, the Lilliputian little-endians. Swift was targeting those who expended disproportionate, not to say unjustified effort on consideration of minutiae of almost vacuous significance, typically religious or political. Adopted into information and communications technology to describe the two conventions relating to ordered constructs according as the initial item in the construct is of major (big-endian) or minor (little-endian) significance. Common examples relate to conventions for components of names, representations of integers by bits in a byte or word in either computer memory or transmission over a communication medium. Typically the choice itself is of vestigial intrinsic or intellectual significance but that it is made and specified is essential. Apart from Swift (1726), the reader is referred to Cohen (1980). In the 1980s, the UK used this convention for its host names, the opposite to that adopted by the Internet.

BIND   Berkeley Internet Name Domain; the program written for Unix at the University of California at Berkeley which is probably the most widespread implementation of the Internet DNS.

BITNET   Originally, 'Because It's There Network', later, 'Because It's Time Network'; club-like voluntary consortium network first established in 1981 and based on a network of adjacent (i.e., directly linked) IBM systems, using IBM protocols, often with communication lines sponsored by IBM; later, non-IBM systems joined in by emulating the IBM protocols. It offered mail (*see* LISTSERV) and file transfer services; ultimately, it merged with CSNET (*q.v.*) in 1989 to form CREN (*q.v.*). CREN support for BITNET services was discontinued in 1996. *See also* EARN, to which the UK subscribed.

Blue Book *See* NIFTP. Also colloquially used in a wider context to refer to Version 1.0 of the *de facto* DIX (*q.v.*) Ethernet standard published in 1980 (revised in 1982 to become the DIX Version 2.0 specification).

BNF   Backus-Naur Form: a formal method (meta-syntax) for specifying (context-free) grammars, most often those of computer programming languages.

BOD   Bandwidth on Demand: in effect, a dynamic version of MBS (*q.v.*) in which transmission capacity may be set up dynamically using some form of signalling; originally used in a packet network context, more recently used in the context of channel- or circuit-oriented networks.

Broadband ISDN Follow-on to ISDN; service developed in 1980s for PTTs, standardised by CCITT, as a multi-service bearer service capable of supporting mixed telephony and data. It never reached full deployment, being more or less superseded by QoS-enabled IP technology.

BSC   Binary Synchronous Control protocol; byte-oriented link layer protocol used, for example, in IBM RJE and related protocols; adopted as an interim link layer for X.25 in some early UK academic networks, e.g., SRCnet.

BSD   Berkeley Software Distribution: the derivative version of Unix developed and distributed by the University of California, Berkeley from 1977 through 1995. The 4.2BSD distribution (strictly, the interim version 4.1c of BSD was the first to contain the network components referred to) of Unix was significant for TCP/IP networking in that it included support for the original TCP/IP protocol stack, including DNS (Bind) and Sendmail, the original versions of both written at Berkeley.

BSI   British Standards Institution. Founded in its current form in 1929; its current name dates from 1930. founder member of ISO (*q.v.*).

BSP   Byte Stream Protocol: stream protocol defined for use over the Cambridge ring. *See* Chapter 5.

BT   British Telecom. Originally a government-owned monopoly company formed from the part of the British GPO which operated the telephone system and was responsible to the DTI. Following deregulation in 1981, it was floated on the stock market and sold off into public ownership in stages.

BUNNIE British UNiversities Network Interface Equipment: term coined in the first Wells Report to the Computer Board in 1973 for a packet switching node for the proposed national academic network.

C

C-PoP Core Point of Presence: location of backbone node in SuperJANET4. There were 8 C-PoPs, located at WorldCom premises in Glasgow, Edinburgh, Leeds, London, Reading, Bristol and Manchester. The node routers were Cisco GSR 12016s.

C2k Classroom 2000: national school-level educational networked services in Northern Ireland.

Caltech California Institute of Technology, Pasadena, California.

CB Occasional acronym for the Computer Board (*q.v.*).

CBR Constant bit rate. Traditional digital communications circuits have this property; packet-based network channels or virtual circuits almost never have; compressed video encodings may have; those intended for use over a digital communications circuit do have. Codecs intended for use in this way at the two ends of a digital circuit are synchronised with the digital transmission circuit and hence with each other.

CCITT Comité Consultatif International de Téléphonique et Télégraphique: a part of the ITU (*q.v.*), it became the ITU-T in 1992 (*see* ITU).

CCLRC Council for the Central Laboratories of the Research Councils: the former council with responsibility for CLRC. Merged with PPARC to form STFC in 2007.

CCTA Central Computer and Telecommunications Agency; a government agency which in the 1970s and 1980s assessed information and communications equipment with a view to providing centralised approval and procurement for government-funded bodies. It was also responsible for defining UK GOSIP (*q.v.*).

CDC Control Data Corporation. Manufacturer of the most powerful scientific computers of the 1960s and early 1970s, designed by Seymour Cray, who later formed his own company, designing and manufacturing a long line of supercomputers.

CDC UT200 CDC remote job entry terminal (and associated protocol), originally for use with 6000-series machines in the 1960s.

CEPT European Conference of Postal and Telecommunications Administrations; a European standards organisation primarily concerned with co-ordinating European representation and contributions at CCITT/ITU-T; created ETSI (*q.v.*) in 1988.

CERN Conseil Européen pour la Recherche Nucléaire (European Council for Nuclear Research), the body set up in 1952 to establish a European fundamental physics organisation. When the European Organisation for Nuclear Research was established in 1954 near Geneva, spanning the border between Switzerland and France (and the Council ceased, having done its job), the new organisation adopted the Council's acronym as its short-form name.

CERT Computer Emergency Response Team; a network of such teams came into being worldwide in the late 1980s as the Internet became an ubiquitous, open-access network, to assist *bona fide* users in defending against malicious, network-born attacks on networks and associated computers. The JANET CERT began service in January 1994. The term CERT has since been superseded by CSIRT (*q.v.*).

CERT/CC   CERT/Co-ordination Center; set up by DARPA in 1988 as a direct result of the Morris worm incident (*see* 'worm').

CETL4HealthNE   Centre for Excellence in Teaching and Learning for Health North East; consortium of five universities (Durham, Newcastle (consortium lead), Northumbria, Sunderland and Teesside), two NHS regional health authorities and three NHS trusts. (See http://www.cetl4healthne.ac.uk/view, accessed March 2010.)

CIDR   Classless Inter-Domain Routing; introduced in 1993, it removed the restriction of the three IPv4 unicast address classes (A, B, C), in which the division of the 32-bit address field into network number and host address could only occur on an octet boundary (1st, 2nd, 3rd, respectively). Its purpose was to slow both the growth of routing tables in routers and the consumption of address space.

circuit   An end-to-end communications channel. (*see also* virtual circuit).

circuit switch   A device with at least two input and two output ports (channels) which can change state to connect any input port to any output port, providing that the output port is not already occupied by a connection from another input port; in the latter case, the switch is said to be (partially) blocked.

CLNS   Connectionless Network Service; term defined by ISO to refer to a datagram style of network operation in OSI.

CLRC   Former Central Laboratories of the Research Councils, formed at the same time as CCLRC, providing merged management of Chilbolton Observatory, DL and RAL. The name CLRC ceased to be used in 2007, as the management of these institutions passed to STFC.

CMU   Carnegie Mellon University.

CoD   Conferencing on Demand: pilot videoconferencing service without the need to book conference bridge resources; mounted over JANET in September 2004, it became a service – called Instant Videoconferencing, IVC – in April 2005. IVC is a component of the JVCS portfolio.

codec   Coder-decoder: term used to refer to the software or device to perform the encoding of an analogue signal into digital representation and vice versa. The function is often referred to as analogue-to-digital (abbreviated d-to-a) conversion (and vice versa, *mutatis mutandis*). The terminology may apply to audio or video or both together.

Coloured Books   Set of documents specifying the original network protocols used by JANET. *See* Appendix E and Chapter 3.

Computer Board   Full title Computer Board of the Universities and Research Councils; created in 1966 with the remit to provide UK universities with large-scale computing facilities. During the 1970s it acquired the additional responsibility for academic computer networking. Became the ISC in 1991 and subsequently the JISC (*q.v.*).

congestion   Situation pertaining in a packet network node when packets are being received at a switching node faster than they are being forwarded, the node is full and no more can be accepted.

CONS   Connection-Oriented Network Service; term defined by ISO to refer to a virtual circuit style of operation in OSI.

COSINE   Cooperation for OSI Networking in Europe; European networking initiatives coordinated centrally by RARE and later by DANTE, established as one of the first EUREKA (*q.v.*) projects in late 1985. Funding was provided by the European Commission and EUREKA members. RARE undertook the COSINE Specification Phase under contract to the European Commission. The first project of the COSINE Implementation Phase was the pan-European network IXI (*q.v.*).

COST   European organisation for Co-Operation in the field of Scientific and Technical research. Supports co-operation and harmonisation, funded by the European Commission and currently organised through the European Science Foundation. Funded EIN (*q.v.*) in 1971.

CPSE   Campus PSE; a small version of a PSE (*q.v.*), suitable for use on a university or laboratory campus. Those for the original JANET were supplied by GEC Computers Ltd, based on the smaller models of 4000 series systems. Subsequently, Camtec developed a much smaller CPSE, a derivative of its X.25 PAD.

CR82   Cambridge Ring 1982 standard; *see* Orange Book.

CREN   Formed in 1989 by the merger of CSNET and BITNET. Support for CSNET services was discontinued in 1991, that for BITNET in 1996. CREN ceased operation in 2003.

CRT   cathode ray tube: display device as used in televisions and computer monitors before the gradual introduction of flat-panel displays beginning in the 1990s.

CSCW   Computer-Supported Co-operative Working; original name for work begun in the 1970s which led to collaborative environments used, for example, in business, teaching and research. (*See also* VLE.)

CSIRT   Computer Security Incident Response Team (preferred term to CERT).

CSMA/CD   Carrier Sense Multiple Access with Collision Detect: technically descriptive term used in IEEE standards and elsewhere to refer to the Ethernet LAN technology developed at Xerox PARC in 1973 (Metcalfe and Boggs, 1976), subsequently developed and published as the DIX specifications and then standardised in IEEE 802.3. Briefly, in the original form of the protocol, a station attached to the shared cable medium senses the carrier to determine when the medium is free for it to transmit a frame addressed to one or more stations; the shared medium provides multiple access; stations listen to the medium continuously, both to receive frames from other stations and to detect collisions (own transmission corrupted); and if two stations transmit at near enough the same time to corrupt each others transmissions, both detect the 'collision' (and stop and retry). Subsequently developed to become switched Ethernet: *see* Chapter 5.

CSNET   Computer Science Network; US network originally funded by NSF to support computer science departments with no access to ARPANET. Proposed in 1979, it was initially established in 1981 as a phone-relay mail network but then, as ARPANET adopted TCP/IP, it connected to ARPANET and also used TCP/IP over X.25 to extend coverage. Eventually, it merged with BITNET in 1989 to form CREN (*q.v.*). The CSNET services were discontinued in 1991 because the continued growth of the Internet had by then rendered them superfluous.

D

DANTE   Not-for-profit company created by COSINE in 1993 and initially owned by RARE, as the permanent organisation responsible for international European networking for the academic, research and education sectors, beginning with EuropaNet. Ownership transferred to European NRENs in 1994. At the time of writing, it has also garnered responsibility for a number of international and intercontinental networks worldwide. The name of the company is derived from the acronym, Delivery of Advanced Networking Technology to Europe. *See* http://www.dante.net/, the DANTE website (accessed June 2010).

DARPA   US Defense Advanced Research Project Agency; formerly referred to as ARPA; agency which created ARPANET.

DAT   Digital Audio Tape; cassette magnetic tape packaged similarly to domestic analogue cassette tape, introduced in 1980s for computer use such as backup, with a capacity of around 60 MB.

DCPU   Data Communications Protocol Unit: created in 1978 by the DTI to promote OSI and to co-ordinate network protocol standards development in the UK; it ceased operation in January 1982.

DCT   Discrete Cosine Transform: mathematical analysis technique closely related to Fourier analysis, used in most of the widely deployed image (including motion picture) compression techniques.

DDoS   Distributed Denial of Service; a particular form of DoS (*q.v.*) achieved by orchestrating traffic or request generation from a (large) number of network hosts.

DEC   Digital Equipment Corporation; founded in 1957 and began making computers in 1960; later renamed Digital and subsequently acquired by Compaq in 1998. Manufacturer of PDP range, DEC10 (of which Tenex was a modified version) and VAX.

DECnet   DEC's suite of network protocols and accompanying implementations.

DELNI   Department for Employment and Learning, Northern Ireland: responsible for FE and HE in Northern Ireland. (Originally the Department for Further and Higher Education, Training and Development, Northern Ireland.)

DENI   Department of Education, Northern Ireland: responsible for school education in Northern Ireland, via five regional Education and Library Boards.

DERA   Defence Evaluation and Research Agency; *see* RSRE.

DES   1. Data Encryption Standard; developed by IBM in the early 1970s; selected and published by NBS in 1975; adopted as a FIPS in 1976. It was superseded for US Government use in 2002 by the Advanced Encryption Standard published by NIST in 2001. There was widespread unsubstantiated speculation at the time that the key used in DES was shortened at the behest of the US National Security Agency (NSA).

2. Department of Education and Science. Responsible for funding universities and the Research Councils at the time of the early days of JANET during the 1970s and 1980s.

DESY   Deutsches Elektronen-Synchrotron. Particle physics laboratory, located in Hamburg (accelerators) and Zeuthen in Brandenburg (neutrino studies). It supports international particle physics experiments.

DfES   Department for Education and Skills.

DIAL   Dual Interconnect Across London; the concept of providing dual paths (links and interconnection points) across London between JANET and its peer networks.

DiffServ   Differentiated Services: used to refer to both the IETF working group and the alternative model to IntServ (*q.v.*) for provision of multiple services with a range of QoS (*q.v.*).

DIX   DEC, Intel and Xerox: consortium, each with an interest in Ethernet and its exploitation, which published two versions of a *de facto* Ethernet standard, the second commonly referred to as the DIX standard. Ethernet had been invented at Xerox PARC; Intel had developed VLSI chip support; and DEC was committed to Ethernet for its VAX range of computers. (Radia Perlman, while at DEC, would subsequently invent the bridge spanning tree algorithm.)

DL   Daresbury Laboratory, situated near Runcorn, Cheshire. Formed in 1962 as part of NIRNS (*q.v.*). Operated as part of STFC (*q.v.*) since 2007. For the parts of its organisational history shared with RAL, *see* RAL.

DNS   Domain Name Service: developed at the end of 1970s by the ARPANET community, it came into widespread use in the early 1980s as the architecture evolved to that of the Internet. Also used to mean Domain Name Server, an instance of the set of distributed servers which implement the service; or Domain Name System, both the system concept and the worldwide system implementation.

DoD   US Department of Defense; funded DARPA.

DoDAG   Department of Defense Advisory Group; curiously named group which advised on the feasibility of JANET mounting an IP pilot network over X.25 in 1990.

DOS   Disk Operating System.

DoS   Denial of Service: term used to describe malicious network activity with intent to prevent service to others from the network or one or more hosts, usually by flooding the network with traffic or a service with requests.

DQDB   Distributed Queue Dual Bus; IEEE 802.6 name for QPSX.

DRA   Defence Research Agency; *see* RSRE.

DRAL   Daresbury and Rutherford Appleton Laboratory. *See* RAL.

DREN   Defense Research and Engineering Network: a US Department of Defense network; a participant in StarLight (*q.v.*).

Dstl   Defence Science and Technology Laboratory; *see* RSRE.

DTI   Department of Trade and Industry.

duopoly   In 1983 the British Government ended BT's UK telecommunications monopoly, which had been inherited from the former Post Office. Competition was introduced but only from one other company, Mercury Telecommunications. This situation, which came to be referred to as a 'duopoly', lasted for 7 years, after which it was reviewed in 1990 and formally ended in 1991

duplex   Form of point-to-point transmission, channel or circuit in which signals or data may be sent in either direction.

DVD   Digital Versatile Disc: digital video recording format.

DWDM   Dense Wave Division Multiplexing; form of WDM in which a large number of wavelengths are transmitted over a fibre (several hundred at the time of writing). This requires the bandwidth of each signal to be narrow so that the wavelength separation between each can also be small with acceptably low interference – hence the terminology. This all requires higher performance equipment than for, say, 10 wavelengths; it took correspondingly longer to develop and costs more.

E

EARN   Former European Academic Research Network (1983–1994); like BITNET in USA, to which it connected, but managed and funded separately. Like BITNET, it operated using the proprietary protocols and subsystems (RSCS, NJE and JES) originally associated with job submission and retrieval used by IBM and IBM-compatible systems, but with expanded functionality to include email and file transfer; lines were initially sponsored by IBM. The UK was part of EARN, connected via the gateway at RAL. As it ceased to operate as a network, the organisation merged with RARE to form TERENA in 1994.

EBCDIC   Extended Binary Coded Decimal Interchange Code: character set definition and encoding promulgated by IBM in the early 1960s, and also used by ICL in the UK. The code had many variations including amongst different peripherals from IBM. The code was originally derived from the codes used for 80-column computer punched cards.

EBONE   European Backbone: IP backbone which came into operation in September 1992 (and ultimately closed in 2002, following the collapse of KPNWest which had taken it over).

ECMA   European Computer Manufacturers Association.

ECN   Engineering Computing Newsletter, published by RAL on behalf of SERC/EPSRC, mid-1970s to March 1996 (Issue 60) (accessed at http://www.chilton-computing.org.uk/acd/literature/newsletters/p001.htm, March 2010).

EDFA   Erbium Doped Fibre Amplifier.

EDINA   National data and information repository service to UK tertiary education, funded by JISC and based at the University of Edinburgh Information Services.

EEC   European Economic Community, predecessor of the European Union.

EIN   European Informatics Network. Initial agreement to proceed 1971; originally referred to as the 'COST 11 Project'; intergovernment trial network in Europe; began operation in 1976; Director, Derek Barber (NPL).

ELB Education and Library Board: school education in Northern Ireland is divided into five regions, each with an ELB responsible to DENI for provision in its region.

EMAS Edinburgh Multiple Access System: an interactive multi-access timesharing operating system developed at the University of Edinburgh in the early 1970s.

EMPB European Multi-Protocol Backbone; 2Mbps successor to 64Kbps IXI, offering both IP and X.25 service from September 1992. See IXI.

ENISA European Network and Information Security Agency; an EU body created in 2004 to advise member states, governments and the European Commission on information security.

EPSRC Engineering and Physical Sciences Research Council. Originally formed as the Science Research Council (SRC) by Royal Charter in 1965, taking over from the Department for Scientific and Industrial Research (DSIR) the physical sciences research part of its remit, and with it NIRNS (*q.v.*). With the broadening of the SRC remit in 1981 it became the Science and Engineering Research Council (SERC). In 1994 there was a wider rearrangement of the UK Research Councils and their remits. In particular particle physics, along with astrophysics and astronomy, were split off from SERC to form PPARC (*q.v.*), and SERC became EPSRC. A constituent of RCUK.

EPSS Experimental Packet Switched Service; operated by the British Post Office; the intention to mount such an experiment was announced in 1971; the definition and scope of the experiment was defined in 1975 (PO Technical Guide 17) when participation was solicited; it became operational in 1977; and it closed in 1981 after PSS (*q.v.*) became fully operational.

ERCC Edinburgh Regional Computer Centre, Edinburgh University.

ERCIM The European Research Consortium for Informatics and Mathematics. Generally consisted of a single member laboratory from each nation, though in at least one case this was recursive, in that the national member was itself a consortium. ERCIM was involved in QUANTUM.

ESIG European SMDS Interest Group; formed in October 1991 to promulgate standards and interworking for European SMDS.

ESnet Energy Sciences Network: US Department of Energy network; a participant in StarLight (*q.v.*).

ESRC Economic and Social Research Council, originally formed in 1965 as the Social Sciences Research Council (SSRC), it was renamed the Economic and Social Research Council in 1984 to reflect its expanded remit. A constituent of RCUK.

ETH Eidgenössische Technische Hochschule, Zürich.

Ethernet *See* CSMA/CD and Chapter 5.

ETSI European Telecommunications Standards Institute; formed in 1988 by CEPT (*q.v.*), it undertakes European-oriented standards work, particularly in the mobile and cordless telephone areas. Early work also included use of European PDH for SMDS and ATM.

EU European Union.

EUCS Edinburgh University Computing Services.

EUNET European arm of USENET (*q.v.*).

EUREKA European intergovernmental initiative consisting of 39 full member countries at the time of writing; founded in 1985, it 'is a pan-European framework for research and development through which industry and research organisations from European countries and the European Union develop and exploit the technologies crucial to global competitiveness' (from EC Cordis archive). It was the framework vehicle for the European networking project COSINE (*q.v.*). (The name is not an acronym: simply Archimedes' famous exclamation 'I found it!', on the occasion of discovering a method of assessing the purity of gold.)

EuropaNet Eventual name, in 1992, for IXI (*q.v.*). *See also* DANTE.

EVN European VLBI (*q.v.*) Network.

EVO Enabling Virtual Organisations; developed at Caltech and released in 2007 as successor to VRVS (*q.v.*).

F

Fawn Book *See* SSMP.

FCC Former Facilities Committee for Computing: joint Research Council committee (under the aegis of SERC) with responsibility for provision and allocation of major computing facilities (primarily those at DL and RL) in support of the Research Councils' programmes. Met jointly with the Computer Board regarding the formation of JANET. (Also, more commonly, the Federal Communications Commission, the US federal agency responsible for US communication regulation.)

FCFS First Come, First Served: a self-explanatory form of scheduling used in packet network switching or routing, in which packets are transmitted onto an output link in the order in which they arrive. (Also known as FIFO scheduling.)

FDDI Fiber Distributed Data Interface: 100Mbps token ring LAN.

FDDI-II Successor to FDDI which added isochronous circuit facilities to FDDI; partially standardised but never manufactured.

FE Further Education.

FEFCW Further Education Funding Council for Wales; it was responsible for funding Further Education and Sixth Form Colleges in Wales from 1992 until 2001.

FFOL Fiber Follow-On LAN; projected successor to FDDI-II; never developed.

FIFO First In, First Out; alternative name for FCFS (*q.v.*) scheduling discipline.

file transfer Activity of or ability for transferring a file from one system to another. Typically applied to software to do this, a service offered by a host system, or a network protocol to achieve it.

FIPS Federal Information Processing Standards: collection of standards developed or selected by the US NBS, subsequently NIST, including in particular a set of recommended (international) networking standards.

FIRST (World-wide) Forum for Incident Response Teams.

freeware Software free to download off the Internet; includes shareware and free versions of software for which chargeable versions are available that typically include more facilities as well as support. The Linux version of Unix, several anti-virus programs, Apache Web server and Sendmail mail transfer agent are examples at the time of writing.

FTAM ISO File Transfer and Access Management. NIFTP and its associated experience formed input to the development of FTAM in the early 1980s. FTAM encompassed both file transfer and remote file access. However, publication by Sun of its NFS (*q.v.*), coupled with adoption of IETF protocols (which included FTP), both in the early 1980s, would eventually eclipse FTAM.

FTP File Transfer Protocol; application protocol to enable files to be transferred across a data network, typically based on the existence of a supporting reliable transport protocol. The protocol might also include facilities for file management and the transfer of whole directories (folders) of files. (Also used to refer specifically to ARPANET/Internet FTP.)

full-duplex Form of duplex (*q.v.*) channel in which data may be transmitted in both directions simultaneously.

G

gateway Generally, an interconnection between two networks, probably with some element of control, such as authorisation. Historically, also used in the US to mean a router (*q.v.*), and in the UK to mean a (transport service) relay (*q.v.*).

GCHQ Government Communications Headquarters, Cheltenham, UK.

GÉANT Gigabit European Academic Network: the European NRENs international interconnect network managed by DANTE. The original began service in December 2001; it was succeeded by GÉANT2 in 2004; and the third incarnation obtained funding approval in 2008 and was launched in 2009. All three have been funded jointly by the European Commission (under successive Framework programmes: V, 6 and 7) and the European NRENs. (The acute accent in the spelling of the acronym derives from the French for 'giant'.)

GÉANT2 *See* GÉANT.

GEC General Electric Company; by the 1970s, GEC was a substantial holding company in the UK, which had been formed by mergers and acquisitions of companies in the electrical, electronic and associated engineering industries, including Associated Electrical Industries in 1967 and English Electric (which included Elliot Brothers and Marconi) in 1968. Of relevance to JANET was GEC Computers (formerly Elliot Automation) which developed the 4000 series of minicomputers in the 1970s, one of the systems used as an ICF MUM and then as the first JANET PSE. GEC subsequently acquired ownership of Plessey (with Siemens) and later formed GPT (*q.v.*).

Glow Scottish schools nationwide education infrastructure, based upon 32 local authority schools networks linked together via JANET with each other and 5 administrative and educational content provision sites. Glow is a combination of application overlay and upgraded version of the previous facility known as SSDN (*q.v.*), which it superseded in 2006. The basic interconnect infrastructure is undergoing upgrading (known as Interconnect 2.0) as of early 2010. The education application overlay is being provided by Research Machines, RM, on behalf of Learning and Teaching Scotland.

GMPLS Generalised MPLS; partially defined technique for switching packets according to some tag or label which associates a packet with a concatenated sequence of channels across a TDM or WDM network. Includes an associated signalling protocol. *See* MPLS.

GOSIP Government Open Systems Interconnection Profile: profiles specifying use of OSI standards; specifications were developed in the UK by the CCTA during 1986–88; the US GOSIP was first published by NIST (*q.v.*) in 1990 (as FIPS 146-1) and was mandatory for US Federal Government purchases until 1995.

GPO General Post Office, the public organisation in the UK which operated the postal and telephone services until deregulation in 1980.

GPT Originally GEC (*q.v.*) Plessey Telecommunications; created in 1988 by GEC & Plessey merging their telecommunications operations; in 1989, GEC and Siemens acquired Plessey, and GPT ceased to be an acronym (as Plessey no longer existed) and became the name of the company. Of relevant to JANET, GPT took over manufacture of the GEC X.25 switches during the 1980s and supplied JANET with H.261 codecs and MCUs during the early stages of SuperJANET.

Green Book Document specifying terminal protocols triple X and TS29 adopted for use over JANET; *see* Chapter 3 and Appendix E.

Grey Book JNT mail protocol; *see* Chapter 3 and Appendix E.

H

H.261 Motion picture or video compression standard published by CCITT in 1990.

H.263 Successor to H.261, with improved features, including being more suitable for use over networks (such as packet networks) with greater variation in delay than that offered by circuit networks.

H.323 Umbrella standard for real-time interactive working; includes session control (signalling), IETF RTP (*q.v.*) and use of specific video and audio compression standards. Effectively ITU-T/ISO competitor to IETF SIP.

hacker A person who gains unauthorised entry to a computer system or network. Although in use earlier – with various meanings – by the 1970s, in computer circles, 'hacking' was most often used to describe altering existing program code to perform better in some way, to add features or to create a new program. By the early 1980s, with the widespread appearance of networked systems, the challenge of accessing a system while avoiding access control evolved from a sport for the curious, through mischief-making, malicious damage to systems and data, to outright criminal activity – and acquired its pejorative flavour. Whether the term owes its origin to perpetrators having intimate knowledge of software, harking back to its earlier meaning, perhaps coupled with the metaphor of penetrating or cutting through defences, or to some other usage, its current meaning was well-established by the mid-1980s.

half-duplex Form of duplex (*q.v.*) channel in which data may be transmitted in only one direction at a time.

HASP Houston Automatic Spooling Program; job queue management subsystem on IBM OS/360 systems, which also handled RJE using a proprietary multiplexing (termed 'multileaving' in IBM jargon of the time) application protocol over a BSC link layer. The name derives from the program having been created by an IBM group specifically for its customer at NASA Houston in 1965 (Braden, 1973).

HE Higher Education.

HEAnet Ireland's NREN; established in 1983 through collaboration of Irish HE institutions with support from HEA, the Higher Education Authority. Since 1997 it has been a company limited by guarantee and now serves the whole of the Irish education and research sector.

HEFC Higher Education Funding Council. There are three such, one each for England, Wales and Scotland, which took over from the UFC and PCFC responsibility for funding universities at the time when polytechnics had become universities. *See* HEFCE, HEFCW, and SHEFC/SFC.

HEFCE Higher Education Funding Council for England: responsible for funding higher education in England since 1993.

HEFCW Higher Education Funding Council for Wales: responsible for funding higher education in Wales since 1993.

HEI Higher Education Institute: generic term for a university, polytechnic, Research Council laboratory, or any other establishment primarily associated with higher education.

HEP High energy physics; synonym for particle physics; constituent of the acronym HEPnet (*q.v.*). The term arises from the fact that although the particles concerned are minuscule, performing experiments with them to find out more about them requires considerable amounts of energy.

HEPnet Not so much a network as an international community, the high energy or particle physics community. The need to collaborate on large experiments conducted at a few international facilities around the world ensured that networking was important to the community as soon as it was available, as a result of which it has frequently had a strong presence in various forums to ensure some means were available to meet the requirements of its programme. Two developments of much wider significance have their origins in this community: the World Wide Web and Grid computing.

HSSI High Speed Serial Interface; industry standard of the 1990s used by routers to access, for example, PNO PDH links operating at T3 (45Mbps) and E3 (34Mbps), and SMDS.

HUMBUL Humanities Bulletin Board system, originally developed and provided as a service by the University of Leicester, using USERBUL, the software created by Leicester for managing bulletin boards.

I

IANA Internet Assigned Number Authority; originally Jon Postel at University of Southern California Information Sciences Institute, now an independent not-for-profit organisation.

IBM International Business Machines. Manufacturer of large, medium and small computers primarily aimed at business data processing, including in the early 1980s the original PC (personal computer). The 1960s saw the development of the 360 architecture by Gene Amdahl, who later formed his own company to continue development and production of this architecture in competition with IBM.

IBM 1130 1960s computer having 8, 16 or 32Kbytes of core memory, with a simple disk operating system; could be used as a stand-alone computer or as an RJE station for an IBM 360 system.

IBM 2780 Basic RJE station, consisting of card reader and lineprinter for use with an IBM 360 system.

ICF Former Interactive Computing Facility: it provided interactive computing facilities to support engineering, especially computer-aided design (CAD), by means of centralised facilities based on DEC10 systems at Manchester and Edinburgh and localised facilities at selected universities. It included network support via SRCnet of the multi-user mini-computers (colloquially known as multi-user minis or MUMs: GEC 4000 series machines and Primes) out of which the localised facilities were constructed.

ICFC Interactive Computing Facility Committee: a former SRC committee with responsibility for funding ICF provision.

ICL International Computers Ltd: UK manufacturer of computers; supplied many of those used in UK universities in 1960s, 1970s, and 1980s, particularly its 1900 and 2900 ranges.

ICL 7020 RJE terminal originally for use with ICL's 1900 series systems.

ICMP Internet Control and Management Protocol.

ICS Abbreviated acronym for ULICS (*q.v.*).

ICT Information and Communication Technology; has partially overtaken usage of the term 'IT' in recent years.

IEEE Institute of Electrical and Electronics Engineers; responsible in the USA for many standards in this sector of the engineering industry; the IEEE 802 committee is responsible for standardising most LAN-related communications technology; many of the more successful have become ISO standards with (virtually) no change (except for a small change in the reference number). Effectively an international standards body in the 802 context and it correspondingly has international participation. Possibly the most notable exception in the LAN standards context was FDDI, which was an ANSI standard (before becoming an ISO one); slotted rings were also an exception, typically going directly to ISO.

IETF Internet Engineering Task Force: the organisation responsible for developing and specifying the protocols and engineering of the Internet.

IGMP   Internet Group Management Protocol: the control and management protocol for IPv4 multicast; separate from ICMP in IPv4 because multicast was added afterwards, it was incorporated into ICMP for IPv6 in 1998.

IMP   Interface Message Processor: network switching node in the ARPANET. It interfaced with other IMPs to implement the packet network, as well as providing host attachment to the network.

INRIA   Institut National de Recherche en Informatique et en Automatique; leading institution for computer science and network research in France; formed from IRIA in 1979.

Internet2   Originally title of UCAID (*q.v.*) project to develop next generation Internet technology, begun in the mid-1990s; subsequently, the name replaced UCAID as the name of the organisation.

IntServ   Integrated Services: refers to both the working group of the IETF and the Internet model architecture for integrated services published in 1994.

IP   Internet Protocol: the fundamental network datagram packet protocol of the Internet; specified by the IETF. It currently exists in two versions: IPv4 and IPv6.

IPv4   The version of IP defined in 1978 and adopted by ARPANET in 1982 to become the familiar protocol of the Internet; currently the major version in use on the Internet, though, as of 2010, IPv4 address space exhaustion is anticipated in the next couple of years or so. *See*, for example, http://www.potaroo.net/tools/ipv4/ (accessed June 2010).

IPv6   The version of IP ratified in 1998 (RFC 2460) as the future IP standard for the Internet, the major stimulus for its definition and introduction having been the looming exhaustion of IPv4 address space. Currently supported, together with IPv4, by the majority of routers and operating systems in common use; as of 2008, an estimate made by Google suggests that penetration of IPv6 is less than one percent (Gunderson, 2008).

IPSS   International Packet Switched Service; international X.25 packet network service; originally created in 1978 by a collaboration of the British Post Office, Western Union and Tymnet. Once PSS was available in the UK, it became the normal means of access to IPSS.

IPTV   Internet protocol television. Portmanteau term typically used to cover delivery of TV-style material over an IP-based network, whether the public Internet, a community network such as JANET or a private enterprise network. May include delivery of a spectrum of material, from real-time to on-demand delivery of near-real-time or archived material.

IRIA   Institut de Recherche en Informatique et Automatique, Rocquencourt, Paris; created in 1967, it became INRIA (*q.v.*) in December 1979.

IRTF   Internet Research Task Force.

ISC   Former Information Systems Committee: created in 1991 from the former Computer Board. *See* JISC.

ISDN   Integrated Services Digital Network; initial digital subscriber service offered by PTTs (standardised by CCITT) following digitisation of the trunk telephone network.

ISO   International Organisation for Standardisation.

ISP	Internet Service Provider.

IT	Information Technology; *see also* ICT.

ITAB	Information Technology Advisory Board; the senior committee responsible for coordinating joint funding of information technology research by DTI and SERC during the early 1990s.

ITP	Interactive Terminal Protocol; *see* Chapter 3 and Appendix E.

ITSU	Information Technology (or IT) Standards Unit; set up by the DTI, shortly after the DCPU ceased, to support and promulgate ISO, particularly OSI, standards in the UK.

ITT	Invitation to Tender: constituent stage of a formal procurement operation, inviting binding, costed proposals of goods and services to be supplied, to be submitted by a specified closing date in response to an OR (*q.v.*)..

ITU	International Telecommunications Union: organisation for harmonising the telecommunications industry; it is a specialised agency of the United Nations. Its standardisation activities are divided into sectors: that for telecommunications is known as ITU-T, which superseded the CCITT in 1992.

IVC	Instant Videoconferencing: a component of JVCS for which no booking is required; introduced in April 2005; *see also* CoD.

IWF	Internet Watch Foundation; UK industry consortium agency founded in 1996, with the primary mission of listing offensive and obscene material on the Internet, particularly in the context of children but also including incitement to racial hatred. The list is used for blocking sites by many UK ISPs and also the JANET Web Filtering Service. IWF was granted charitable status in 2004.

IXI	International X.25 Interconnect (sometimes given as Infrastructure); the European international interconnect facility, planned and organised by RARE under contract to the European Commission on behalf of COSINE. The service was initially provided by PTT Telecom (The Netherlands) in 1990 (pilot phase). Subsequently renamed EMPB (1991) and then EuropaNet (1992), it became the responsibility of DANTE in 1993.

J

JAMES	Joint ATM Experiment on European Services: two-year European PNO precompetitive market trial begun in April 1996, following the conclusion of the Pan-European ATM technology pilot (*see* Chapter 11, Section 11.2). The trial ended in March 1998.

JANET	Originally, Joint Academic[1] Network, named for its origin as the UK academic – that is, university or higher education network – and synergistically for the name of the first secretary of the JNT. Now simply the name of the UK network for the whole of the education and research sector.

JANET(UK)	Organisation responsible for the operation and development of JANET; trading name of The JNT Association since June 2007 (replacing the original trading name UKERNA).

---

1   Academy: from the name, Academos, of the landowner of the groves in which the original Plato's academy met for around 900 years.

JCL Job control language: set of commands or directives originally for use in specifying and controlling the execution of batch jobs on the large mainframe computers of the 1960s and 1970s.

JCN Former JISC Committee for Networking: *see* ACN. The JCN ceased in July 2009, its role having been subsumed into the new JISC Infrastructure and Resources (JIR) committee.

JCSR JISC Committee for the Support of Research; created as a result of the Follett Report (2000), at the time of the UK e-Science programme.

JES Job Entry Subsystem: job management and RJE subsystem on later IBM System 370 systems.

JIPS JANET IP Service; originally named for the alternative to the main JANET service based on X.25.

JIR JISC Infrastructure and Resources committee; formed in 2009, it took over the remit of the JCN (*q.v.*) among others; it has responsibility for a wide range of shared information and communication services.

JISC Joint Information Systems Committee: created in 1993, subsuming the ISC and its function; responsible to HEFCE, HEFCW, LSC, SFC and DENI for advising on and providing information services to the further and higher education sectors of the UK.

JIVE Joint Institute for VLBI in Europe; situated in Dwingeloo, The Netherlands.

JNT Joint Network Team: formed in 1979; replaced by The JNT Association in 1994; responsible initially for realising JANET and subsequently for its organisation, development and operation following its formal inception in 1984.

JNT Association, The Formed in 1994; not-for-profit company limited by guarantee which is responsible for JANET, having taken over the role from the JNT; trading originally as UKERNA, trading name re-branded JANET(UK) (*q.v.*) in June 2007.

JPEG Joint Photographic Experts Group: joint ISO and ITU-T working group responsible for standardising still image compression.

JTM ISO Job Transfer and Management; the specification of the protocol and services was substantially based on the UK JTMP work.

JTMP Job Transfer and Manipulation Protocol; *see* Chapter 3 and Appendix E.

JVCS JANET Videoconferencing Service: general name used to refer to the service from around 1998 (as SJVCN was decommissioned and replaced first by ISDN, and then IP-based H.323).

K

KPMG Professional services consultancy; commissioned in 1990 by the Computer Board to evaluate and advise on the possible forms of legal entity which the JNT Association might take. (The firm was formed in 1987 by the merger of Peat Marwick International and Klynveld Main Goerdeler; the name encapsulates the names of its historical principal founders: Klynveld, Peat, Marwick and Goerdeler.)

L

LAN Local Area Network. Term originally coined in the 1970s to describe shared media packet network technologies which were confined to working in a limited area by imposition of timing relationships arising from signal propagation speed and physical dimension of the medium necessary to operation and performance of the medium access protocol. Examples include bus- and ring-based topologies using random, slotted or token access schemes, such as original Ethernet, token ring, FDDI, ATM ring, etc. Most LANs at the time of writing are based on the (full-duplex) switched form of Ethernet in which contention has been eliminated, which thereby also eliminates the original distance limitation.

LANE LAN Emulation: technology developed to provide IEEE 802-style LAN service over an ATM network.

LDP Label Distribution Protocol; *see* MPLS.

LBL Lawrence Berkeley Laboratory (also known as LBNL).

LBNL Lawrence Berkeley National Laboratory (also known as LBL).

LEA Local Education Authority; that part of local government responsible for schools in its area.

LED light-emitting diode.

LES LAN Extension Service; proprietary service developed in the late 1990s by some PNOs to provide point-to-point, wide-area Ethernet-like service. Never standardised: the precise definition of the service varied with provider; quite widely used by JANET regional networks from the late 1990s onwards; subsequently generally referred to as SHDS (*q.v.*).

LHC Large Hadron Collider, situated at CERN, near Geneva. The latest in a succession of accelerators to explore the behaviour of subatomic particles when interacting with each other at high energy – processes also relevant to the development of the universe immediately after its big bang creation.

LIN Location Independent Networking; in the context of JANET, the name given during development to the prototype service subsequently released in 2006 as the JANET Roaming Service.

LINX London Internet Exchange: organisation providing national and international interconnection amongst ISPs; it operates a number of exchanges situated at premises in London's Docklands, including Telehouse and Telecity. It was created in 1994, the outcome of an initiative by a number of ISPs including PIPEX, UKnet and Demon, together with UKERNA and BT, and operates as a non-profit company limited by guarantee, owned by its members.

LISTSERV In the mid-1980s, two systems for managing mailing lists were developed independently. The first was associated with BITNET and was developed in 1984. The second was developed by Eric Thomas in 1986 and it shortly replaced the system already in use on BITNET. L-Soft was formed in 1994 to market LISTSERV. It was chosen as the system for JISCmail in 2000, the service at RAL which succeeded the Mailbase service at Newcastle.

little-endian *See* 'big-endian'.

LLNW Lifelong Learning Network for Wales, serving schools, libraries and lifelong learning centres; formed the initial basis for the development of the PSBA Network for Wales, which superseded LLNW in 2008.

LMN London Metropolitan Network.

LNT London Network Team.

LSC Learning and Skills Council: main source of funds for further education in England.

LSE London School of Economics, one of the constituent institutions of the University of London.

LTS Learning and Teaching Scotland: responsible for Glow.

Lumen From the Latin for 'light', a unit of light intensity. In March 2007, JANET(UK) moved into new premises: the name of the building, Lumen House, is a reference to high-capacity communication being increasingly optical.

M

MAC Media ACcess; as in 'MAC protocol', the sub-layer of the link layer protocol originally associated with shared-media networks, often local wired or wireless, but also satellite-based; and 'MAC address', the address used in a MAC protocol.

malware malicious software.

MAN Metropolitan Area Network. Used in the JANET context to refer to a regional network; in this context, the usage is unrelated technically to the IEEE 802.6 standard of the same name.

MBone Multicast Backbone: IP overlay network implementing the IP multicast extensions of Deering (1989); initially constructed in 1991/92. The need for MBone support gradually reduced during SuperJANET4, as the native multicast service became more reliable.

MBS Managed Bandwidth Service; originally introduced in the context of packet networks, this typically referred to an allocation of transmission capacity over an end-to-end path; the extent to which it might be protected from other users infringing the capacity (or vice versa) depended upon the specifics of an individual service. A variety of techniques related to QoS and MPLS have been used to provide such services. Typically, such a service might be set up on a semi-permanent basis. More recently the term has also sometimes been applied to channel-based services. *See also* BOD.

MCU Multipoint Control Unit: video switch used in some videoconference architectures when more than two sites participate; coined at the time when circuit-oriented H.261 codecs were being deployed for videoconferencing services (public and private) over the public switched network.

MECCANO Multimedia Education and Conferencing Collaboration over ATM Networks and Others; UCL-led Framework V (Telematics for Research) project (1998–2000), third of the MICE (*q.v.*) trilogy developing desktop multimedia videoconferencing technology, parts of which formed the basis of Access Grid.

MERCI Multimedia European Research Conferencing Integration; UCL-led European project (1995–1997), second of the MICE (*q.v.*) trilogy developing desktop multimedia videoconferencing technology, parts of which formed the basis of Access Grid.

MHS Message Handling System; *see* X.400.

MICE Multimedia Integrated Conferencing for Europe; UCL-led project (1992–1995), first of a trilogy of European experiments developing desktop multimedia videoconferencing technology, initially based on the suite of programs (Vic, VAT, WB and SD), commonly referred to as the 'Mbone tools', originally developed at LBL; RAT and SDR, developed at UCL, subsequently replaced VAT and SD (respectively). Parts of the suite subsequently formed the basis of Access Grid.

MIT Massachusetts Institute of Technology, Cambridge. Home of Lincoln Laboratories, where much of the original ARPANET work was centred – though none of the first four nodes in 1969.

MJPEG Motion JPEG (*q.v.*).

MoD Ministry of Defence.

modem Modulator-demodulator: device which encodes a digital signal into an analogue one for sending over an analogue transmission path, and the reverse. An acoustic modem is used for operation over an audio transmission path. Apart from initial domestic broadband modems dating from before *c.*2005, most computer communication modems at the time of writing are built into equipment, usually as a single chip.

MPEG Moving Picture Experts Group: ISO working group responsible for motion picture or video compression.

MPLS Multiprotocol Label Switching; technique (stimulated in part by ATM switching) in which a small (4-byte) label header is prepended to an IP packet to enable the packet to be switched, on the basis of the label alone, along a predefined path through a series of label-switching routers. In effect, a point-to-point unidirectional tunnel or virtual path is implemented which need not follow the same routing between its end points as that defined by the routing tables in force in routers at any particular time. A signalling protocol is needed to manage (allocate, distribute and reclaim) the labels and associated paths: a specific Label Distribution Protocol (LDP) or RSVP (*q.v.*) have both been used for this purpose. More recently, the technique has been separated from its use in connection with IP networks and used in more general provision of channel-oriented packet (or packet-oriented channel) services.

MRC Medical Research Council, responsible for funding academic research in this sector. (The Medical Research Committee was established in 1913; it became the Medical Research Council by Royal Charter in 1920, with essentially the remit it enjoys currently.) A constituent of RCUK.

MRI Magnetic resonance imaging.

MS-DOS Microsoft's version of DOS, introduced for use on the IBM PC (*q.v.*) and compatible systems.

MTA Mail Transfer Agent; a system responsible for forwarding email.

MTS Michigan Terminal System: interactive multi-access timesharing system for IBM System/360 and 370 systems developed by the University of Michigan in the late 1960s; ran on various NUMAC IBM systems at Newcastle and Durham Universities.

multimode fibre Family of glass (or plastic) fibres for use in telecommunications transmission. Solving Maxwell's equations for electromagnetic propagation in a cylinder leads to a family of solutions, dependent on the geometry of the cylinder and the wavelength of the radiation. If the diameter of a glass fibre, compared to the wavelength of the light being used for transmission, is sufficiently small, there will be only one mode in which case the term single mode or monomode fibre is used. Modal distortion (or dispersion) occurs in multimode transmission because the signal being propagated is represented by a number of modes but the modes do not propagate at the same speed; as a consequence, the signal becomes increasingly spread out or distorted with distance. Glass fibre used for transmission is made of concentric cylinders of glass having different refractive indices; transmission takes place (predominantly) along the inner cylinder of glass, and is confined to the fibre by total internal reflection. Typical fibre has an outside diameter of 125μm; multimode fibre has an inner core diameter of typically 50 or 62.5μm; single-mode, less than about 10μm. (The wavelength of light used for transmission is typically in the range 600–1600nm, though use of slightly longer wavelengths is under development.)

MUM Multi-User Mini; term coined at the time of the SRC ICF in the mid-1970s to denote a small computer (by comparison with the mainframes of the time) which provided interactive multi-access facilities to a relatively small community of users, such as a research group or small department in a university.

N

N3 NHS network extant at the time of writing, provided under contract by BT since 2004 (2005 for NHS Scotland).

NAC Network Advisory Committee, formed in 1983 out of the NMC to advise the CB; renamed ACN (*q.v.*) in 1994.

NASA US National Aeronautics and Space Administration.

NASG Network Association Steering Group; formed in 1990 to advise the Computer Board (and subsequently the ISC) on the creation of the JNT Association.

NAT Network Address Translation; the process of remapping part of the IPv4 address space, most often into a single address. It enables address re-use and in this context has contributed to slowing the consumption of IPv4 addresses. It may also be used to hide a complete subnet from public access over the Internet, in which guise it has become popular in contributing to security as a form of firewall mechanism. As is evident, from a network architecture perspective it is non-transparent, so destroying a number of the properties of the IP architecture.

NBMA Non-broadcast multi-access: term used to describe a network used as layer 2 in support of IP which has no broadcast feature (unlike a shared-media network) and so cannot support ARP; example networks include ATM, X.25 and Frame Relay. IP routing over such a network also presents additional problems. (SMDS supports multicast and this may be used to support ARP, as exploited for JANET.) *See also* ROLC.

NBS US National Bureau of Standards: *see* NIST.

NCAR National Corporation for Atmospheric Research; situated in Boulder, Colorado, it was one of five supercomputer centres funded by NSF in the mid-1980s; though the NCAR supercomputer was for atmospheric research, access to all five centres was one of the important requirements which contributed to the establishment of the NSFNET backbone network during 1985/86.

NCC Network Control Centre, established at RAL in 1984 as part of the original operating structure for JANET; continued until JANET X.25 service closed in 1997, though increasingly vestigial once the JANET IP service (for which the NOSC (*q.v.*) had operational responsibility) began in 1991.

NCCN National Committee on Computer Networks, set up by the Department of Industry in summer 1976, chaired by Jack Howlett (Director, Atlas Laboratory), reported October 1978.

NCP Network Control Program: ARPANET host software which provided end-to-end, host-host network service, as well as implementing the host-IMP interface.

NERC Natural Environment Research Council, formed in 1965, responsible for funding academic research in every aspect of the environment, including Earth System Science. A constituent of RCUK.

NFS Network File System. This was a proprietary system developed by Sun for use with its range of LAN-based workstations, some of which were diskless. Although proprietary, the set of network protocols to support the system was published in the early 1980s and became the first and probably most widely adopted of such systems. Apart from its original Sun Unix derivative of BSD, it became widely available on other flavours of Unix and the client side was also available later for MS-DOS, the system to be found on early PCs.

NHS National Health Service. (SNHS, the Scottish National Health Service, is technically separate, sharing services with the NHS, but this distinction is not significant for JANET and is not made in the text.)

NHSnet NHS network prior to N3 (*q.v.*); provided by BT and Cable & Wireless.

NIFTP Network Independent File Transfer Protocol; *see* Chapter 3 and Appendix E.

NIRNS National Institute for Research in Nuclear Science. Formed in 1957, subsumed as part of SRC (*q.v.*) in 1965. *See also* DL and RAL.

NIS Network Information System: directory system developed by Sun (originally under the name Yellow Pages, subsequently changed to NIS to avoid infringing BT's rights in regard to its telephone directory product). Apart from providing name-to-address translation in support of a set of networked Sun workstations, it also provided some distributed system configuration information.

NISP Networked Information Services Project; funded at the University of Newcastle by the Computer Board to investigate the provision of information services over JANET.

NISS National Information on Software and Services; originally a joint project between the Universities of Bath and Southampton, begun in 1987, funded by the Computer Board, to catalogue and publish information on IT resources available in the academic community. Later, it developed Athens (*q.v.*). Some historical information on NISS is available via archives at the University of Kent: http://www.mu.jisc.ac.uk/servicedata/niss/ accessed February 2010.

GLOSSARY

NIST   US National Institute of Standards and Technology, a federal agency of the US Department of Commerce. Founded in 1901, it was known as the National Bureau of Standards until 1988.

NITS   Network Independent Transport Service: *see* Chapter 3 and Appendix E.

NJE   Network Job Entry. A host-host (peer-to-peer) version of the IBM RJE protocol.

NLR   National LambdaRail: US optical backbone network, offering wavelength, frame and packet services to researchers. NLR consortium formed in 2003.

NMC   Network Management Committee, formed in 1981 with David Hartley as chair, to advise the Computer Board and FCC on transition and initial organisation of JANET; in 1983 it became the (smaller) NAC advising the Computer Board.

NNTP   Network News Transfer Protocol.

NOC   Network Operations Centre. Ten of these were originally created and contracted in 1984 to operate the ten PSEs forming the initial JANET configuration. The PSEs at Swindon and Bidston were shortly thereafter discontinued; the remainder continued until the transition to IP. Thereafter they were effectively absorbed into the regional networks, the specific role in regard to X.25 disappearing as X.25 was discontinued.

NORDUnet   Nordic universities network; originated through discussions beginning in 1980; began operation 1988; serves Denmark, Finland, Iceland, Norway and Sweden. Its story is told in the book by Lehtisalo (2005).

NOSC   Network Operations and Service Centre, initially part of ULCC; created in 1990 at the time when the Shoestring project had demonstrated that mounting an IP service over the JANET X.25 infrastructure was feasible; responsible for the operation of the JANET IP Service, JIPS; evolved subsequently to take overall responsibility for the operation of JANET. Became part of UKERNA in 2006.

NPL   National Physical Laboratory, Teddington, England.

NREN   1. National Research and Education Network; generic term used to designate a national network primarily dedicated to the provision of communication infrastructure and services to the research and education sector(s) – like JANET. The term NRN (*q.v.*) is also sometimes used with the same meaning.

   2. NASA Research and Engineering Network.

NRN   National Research Network; *see* NREN.

NRS   Name Registration Scheme: eventually became the portmanteau term used when referring not only to the JANET Name Registration Scheme but also the system supporting it; the naming authority remained the JNT.

NSA   US National Security Agency.

NSF   National Science Foundation; primary US agency for funding academic science research; provided seed funding for CSNET and NSFnet.

NSFnet   National Science Foundation network; in 1984, soon after the ARPANET MILNET split, discussion began about the provision of a national IP backbone to support research generally in the USA (for which DARPA had no remit). The result was the NSFnet backbone which began operation in 1986. It was effectively superseded by Internet2 (*q.v.*) in 1998. See also **UCAID**.

NUMAC Northumbrian Universities Multiple Access Computer(s): joint computer service for Durham University and Newcastle University; originally provided by an IBM 360/67 at Newcastle, later various 370-compatible systems at both Durham and Newcastle, running MTS (*q.v.*), linked by NUNET.

NUNET Northumbrian Universities Network; early regional network, primarily linking Durham and Newcastle universities but also linked to RCOnet.

O

OASIS Organization for the Advancement of Structured Information Standards; industry consortium originally formed in 1993 called SGML Open, an association of vendors of tools based on SGML; changed to current name in 1998 as SGML was overtaken by XML developments.

open-source Term referring to software for which the source is published and freely available; *see also* shareware.

OR Operational Requirement: a constituent of a formal procurement, consisting of a precise statement of requirements, typically distinguishing mandatory, desirable, optional and informational elements, which forms part of an Invitation to Tender.

Orange Book UK standardised version of Cambridge ring protocols. Original was TSBSP (Ian Dallas, University of Kent); subsequently, in CR82 (John Larmouth, Salford University) the scope was widened to include the whole technology of the Cambridge Ring LAN, which was defined in ISO style and eventually published as ISO 8802-7; *see* Chapter 5 and Appendix E.

OSCRL Operating System (OS) Command and Response Language; ISO work item in the 1970s on the topic of standardised JCL for batch operating systems.

OSI Open Systems Interconnection: ISO framework for open (i.e., non-proprietary) standard network protocols and services.

OSI CR A Functional Standard for the use of OSI Protocols over Slotted Rings. *See* Appendix E.

OST Office of Science and Technology; responsible for Research Council funding since the early 1980s, initially as part of the Cabinet Office, currently as part of the DTI.

P

PA Pre-Allocated: DQDB managed slot allocation scheme used for isochronous circuit service.

PABX Private Automatic Branch Exchange: a telephone switch or exchange situated on an organisation's premises providing its internal telephone service and telephone connections to the public telephone network. (The less automated version which needed an operator was referred to as a Private Branch Exchange or PBX.)

packet Contiguous finite sequence of bytes, including control and addressing or routing information, transmitted as a single unit between two computer devices attached to a communications network; typical packet sizes range from a few bytes to a few thousand bytes.

packet switch Device with at least two input and two output channels which directs a packet received on any input channel to the output channel determined by the addressing or routing information contained in the packet. (Term coined by Donald Davies at NPL in 1965.)

PACX Private Automated Computer Exchange: a device somewhat similar to a PABX which provided an internal (circuit) switching function to enable interactive terminals in an organisation to select which of the organisation's computers to connect to. Since computers had multiple ports, this typically included contention resolution for a free port.

PAD Packet Assembler Disassembler. A device or program used to interface character-mode interactive terminals to a packet network. Most often used in the context of terminals operating to the X.3, X.28 and X.29 specifications in the context of an X.25 network.

PARC *See* Xerox PARC.

PBB Provider Backbone Bridge: form of Ethernet switch for use in a public carrier network; based on VLAN technology using VLAN MAC encapsulation to interconnect switches over the wide area; defined in IEEE802.1ah-2008.

PBB-TE Provider Backbone Bridge Traffic Engineering: form of Ethernet switch for use in a public carrier network, similar to PBB but omitting flooding, the learning bridge algorithm and the spanning tree algorithm. Although this results in a switch in which VLAN connectivity has to be configured from outside the switch, manually or otherwise, the configuration is predictable and static; in the absence of spanning-tree exploitation of alternative paths, resilience has to be provided by other means. Defined in IEEE802.1Qay-2009. Some of its connection-oriented features, as well as operation and management approach, are similar to SONET/SDH.

PC Personal Computer. Term originally used in describing the Xerox Alto in 1972; subsequently used to refer to microcomputers for single-user personal use and this use continues; since 1981, also used to refer to any clone of the microcomputer originally developed and introduced by IBM in that year. Originally based on 8-bit 8088 microprocessor, it was subsequently based on Intel 8086 and derivative 16-bit (x86) architectures; PCs of the latter variety are still based on processors conforming to the basics of this architecture.

PCFC Former Polytechnics and Colleges Funding Council, created in 1988. The remit for funding polytechnics was effectively subsumed into the various higher education funding councils when the latter came into being, replacing the UFC in 1993. Polytechnics became universities at the same time.

PDH Plesiochronous Digital Hierarchy; *see* Appendix G.

PDP Programmable Data Processor: name of DEC's first range of computers.

Peach Book UK standardised version of an ISO-oriented transport protocol for use over Cambridge Ring ; *see* Chapter 5 and Appendix E.

PES Public Expenditure Survey; procedure by which, at the time the original SuperJANET was being funded, the Government allocated its Departments their budgets for the next three years: the first year would be the following year's firm allocation; the subsequent years would be estimates, effectively the starting point for the following year's PES round. The PES took place in the autumn; the allocations would subsequently form the starting point for the Government Budget the following spring, the latter being the determination of the amount of public funds needed for the year.

PET   Positron emission tomography.

phishing   Term in field of computer network security for fraudulently attempting to gain personal information by various forms of (usually) email confidence trick. Information sought ranges from passwords and bank account details all the way to identity theft, most commonly for the purpose of financial theft. Origin of term not definitively established but certainly alludes to the expression 'fishing for information' and may be a semi-acronym derivative of 'password harvesting fishing'.

phreaking   Telephone version of phishing (*q.v.*), where a telephone conversation replaces, for example, email.

Pink Book   UK standardised version of an ISO-oriented transport protocol for use over Ethernet (CSMA/CD); *see* Chapter 5 and Appendix E.

PNO   Public Network Operator. A generic term for an organisation offering general communications services, which over time have included telegraph, telephone (including mobile), facsimile, Internet and lower-layer transmission facilities such as microwave, satellite, fibre and shares of these by means of frequency, wave or time division multiplexing. *See also* PTT.

PoS   Packet over SONET; also used to refer to the international version of the interface, Packet over SDH.

PPARC   Former Particle Physics and Astronomy Research Council, formed in 1994 at the same time as EPSRC, with responsibility for funding academic research in these subjects, which it took over from the former SERC. Merged with CCLRC to form STFC in 2007.

protocol   The fundamental procedures by which a network operates. Included are the formats of control messages and data, together with the purpose of each control message, the actions to be performed on its receipt and the state of communication after its transfer. Procedures relating to errors are also included.

PSBA   Public Sector Broadband Aggregation; aggregation, in this context, generally refers to government initiatives or policies to reduce costs by shared provision; more positively, it may also be used to encourage cross-sector collaboration in the public sector. A particular example is the PSBA Network in Wales.

PSE   Packet Switching Exchange: basically a packet switch; also occasionally used to denote the operational entity composed of a packet switch, its operational environment and its supporting staff. The original versions of the PSEs for JANET were based on higher-end models from GEC Computers Ltd range of 4000 series systems.

PSS   Packet Switched Service; BT's packet switched data service, which opened for trial use in 1980 following experience gained with EPSS. It became a full service in 1981. It was later re-branded as Packet Switch Stream, alongside several other 'Stream'-branded services.

PTT   1. Postal, Telegraph and Telephone; a PTT operator was the term typically applied to describe a government controlled organisation which historically provided all three services; the UK GPO was an example.

2. Push-to-Talk. A recent service at the time of writing, being introduced on mobile phones but offering a service akin to citizen band (CB) radio, much used by police, lorry drivers and, increasingly, others because it is much cheaper and the half-duplex mode of operation (from which the name is derived) hardly detracts in the context of the short transactional exchanges for which it is chiefly used.

PUP   PARC Universal Packet; format and, by derivation, architecture of Xerox PARC's Internet architecture, which influenced the concept and definition of IP as well as being the direct inspiration for the Novell Network system.

PVC   Permanent Virtual Circuit; a permanently allocated channel in a packet-switched network; X.25 and ATM networks are typical examples.

Q

QA   1. Queue-Arbitrated: DQDB slot allocation scheme used for asynchronous data service.

2. Quality Assurance: commonly used in a variety of contexts, including in reference to tests relating to quality of venue environment and configuration for sites using JVCS.

QMUL   Queen Mary, University of London. *See* QMW.

QMW   Queen Mary and Westfield, one of the colleges of the University of London. It was originally Queen Mary; subsequently, Westfield College was merged with it to form QMW. Later, the Westfield part in Hampstead was disposed of and the name reverted to Queen Mary, University of London (QMUL).

QoS   Quality of Service; over-worked acronym, but used here to apply specifically to techniques by which a single network may provide multiple transport services (with particular reference to delay, timing and error control) over a single bearer infrastructure.

QPSX   Queued Packet and Synchronous circuit eXchange. Also Australian company named for the technology. *See* under DQDB in Section 5.3.

QUANTUM   QUAlity Network Technology for User-oriented Multimedia; EU Framework IV-V project in 1997-98 which engineered the upgrade of TEN-34 to TEN-155.

QUB   Queen's University Belfast.

R

RA   Registration Authority; in the context of authentication, a legal entity which is trusted by a community to establish the authenticity of other legal entities.

RADIUS   Remote Authentication Dial In User Service; although originally developed in 1991 in the context of telephone dial-in, the protocol and service is applicable in a much more general context and was adopted by the IETF (RFC 2865).

RAE   Royal Aircraft Establishment.

RAL Rutherford Appleton Laboratory, situated near Chilton, originally in Berkshire, it has been in South Oxfordshire since 1974, following substantial rearrangement of county boundaries. Rutherford High Energy Laboratory was formed in 1957 as part of the National Institute for Research in Nuclear Science (NIRNS). It became the Rutherford Laboratory (RL) in 1974, recognising the diversification already under way to include other areas of physics in addition to high energy (particle) physics. The Atlas Laboratory, adjacent to RL, was merged with the latter in 1975. In 1979 the Appleton Laboratory, then situated at Slough, moved and merged with RL to form Rutherford and Appleton Laboratories for a year, before becoming Rutherford Appleton Laboratory in 1980. DL and RAL came under a single management in 1994, at a time of major reorganisation of the Research Councils, and adopted the acronym DRAL (*q.v.*) for a year. In 1995, the organisation negotiated status as a Research Council, the Laboratories becoming the Central Laboratories of the Research Councils (CLRC) under the Council for the Central Laboratories of the Research Councils (CCLRC). Then in 2007, the Laboratories became part of STFC (*q.v.*)

RARE Réseaux Associés pour la Recherche Européenne; association of European academic and research networking organisations, set up in 1986; became TERENA in 1994 when it merged with EARN.

RAT Reliable Audio Tool; audio component of LBL/UCL/MICE videoconferencing tools; written at UCL as successor to Vat, the original LBL component.

RBC Regional Broadband Consortium: regional collaboration of English LEAs, primarily for the purpose of procuring educational ICT support.

RCUK Research Councils UK, a strategic partnership of the UK Research Councils formed in 2002. As of 2010 the constituent councils were AHRC, BBSRC, EPSRC, ESRC, MRC, NERC and STFC. (*See under the individual acronyms for details of each.*)

Red Book *See* JTMP.

relay ISO OSI term relating to interconnection between two networks which use different protocols. Usually associated with a particular layer in the protocol stack at which the protocol is functionally mapped and above which protocols may operate (more or less) transparently across both networks. In the particular case of an 'application relay', none of the protocols operate transparently across both networks: instead, the application itself, for example file transfer, is mapped between the native protocols in each network.

RFC Request for Comments: generic title for a formal IETF document relating to the Internet, more particularly its protocols and operation; the collection of RFCs – approaching 6000 at the time of writing – 'Contains technical and organisational documents about the Internet, including the technical specifications and policy documents produced by the Internet Engineering Task Force', to quote the IETF RFC Editor page. (http://www.rfc-editor.org/, accessed June 2010. *See also* Forty Years of RFCs, RFC 5540, http://www.rfc-editor.org/rfc/rfc5540.txt, accessed June 2010.)

RFI Request for Information: a formal public document which may be used in one of the preliminary stages leading to a procurement.

RFP Request for Proposals: a public document sometimes issued at a preliminary stage in large procurements, to solicit interest from potential suppliers and to gain initial, formal indications of what each might be able to supply. This was used as the first stage in the formal process of procuring the original SuperJANET.

RHBNC Royal Holloway and Bedford New College; formed by the merger in 1985 of Royal Holloway and Bedford Colleges, both institutions of the University of London. The new institution was inaugurated under its new name in 1986. In 1992 the College adopted the shorter, working title of Royal Holloway, University of London, with the corresponding acronym form of RHUL when needed. The earlier acronym survives in JANET history in some of the diagrams of the network dating from the time of its currency

RHEL   Rutherford High Energy Laboratory. *See* RAL.

RIPE   Réseaux IP Européens; a consortium of European IP network operators, formed in 1989, primarily to share operational experience and pursue technical coordination. The RIPE Network Coordination Centre (NCC) was established in 1992, based on the RARE legal framework. Although not part of its initial remit, RIPE NCC took over responsibility for European IP address allocation in 1992.

RJE   Remote Job Entry, an expression originated by IBM to describe the capability for submission of a program to be run on a (large) remote computer (mainframe) and the retrieval of the resultant output remotely from the mainframe. An RJE station was the collection of equipment (card reader, lineprinter, console, modem and communications line) which provided such a service.

RL   Rutherford Laboratory. *See* RAL.

RM   Oxfordshire-based company specialising in ICT systems and services for education and learning worldwide, with a history of supplying the schools sector in the UK since 1977 (when it sold its first computer to a local education authority). Founded by Mike Fischer and Mike O'Regan in 1973 originally supplying hobbyist components; formerly known as Research Machines Ltd.

RN   Regional Network in JANET; originally referred to as a MAN.

roaming   Roaming refers to a network connection which moves. The term is used technically in two contexts, typically referred to as 'internal' and 'external'. A mobile device in operation, signed on to a particular wireless access point, may move sufficiently that the signal from another access point becomes stronger and the device changes its connection to use the more favourable access point: on the assumption that this is a rearrangement of the connection within the network of an organisation (an AS), this is termed internal roaming. A device which disconnects entirely and then reconnects to a different network run by a different operator (a different AS) is said to be roaming externally.

ROLC   Routing Over Large Clouds: name of an IETF Working Group formed in December 1993 to study the issues surrounding IP routing over non-broadcast multi-access (NBMA, *q.v.*) networks, such as SMDS, ATM, Frame Relay and X.25.

router   System which interconnects two or more (sub)nets of the Internet. Primary function is to examine the address of each packet received, decide which outgoing link to forward it on and send it on its way. In support of this, it communicates with other routers to maintain its knowledge of the best link on which to transmit packets with any given address.

routing   Action of deciding from a network address on which of several output links to transmit a packet when forwarding towards its destination. In a datagram or connectionless network, this is done independently for each packet; in a connection-oriented network, it is done once during connection setup and subsequent packets are effectively tagged to use the same route through the network.

RSA   Rivest-Shamir-Adleman algorithm for public key encryption (Rivest *et al.*, 1978).

RSC Regional Support Centre; based on the concept pioneered in Wales for supporting the FE sector, including Sixth Form Colleges, and later adopted by JISC, the RSCs provide support for JANET services to FE colleges in the UK.

RSCS Remote Spooling and Communications System; IBM protocol and subsystem on later 370 systems for supporting its peer-to-peer version of RJE (and superseding HASP, *q.v.*).

RSRE Former Royal Signals and Radar Establishment, Malvern, an establishment of the Ministry of Defence (MoD); it became part of the Defence Research Agency (DRA) in 1991, which in turn became the Defence Evaluation and Research Agency (DERA) in 1995. DERA was split in 2001, one part becoming QnetiQ, the other Dstl, the Defence Science and Technology Laboratory, an agency of the MoD. After forming a public private partnership in 2003 with equity investment from US-based Carlyle Group, QnetiQ was sold on the London stock market in 2006.

RSVP Resource Reservation Protocol: Internet signalling protocol originally defined in the mid-1990s for IntServ.

RTP Real Time Protocol: IETF Internet application protocol defined for transport of real-time audio-video continuous media.

S

ST-II Stream protocol proposed in 1990 as candidate for continuous media transmission over the Internet; sometimes referred to as IPv5 – from the fact that both the values of the packet format version and the protocol identifier are 5 – though it is not a version of IP.

SAML Security Assertion Markup Language; *de facto* standard to express assertions about authentication and authorisation; it is based on XML and was created by OASIS consortium, originating from work begun in 2001.

SAR Segmentation and Reassembly: process of 'chopping' up a packet into smaller segments to enable transport across a lower layer technology with a (maximum) frame, packet or cell size smaller than the packet for transport. For example, SAR schemes are defined for each of IP, IEEE802.6 and B-ISDN (the latter two having been harmonised to be the same).

SD Session Directory: component of original LBL desktop videoconferencing tools.

SDH Synchronous Digital Hierarchy: ITU-T telecommunications TDM standard, originally standardised by CCITT, ITU-T's predecessor, in 1986; similar to US SONET (*q.v.*). *See* Appendix G.

SDR Session directory component of LBL/UCL/MICE videoconferencing tools; written at UCL as successor to SD, the corresponding original LBL component.

SDSS Shibboleth Development and Support Services; project which developed prototype Shibboleth service for JANET community, funded by JISC at EDINA; predecessor to the UK Access Management Federation for Education and Research.

SERC *See* EPSRC.

Glossary 361

SERCnet X.25-based wide area network constructed (initially) by RL (including the Atlas Laboratory) and DL as part of the computing infrastructure to support SRC (later SERC) research programmes. Originally known as SRCnet, it became known as SERCnet at the time of the corresponding renaming of SRC in 1981. It later formed the basis of the wide area segment of JANET.

SFC Scottish Funding Council: full title, Scottish Further and Higher Education Funding Council, responsible for funding further and higher education in Scotland. Formed by merger of SHEFC (*q.v.*) and SFEFC (*q.v.*) in 2005.

SFEFC Former Scottish Further Education Funding Council; merged with SHEFC to form SFC (*q.v.*) in 2005.

SGML Standard Generalised Markup Language; an ISO standard for defining mark-up languages published in 1986; largely superseded by XML.

shareware Community-supported computer software available for downloading on the Internet free of charge, often on the basis of a public licence to protect intellectual property rights, frequently supported by sponsorship and typically open-source. Gnu C compiler, Linux operating system, Apache Web server, Sendmail mailserver, and the Mozilla family of Web browsers and mail clients are some of the better known examples.

Shibboleth A project and an open-source system for supporting identity management and access control for Web-based resources; the project is part of the Internet2 Middleware Initiative. See Chapter 13 and http://shibboleth.internet2.edu/ (accessed July 2010).

SHDS Short Haul Data Service: more recent (at the time of writing) term for the services which include those originally marketed as LES (*q.v.*).

SHEFC Scottish Higher Education Funding Council, formed in 1992 to take over funding of Scottish universities from UFC; merged with SHEFC (*q.v.*) to form SFC (*q.v.*) in 2005.

single-mode fibre also (formerly) known as monomode fibre; *see* 'multimode fibre'.

SIP Session Initiation Protocol; IETF framework protocol for real-time, interactive, unicast or multicast, session control (initiation, modification and termination); competitor to ITU-T H.323.

SJVCN SuperJANET Videoconferencing Network; 14-site, studio-based system, using H.261 codecs and MCUs over 2Mbps ATM circuit emulation, from 1994 until early 1998.

SLA Service Level Agreement; originally, contract-like agreement between non-legal entities; more recently, it has been used for the detailed performance specification in a contract or similar agreement – though this is now often specified in an SLD (*q.v.*).

SLAC Stanford Linear Accelerator Center.

SLD Service Level Description; detailed description of services being provided in a supplier-consumer relationship, whether it be public service entities (legal or informal) or in QoS (*q.v.*) agreements between network ASs (*q.v.*).

SMDS Switched Multi-megabit Data Service: defined by Bellcore in 1989/90 in Technical Advisory document TA-TSY-000772 (Issue 3, October 1989) and a subsequent Supplement 1, December 1990. Subscriber service access specification for a data service compatible with the IEEE 802.6 MAN standard but not requiring the latter technology for its implementation. Capable of operation over a variety of transmission systems.

SMS Short Message Service: the capability to send and receive text messages of up to 160 characters between mobile phones.

SMTP Simple Mail Transfer Protocol

SMVCN Scottish MANs Videoconferencing Network. From 1996 until 2002, the regional networks (MANs) in Scotland exploited ATM technology both as the basis of the regional networks and to form the interconnect amongst them. To support shared teaching, the SMVCN was created over this, using M-JPEG codecs.

sniffing Observing packets, including part or all of the content, traversing a (shared) network; used for traffic or QoS engineering, fault diagnosis, development, teaching and malicious reading of other people's communications.

SONET Synchronous Optical Network; US telecommunications TDM standard, primarily intended for use over optical fibre networks; originally standardised by ANSI in 1985; SDH (*q.v.*) is the corresponding, similar ITU-T international standard. *See* Appendix G.

spam Common term for UBE (*q.v.*), though strictly it includes essentially all electronic methods of distributing unsolicited or unwanted messages. Although descriptive and colourful acronym interpretations abound, perhaps the leading candidate for the explanation of the origin of the term lies in the repeated use in a sketch from the 1970 BBC programme *Monty Python's Flying Circus* of the name SPAM, a form of cheap, tinned luncheon meat ubiquitous in post-war Britain in the 1950s and 60s. (More details may be found, for example, in the entry on electronic spam in Wikipedia: http://en.wikipedia.org/wiki/Electronic_spam (accessed June 2010).)

SPAN Space Physics Analysis Network; DECnet-based network, controlled primarily out of NASA, originated in 1981; in active use throughout the 1980s.

Spark Scottish Schools Broadband Project which initiated SSDN (*q.v.*).

SRC *See* EPSRC.

SRCnet *See* SERCnet.

SRI Stanford Research Institute. Founded in 1946, it later became independent of Stanford University and adopted the name SRI International.

SSDN Former Scottish Schools Digital Network; came into operation in 2003; superseded by GLOW (*q.v.*).

SSMP Simple Screen Management Protocol. *See* Chapter 3 and Appendix E.

SSRC *See* ESRC.

STAR TAP Science, Technology, And Research Transit Access Point: situated in Chicago, funded by NSF, it provides an interconnect facility for international research use of optical networking.

StarLight Collection of US networks participating in worldwide pilot optical networking via the optical interconnect in Chicago (STAR TAP); includes Abilene/Internet2, NLR, and ESnet, amongst others.

Starlink Former interactive computing facilities in support of SRC astrophysics and astronomy research programme, originally based on Digital Equipment Corporation (DEC) VAX systems. Included wide area network support originally based on DECnet, DEC's proprietary equipment and protocol suite.

STFC Science and Technology Facilities Council, formed in 2007 as a result of merging PPARC and CCLRC, responsible for operation of and access to major scientific facilities, nationally and internationally, and for funding research in particle physics, nuclear physics (which it took over from EPSRC) and astrophysics / astronomy. It is a member of RCUK. DL and RAL, laboratories of particular relevance to the history of JANET and previously the responsibility of CCLRC, it became the responsibility of STFC in 2007.

SURFnet The Netherlands' NREN.

SVC Switched Virtual Circuit; a dynamically allocated channel in a packet-switched network, set up using signalling.

SWURCC South West Universities Regional Computer Centre.

T

TACS Terminal Access Conversion Service in JANET which provided conversion between the Internet terminal protocol, TELNET, and X.29.

TAT-8 8th Transatlantic Telephone Cable; constructed by consortium led by AT&T, BT and France Telecom; brought into operation in 1988. The first cable to use optical fibre; decommissioned in 2002.

TAU Former Technical Advisory Unit; unit set up and contracted out to the University of Kent by JISC in 1993 at the time of the creation of UKERNA to provide technical advice to JISC and monitor the performance of UKERNA against its SLA with JISC in respect of JANET services. The TAU drafted the initial version of this SLA on behalf of JISC. Since subsumed as part of the JISC Monitoring Unit (still contracted out to University of Kent as of 2009).

TAXI Transparent Asynchronous Xmitter-Receiver Interface, where 'xmitter' stands for 'transmitter'. Part of the trade name of a chipset, TAXIchip, manufactured by AMD (Advanced Micro Devices), implementing a transmission scheme commonly referred to as 'TAXI', used in FDDI, 100BaseFX (100Mbps fibre Ethernet) and 100Mbps local transmission scheme over fibre for ATM (defined by the ATM Forum).

TCP Transmission Control Protocol: reliable, ordered, byte-stream transport protocol of the Internet; operates over IP.

TDM Time Division Multiplexing; method of sharing a transmission medium amongst a group of synchronous (digital) signal streams by allowing each stream a regular turn, usually for a fixed length of time and in round-robin fashion.

TELNET ARPANET interactive terminal protocol, inherited by Internet. Supported as one of the services provided by the UCL UK-US gateway from 1973 until 1989, when the gateway service for JANET transferred to ULCC. The need for a gateway ceased as JIPS began in 1991, when JANET joined the Internet.

TEN-34 Trans-European Network at 34Mbps; EU ACTS project which provided European international NREN interconnection at a nominal 34Mbps; proposed in 1995, service began in 1997. The service was succeeded by the 155Mbps service, TEN-155, in 1998.

TEN-155 Trans-European Network at 155Mbps; EU Framework V successor to TEN-34 which provided European international NREN interconnection initially at a nominal 155Mbps; proposed in 1997, service began in 1998. The service was subsequently upgraded to 622Mbps in 2000 and succeeded by GÉANT in 2001.

Tenex Operating system developed at BBN during the late 1960s and beginning of the 1970s, ultimately on a particular version of DECsystem-10 hardware which had a paging drum and virtual memory mapping hardware. Stimulated originally by LISP's run-time requirement for large memory. Ultimately drew on Multics and Berkeley timesharing system experience. DEC TOPS20 was a direct descendant. Many of the early hosts on the ARPANET were Tenex systems.

TERENA Trans-European Research and Education Networking Association; formed by a merger of RARE and EARN in 1994. Provides forum for and organisation of international European technical networking discussion relating to standards, interworking, support, education and new services. (*See* http://www.terena.org/, the TERENA website (accessed June 2010).)

TIP Terminal IMP; as well as providing all the services of an ARPANET IMP (*q.v.*), a TIP also provided interactive terminal access to the ARPANET, either directly or via dial-in.

top-slicing The practice of taking a portion of a budget intended for devolved funding of a number of organisations and administering that portion centrally for a specific communal purpose (in the example exemplified by the Computer Board, originally high-end computer facilities, later national network facilities) for joint, communal benefit, rather than devolving all of the budget and leaving it up to the community of organisations to arrange provision (whether jointly or severally).

triple X Portmanteau expression for the set of three specifications, X.3, X.28 and X.29, published by the CCITT, defining how character-mode terminal access to an X.25 network is handled. Also sometimes written XXX.

TS29 Terminal protocol compatible with X.29 for use over Yellow Book Transport Service; defined as part of the Green Book; *see* Chapter 3 and Appendix E.

TSBSP Transport Service over Byte Stream Protocol; *see* Orange Book.

TVN Thames Valley Network: *see* Appendix B.

U

UBE Unsolicited bulk email: the electronic equivalent of paper post 'junk mail', i.e., email distributed to a huge set of recipients, often (mostly historically) with the aid of unsuspecting mail relays, without having been requested; also called spam (*q.v.*). The content and purpose typically includes advertising, offensive material, phishing (*q.v.*) and fraud of one sort or another.

UCAID   University Corporation for Advanced Internet Development; formed in 1996 and took over responsibility for the US academic and research internet backbone in 1998 (from NSFnet); it adopted Internet2 as the title for the project to develop the next generation Internet technology and, subsequently, this title replaced UCAID as the name of the organisation.

UCAS   Universities and Colleges Admissions Service: central organisation responsible for administering admissions to higher education institutions in the UK.

UCL   University College London.

UCCS   University of Cambridge Computing Service.

UDP   User Datagram Protocol: alternative transport protocol to TCP, for use over IP where a packet protocol with neither flow control nor error recovery are required.

UFC   Former Universities Funding Council. Formed in 1988, took over funding of UK universities from the former UGC. Superseded by higher education funding councils in 1993.

UGC   Former University Grants Committee, created in 1919 with responsibility for university funding. Superseded by UFC in 1988.

UKC   University of Kent at Canterbury.

UKERNA   United Kingdom Education and Research Networking Association: former trading name of The JNT Association; rebranded as JANET(UK) (*q.v.*) in June 2007.

ULA   Uncommitted logic array; a 'half-way house' to custom chip design around 1980. The silicon chip had a number of gate 'cells' embedded; however, their interconnection was undefined and could be added at a subsequent stage to order. At the time, such a chip contained less than 1000 gates. Some ULAs, like the Ferranti one referred to in the text, also included some analogue cells which could be used for driving low-level analogue circuitry.

ULCC   University of London Computer Centre.

ULICS   University of London Institute of Computer Science; closed in 1973, with transfer of some of its members to the new Department of Statistics and Computer Science at UCL; became the UCL Department of Computer Science in 1980.

UMIST   University of Manchester Institute of Science and Technology.

UMRCC   University of Manchester Regional Computing Centre.

Unix   Interactive timesharing operating system originally developed in the late 1960s at Bell Labs; since proliferated in many versions, including BSD and Linux.

Usenet   Network news service based on UUCP for mail and news forwarding. *See* Chapter 5.

USERBUL   Software system created by the University of Leicester in the 1980s to manage bulletin board systems. *See also* HUMBUL.

UT200   *See* CDC UT200.

UUCP   Unix-to-Unix CoPy (CP being the Unix command abbreviation for the Copy command): a program and protocol to copy data between directly linked ('adjacent') Unix systems. Also used collectively to denote a set of programs which could perform a number of command-like operations between adjacent Unix systems. Occasionally used to mean UUCPnet. UUCP was also implemented on a few other operating systems of the time.

UUCPnet   Informal network of (mostly Unix) systems supporting UUCP. In particular, UUCP systems would often forward each other's mail, creating a quite widespread informal mail network.

UWA   University of Western Australia, where QPSX was originally developed.

UWIST   University of Wales Institute of Science and Technology; merged with University College, Cardiff, to form the University of Cardiff in 1988.

V

Vat   Audio component of LBL/UCL/MICE videoconferencing tools, originally developed at LBL.

vBNS   Very-high-performance Backbone Network Service in the USA; originally developed by MCI in partnership with NSF for the academic research community in 1995 as an upgrade to the existing NSFNET backbone.

VCAS   Videoconferencing Advisory Service; created in 1998, this service was initially provided for JANET jointly by the University of Newcastle upon Tyne and UCL. The helpdesk was originally provided by Newcastle. *See also* VTAS.

VCEG   Video Coding Experts Group: ITU-T working group responsible for video or motion picture compression.

VCMC   Videoconferencing Management Centre: term adopted from about 1998 for the management service provided by Edinburgh University for JVCS, including booking and interworking (transcoding and switching) between various technologies.

VDU   Visual display unit based on a CRT; typically used in the expression 'VDU terminal' to denote an interactive computer terminal consisting of such a display and a keyboard. Encouraged by the advent of the flat screen display, the term is giving way to 'screen' colloquially.

VHS   Video Home System; analogue video tape recording standard introduced in 1976 for domestic use. There were several variations in format, the best being comparable with domestic analogue TV at the time.

Vic   Original video component of LBL/UCL/MICE videoconferencing tools developed at LBL.

virtual circuit   End-to-end path across a packet-switched network. The flow of data along a virtual circuit is conveyed by a sequence of packets, each containing the same value of virtual circuit identifier while traversing a given link; at each packet switch, the incoming link and virtual circuit identifier identifies the output link on which the packet should be transmitted, as well as a new value of the virtual circuit identifier to be used while traversing that link (so that virtual circuit identifier values are local to a link, avoiding the need for global values unique for every virtual circuit throughout the whole of a large network). The set of mappings defining the virtual circuit through the switches is set up before any data packets are transmitted, either by management (permanent virtual circuit) or dynamically by setup messages (switched virtual circuit); the virtual circuit is cleared in similar fashion.

VLAN Virtual LAN; feature of switched LAN technology which provides several 'virtual' LANs over a single, wired infrastructure, by using port, MAC address or IP address to define a set of ports or hosts forming the virtual LAN to which a host belongs. Enables the logical partitioning of a single physical layer 2 network into a number of virtual layer 2 networks, providing both performance and security advantages, as well as separating the organisational user workstation network groupings from the physical configuration of the network infrastructure.

VLBI Very Long Baseline Interferometry; *see* panel 'Communications for radio astronomy and particle physics' in Section 10.3.

VLE Virtual Learning Environment. A particular outcome of work in CSCW (*q.v.*). A VLE may include facilities such as managed discussion threads mediated by e-mail; automated multiple-choice quizzes as aids to revision, partial assessment, etc; managed submission of coursework; access to learning materials, including recorded lectures and demonstration sessions for practical skills; and real-time attendance at the latter via videoconferencing tools.

VoIP Voice over IP: Internet interactive speech application, similar to telephony.

VRVS Virtual Rooms Videoconferencing System; created at Caltech in 1997; multiple window videoconferencing system based on paradigm of shared (virtual) meeting space ('room'). Originally derivative from the LBL/UCL/MICE system. Exploits application level multicast using reflector nodes (similar in architecture to earlier CUSeeMe system developed at Cornell in 1992). (*See also* EVO and Access Grid.)

VSAT Very Small Aperture Terminal: a satellite terminal using a dish antenna of less than 3 metres, based on technology pioneered in the 1970s using geostationary satellites; came into use during the early 1980s, often for videoconferencing systems, as well as providing digital data links.

VTAS Video Technology Advisory Service: in 2000, VCAS (*q.v.*) changed its name to reflect the broader service it was providing with the growth of video content delivery.

VTP Virtual Terminal Protocol.

W

WAN Wide area network. JANET is an example.

WB White Board: interactive whiteboard component of LBL/UCL/MICE videoconferencing tools developed at LBL.

WDM Wave Division Multiplexing: method of sharing transmission capacity of optical fibre by transmitting signals on different wavelengths (frequencies, colours).

White Book Specification document for the proposed transition of JANET to OSI protocols. Not strictly part of the Coloured Book collection. *See* Appendix E.

Wi-Fi Also commonly written WiFi; popular term used to refer to IEEE802.11 wireless LAN technology. Origin stems from brand name, Wi-Fi Alliance, created in 1999, for trade group Wireless Ethernet Compatibility Alliance. (Wi-Fi has no technical meaning, nor is it an acronym or abbreviation – like 'hi-fi' for 'high fidelity' in relation to audio reproduction.)

WiMAX Commercial name, denoting Worldwide Interoperability for Microwave Access, incorporated into name of industry marketing consortium, WiMAX Alliance, and referring to IEEE802.16 wireless MAN technology.

WLAN Wireless LAN (local area network).

worm A program which executes on a network host to exploit incorrect system design or implementation (a 'vulnerability') in another host to enable its penetration from the network; it then transports itself to the new host and repeats the process on further systems. Apart from achieving penetration and propagation, execution of the program usually also achieves additional more or less malicious effects. The original Morris worm affected about 6000 Unix systems on the Internet, about a tenth of the total number of hosts then connected, and was estimated to have caused damage worth up to a $100 million. It also produced widespread indiscriminate denial of service through consumption of resources.

WREN Welsh Research and Education Network.

WVN Welsh Video Network; the successor to WelshNet, originally funded by HEFCW in 1998 and EU (2000). Initially (like its predecessor) based on ISDN, it began service in 2001 serving over 80 studios, including FE and HE; thereafter, increasingly based on IP using H.323.

WWW World Wide Web.

X

X.25 International network packet (interface) protocol originally used in JANET; first provisionally approved in October 1977, in accordance with Resolution No.2 of the VIth Plenary Assembly of the (former) CCITT (Geneva, 1976); first published as Provisional Recommendations (along with X.3, X.28 and X.28) in 1978 (ISBN 92-61-00591-8). JANET ceased all use of X.25 in 1997.

X.28 *See* triple X.

X.29 *See* triple X.

X.3 *See* triple X.

X.400 Suite of standards developed by CCITT in collaboration with ISO (where it is known as ISO 10021) for Message Handling Systems, the most common application of which is for email. Initial version was published in 1984, with substantial subsequent revision in 1988. For email, it never really competed successfully with the earlier Internet standard – probably not helped by the clumsy, keyword-oriented syntax in which a user's email address was expressed. It has seen use in military and Electronic Data Interchange (EDI) applications.

X.509 ITU-T standard relating to infrastructures based on public keys for privilege management and single sign-on; defines certificate and concept of certificate authority for use in this context. Original version of the standard published by CCITT in 1988.

Xerox PARC Xerox Palo Alto Research Centre, Stanford, California.

XML Extensible Markup Language, for encoding electronic documents. Originally conceived in the context of the World Wide Web, its extensibility has enabled the creation of a myriad of specialised mark-up languages, in contexts not necessarily related to Web usage.

XOT X.25 over TCP/IP; router software support for this overlay.

XXX *See* triple X.

Y

YBTS Yellow Book Transport Service: *see* NITS.

Yellow Book *See* NITS.

Sources

Listed here are the major sources used in compiling this history. In addition, a number of working papers, reports etc. were also consulted; where these were major reports and received substantial circulation at the time, they have been referenced in the normal way. There is a wealth of background information available on the Internet, usually enough to cross-check most topics with several sites. Only a selection of the more complete or authoritative is included – here and in the footnotes and references. I am grateful to JANET(UK) for access to a number of papers in its collection.

- **JANET News (#1, September 2007, continuing).** Published by JANET(UK), available from it or online at http://www.ja.net.

- **Network News (#1, May 1977 to #44, June 1995).** I am grateful to Roland Rosner for the loan of his nearly complete set; and to Peter Linington and Shirley Wood for filling in most of the gaps.

- **Networkshop Reports (September 1977, continuing).** Many of the more recent are available online at http://www.ja.net. I am grateful to Phil Harrison and Shirley Wood for excerpts from some older ones.

- **UKERNA News (#1, October 1996 to #39, June 2007).** Accessed at http://www.ukerna.ac.uk on or before June 2006.

- **NATO Advanced Study Institutes at Bonas in 1978 and 1981.** These have a number of papers which reflect the state of networking at the time. Individual papers have been referenced in the normal way; bibliographic details of the proceedings are included here. Again, I am indebted to Roland Rosner for the loan of his copies.

- *Interlinking of Computer Networks: Proceedings of the NATO Advanced Study Institute, held at Bonas, France, August 28 – September 8, 1978.* K.G. Beauchamp (ed.). D. Reidel Publishing Co. (1979) ISBN 90-277-0979-3.

- *New Advances in Distributed Computer Systems: Proceedings of the NATO Advanced Study Institute, held at Bonas, France, June 15–26, 1981.* K.G. Beauchamp (ed.). D. Reidel Publishing Co. (1982) ISBN 90-277-1368-5.

Internet pages

Funding bodies

- Apart from the sites associated with each of the higher education funding councils and Research Councils, the National Archives at Kew (http://www.nationalarchives.gov.uk/) and the HERO site (http://www.hero.ac.uk/uk/home/index.cfm) also proved useful in checking dates of background events.

- The Chilton computing site (http://www.chilton-computing.org.uk/) has information relating to the early days, including the Flowers Report, the Rutherford and Atlas Laboratories, the ICF, SRC, SRCnet.

- *Post Office / BT:* the BT site (http://www.btplc.com/Thegroup/BTsHistory/History.htm) includes an account of the privatisation of the Post Office, including early regulatory history.

REFERENCES

Abramson, N. (1973). The Aloha System. *Computer-Communication Networks*, Norman Abramson & Franklin F. Kuo (eds), 501–17. Prentice-Hall. ISBN 0-13-165431-4.

Barham, P., Dragovic, B., Fraser, K., Hand, S., Harris, T., Ho, A., Neugebauer, R., Pratt, I., and Warfield, A. (2003). Xen and the Art of Virtualization. *Proceedings of ACM SOSP'03, New York (October)*. (Accessed at http://www.cl.cam.ac.uk/research/srg/netos/papers/2003-xensosp.pdf, January 2010.)

Bartlett, K., Davies, D., Scantlebury, S., and Wilkinson, P. (1967). A digital communication network for computers giving rapid response to remote terminals. *ACM Symposium on Operating System Principles* (October).

Bellcore (1989). *Generic System Requirements in support of Switched Multi-megabit Data Service.* Technical Advisory TA-TSY-000772, Issue 3, Bell Communications Research (October).

Bellcore (1990). *Generic System Requirements in support of SMDS Service.* Technical Advisory TA-TSY-000772, Issue 3, October 1989, Supplement 1, Bell Communications Research (December).

Birrell, A.D., Levin, R., Schroeder, M.D., and Needham, R.M. (1982). Grapevine: an exercise in distributed computing. *Communications of the ACM*, **25**, No.4, 260–74 (April).

Bittau, A., Handley, M., and Lackey, J. (2006). The Final Nail in WEP's Coffin. *Proceedings of the 2006 IEEE Symposium on Security and Privicy*, pp.386–400. (Accessed via http://www.cs.ucl.ac.uk/staff/M.Handley/papers/ March 2009.)

Blake, S., Black, D., Carlson, M., Davies, E., Wang, Z., and Weiss, W. (1998). An Architecture for Differentiated Services. RFC 2475, IETF (December). (Accessed via http://www.ietf.org/rfc.html March 2009.)

Braden, R., Clark, D., and Shenker, S. (1994). Integrated Services in the Internet Architecture: an Overview. RFC 1633, IETF (June). (Accessed via http://www.ietf.org/rfc.html March 2009.)

Braden, R., Zhang, L., Berson, S., Herzog, S., and Jamin, S. (1997). Resource ReSerVation Protocol (RSVP) – Version 1 Functional Specification. RFC 2205, IETF (September). (Accessed via http://www.ietf.org/rfc.html March 2009.)

Brandon, J.P. (1978). METRONET. *Regional Network Activities*, session in *Networkshop 2*, 51–2, Liverpool (April).

Bryant, P.E. (1996). The Rise and Rise of SRCnet. *Engineering Computing Newsletter: Final Issue (60)* (ed. M.R.Jane). (Accessed at http://www.chilton-computing.org.uk/acd/literature/newsletters/p001.htm August 2008.)

Burr, W.E. (2001). Data Encryption Standard. *A Century of Excellence in Measurements, Standards, and Technology; A Chronicle of Selected Publications of NBS/NIST, 1901–2000*, (ed. D.R. Lide), 250–53. NIST SP 958. (Accessed at http://nvl.nist.gov/pub/nistpubs/sp958-lide/250-253.pdf November 2009.)

Burren, J.W., and Cooper, C.S. (1989). *Project UNIVERSE: An Experiment in High-Speed Computer Networking.* Oxford University Press. ISBN 0-19853723-9.

Carroll, L. (1871). *Through the Looking-Glass, and What Alice Found There.* Macmillan.

Cerf, V. (1978). *The Catenet Model for Internetworking.* Information Processing Techniques Office, Defense Advanced Research Projects Agency, IEN 48 (July). (Accessed at http://www.postel.org/ien/txt/ien48.txt January 2009.)

Cerf, V., and Kahn, R. (1974). A protocol for packet network internetworking. *IEEE Trans. Commun.*, **22**, 627–41 (May).

Cerf, V.G., and Kirstein, P.T. (1978). Issues in Packet-Network Interconnection (*Invited Paper*). *Proceedings of the IEEE*, **66**, No.11, 1386–1408 (November).

Chuang, S., Crowcroft, J., Hailes, S., Handley, M., Ismail, N., Lewis, D., and Wakeman, I. (1993). Multimedia application requirements for multicast communications services. *Proceedings of INET 93* (August).

Close, F. (2009). *Antimatter.* Oxford University Press. ISBN 978-0-19-957887-0.

CNRI (1996). *The Gigabit Testbed Initiative Final Report.* Corporation for National Research Initiatives, Cooperation Agreement NCR8919038 (December). (Accessed at http://www.cnri.reston.va.us/gigafr/index.html and http://www.cnri.reston.va.us/gigafr/Gigabit_Final_Rpt.pdf April 2009.)

Cohen, D. (1980). On Holy Wars and a Plea for Peace. IEN 137 (IETF). (Accessed at http://www.ietf.org/rfc/ien/ien137.txt December 2008.) *(An edited version of this note appeared in the October 1981 issue of the IEEE Computer Magazine.)*

Computer Board (1970). *Report of the Computer Board for the period 1st November, 1968 – 31st March, 1970.* Cmnd 4488, HMSO (February).

Computer Board (1973a). *Report of the Computer Board for the period 1st April, 1970 – 31st March 1972.* Cmnd 5220, HMSO (February).

Computer Board (1973b). *Computer Board visit to North America March/April 1973 – Conclusions and Recommendations.* Circulated with IUCC Newsletter **4** (June).

Computer Board (1991). *Structure and service proposals for the creation of a United Kingdom Education and Research Networking Association.* CB(91) REV4.1, 29/30 January 1991.

Conti, G. (2008). *Security Data Visualization.* No Starch Press. ISBN 978-1-59327-143-5.

Cook, M.S. (2008). Virtualisation and JANET. *JANET News*, No.3, 14 (March). (Accessed at http://www.ja.net/documents/publications/news/news-3.pdf January 2010.)

Cope, A., and Southard, D. (2009). A2. Location Accuracy and Awareness. *JANET Investigations in Location Awareness.* (Accessed at http://www.ja.net/documents/development/network-access/location-awareness/investigations/A2-location-accuracy-and-security.pdf January 2010.)

Cooper, C.S. (1991). High-speed networks: the emergence of technologies for multiservice support. *Computer Communications*, **14**, No.1, 27–43 (January/February).

Cooper, R. (1990). *From JANET to SuperJANET: The Development of a High Performance Network to Support UK Higher Education and Research.* Paper to the Computer Board for Universities and Research Councils.

Cormack, A. (2002). NHS/JANET Architectures – a discussion paper for UKERNA and NHSnet. PB/CERT/002 (02/11). (Accessed at http://www.webarchive.ja.net/services/publications/archive/reports/nhs-janet-architectures.pdf December 2009.)

COST (1971). Agreement on the establishment of a European informatics network. Brussels, November 1971. *United Nations Treaty Series*, Vol.1262, No.I–20750.

Davies, D.W. (2001). An Historical Study of the Beginnings of Packet Switching. *The Computer Journal*, **44**, No.3, 152–62. *(This paper was published posthumously at the author's request—see the* Introduction to Davies's paper *in the same issue of* The Computer Journal.*)*

Davies, D.W., and Barber, D.L.A. (1973). *Communication Networks for Computers.* John Wiley & Sons. ISBN 0-471-19874-9.

Davies, H., and Bressan, B. (*eds*) **(2010).** *A History of International Research Networking.* Wiley-Blackwell. ISBN 978-3-527-32710-2.

Davies, J.I. (1978). RCO. *Regional Network Activities,* session in *Networkshop 2,* 53–5, Liverpool (April).

Day, R.A. (1992). The JANET IP Service. *Networkshop 20,* Leeds (March).

De Arce, J.M., Cooper, C., Pallares, M., and Perkins, C. (1999). *TEN-155 Managed Bandwidth Service: Alpha Test Report.* QUANTUM MBS Alpha Test Report.

Deering, S. (1989). *Host Extensions for IP Multicasting.* RFC 1112, IETF (August).

Deibert, R.J., Palfrey, J.G., Rohozinski, R., and Zittrain, J. (2008). *Access Denied.* The MIT Press. ISBN 978-0-262-54196-1.

Dewis, I.G. (1978). The British Steel Corporation Network. *Interlinking of Computer Networks,* Proceedings of the NATO Advanced Study Institute, Bonas, France, 345–52. D. Reidel (1979). ISBN 90-277-0979-3.

Diffie, W., and Hellman, M.E. (1976). New Directions in Cryptography. *Communcations of the ACM,* **21**, No.2, pp.120–26. (Accessed at http://www.cs.rutgers.edu/~tdnguyen/classes/cs671/presentations/Arvind-NEWDIRS.pdf October 2009.)

Disney, M.J., and Wallace, P.T. (1982). Starlink. *Q. Jl. R. Astr. Soc,* **23**, 485–504. (Accessed at http://adsabs.harvard.edu/full/1982QJRAS..23..485D August 2008.)

DTI (1990). *Competition and Choice: Telecommunications Policy for the 1990s: A Consultative Document.* Presented to Parliamant by the Secretary of State for Trade and Industry by Command of Her Majesty (November). Cm 1303, London: HMSO.

Dumas, A. (1884). *The Count of Monte Cristo.*

Edwards, C., and Mackay, M. (2008). *QoS on JANET: The JANET QoS Development Project: Technical Guide.* (Accessed at http://www.ja.net/documents/publications/technical-guides/qos-tg.pdf January 2010.)

Engelbart, D.C., and English, W.K. (1968). A Research Center for Augmenting Human Intellect. *AFIPS Conference Proceedings of the 1968 Fall Joint Computer Conference,* San Francisco, CA, **33** (December), 395–410.

Flowers, B.H. (1966). A report of a Joint Working Group on Computers for Research. *Chaired by Professor B.H. Flowers on behalf of the Council for Scientific Policy and the University Grants Committee.* HMSO (January). (Accessed at http://www.chilton-computing.org.uk/acl/literature/manuals/flowers/foreword.htm April 2008.)

Foster, I., and Kesselman, C. (1999). *The Grid: Blueprint for a New Computing Infrastructure.* Morgan-Kaufmann.

Foster, I., and Kesselman, C. (2004). *The Grid 2: Blueprint for a New Computing Infrastructure (2nd edition).* Morgan-Kaufmann. ISBN 1-55860-933-4.

Forster, J., Satz, G., Glick, G., and Day, R. (1994). Cisco Systems X.25 over TCP. RFC 1613, IETF (May). (Accessed via http://www.ietf.org/rfc.html June 2009.)

Girling, C.G. (1983). *Representation and Authentication on Computer Networks.* University of Cambridge Computer Laboratory Technical Report (PhD Thesis), UCAM-CL-TR-37.

Gray, R.M. (2007). Packet speech on the Arpanet: A history of early LPC speech and its accidental impact on the Internet Protocol. *A historical overview given at Research Triangle Park (June).* (Accessed at http://ee.stanford.edu/~gray/lpcip/nclpcip.pdf April 2009.)

Griffin, T.G., and Sobrinho, J.L. (2005). Metarouting. *Proceedings of ACM SIGCOMM 2005, Philadelphia (August).* (Accessed at http://conferences.sigcomm.org/sigcomm/2005/paper-GriSob.pdf January 2010.)

Gunderson, S.H. (2008). Global IPv6 Statistics - Measuring the current state of IPv6 for ordinary users. RIPE 57, Dubai (October). (Accessed at http://www.ripe.net/ripe/meetings/ripe-57/presentations/Colitti-Global_IPv6_statistics_-_Measuring_the_current_state_of_IPv6_for_ordinary_users_.7gzD.pdf January 2010.)

Handley, M., Kirstein, P., and Sasse, M.A. (1993). Multimedia integrated conferencing for European researchers (MICE): piloting activities and the conference management and multiplexing centre. *Computer Networks and ISDN Systems*, **26**, 275–90.

Handley, M. (2003). The Internet: the last 30 years and the next 30 years. *Presentation at the celebrations of the 30th anniversary of the first international Internet link.* (Accessed via http://www.cs.ucl.ac.uk/staff/M.Handley/slides/ April 2009.)

Harrison, P. (1979). Networking in the Midlands universities. *Networkshop 5*, 42–54, Kent (September).

Helms, H.J. (1978). Organization and Technical Problems of the European Informatics Network. *Interlinking of Computer Networks*, Proceedings of the NATO Advanced Study Institute, Bonas, France, 45–66. D. Reidel (1979). ISBN 90-277-0979-3.

Hopper, A. (1978). *Local Area Computer Communcation Networks.* PhD Thesis; Technical Report No 7, University of Cambridge Computer Laboratory.

Houlder, P. (2007). Starting the Commercial Internet in the UK. *6th UK Network Operators' Forum meeting, 19 January 2007, University of Southampton.* (Accessed via h http://www.uknof.org.uk/uknof6/ January 2010.)

Howlett, J. (1978). Users' view of a national network: a survey by the UK National Committee on Computer Networks. *Interlinking of Computer Networks*, Proceedings of the NATO Advanced Study Institute, Bonas, France, 157–60. D. Reidel (1979). ISBN 90-277-0979-3.

Huitema, C. (1996). *IPv6: The New Internet Protocol.* Prentice-Hall. ISBN 0-13-241936-X.

Induruwa, A.S., Linington, P.F., and Slater, J.B. (1999). Quality-of-Service Measurements on SuperJANET, the U.K. Academic Information Highway. *Proc. INET'99, San Jose, CA.* (Accessed at http://www.isoc.org/inet99/proceedings/4e/4e_2.htm February 2009.)

ITU (2000). H.323: *Packet-Based Multimedia Communications Systems.* Series H: Audiovisual and Multimedia Systems. (Prepublication – unedited – version of approved recommendation was released in November.)

Jain, R. (1990). Performance analysis of FDDI token ring networks: effect of parameters and guidelines for setting TTRT. *Proc. ACM SIGCOMM'90, Philadelphia, PA*, 264–75. Also in *IEEE LTS Magazine*, May 1991; and in *Digital Technical Journal*, **3**, No.3, 78–88.

Jain, R. (1993). FDDI: Current Issues and Future Plans. *IEEE Communications Magazine*, September, 98–105.

Jain, R. (1994). *FDDI Handbook: High-Speed Networking Using Fiber and Other Media.* Addison-Wesley. ISBN 0-201-56376-2.

Johnson, M. (1987). Proof that timing requirements of the FDDI token ring protocol are satisfied. *IEEE Trans. Commun.*, **COM-35**, No.6, 620–25.

Johnson, M. (1988). Performance analysis of FDDI. *Proc. EFOC/LAN'88, Amsterdam, Netherlands.* Information Gatekeepers, Boston, MA, 295–300.

Kay, A.C. (1972). *A Personal Computer for Children of All Ages.* Xerox PARC (August). (Available at a number of sites; accessed at http://www.mprove.de/diplom/gui/Kay72a.pdf March 2009.)

Kay, A.C. (1977). Microelectronics and the Personal Computer. *Scientific American,* **237**, No.3, 230–44 (September).

Kirstein, P.T. (1998). Early Experiences with the ARPANET and INTERNET in the UK. (July) (Accessed at http://nrg.cs.ucl.ac.uk/mjh/kirstein-arpanet.pdf April 2009.)

Kirstein, P.T. (2004). European International Academic Networking: A 20 Year Perspective. *Selected Papers from the TERENA Networking Conference (2004).* University of the Aegean, 7–10 June, Rhodes, Greece. ISBN 90-77559-04-3. (Accessed via http://www.terena.org/publications/tnc2004-proceedings/ July 2009.)

KPMG (1991). *Review of Four Models of a Networking Association.* CB(91) REV4.1 Annex 5, 29/30 January 1991.

Larmouth, J., and Rosner, R.A. (1982). Networking Protocols in the UK Academic Community. *(This paper was widely circulated in the academic community at the time but its date is uncertain: it was most likely late 1981 or early 1982.)*

Lee, S. (1994). HUMBUL – The Humanities Bulletin Board (Report). *Humanist Discussion Group,* **8**, No.0101 (July). (Accessed as archive thread item at http://www.digitalhumanities.org/humanist/Archives/Virginia/v08/0101.html February 2010.)

Lehtisalo, K. (2005). *The History of NORDUnet: Twenty-Five Years of Networking Cooperation in the Nordic Countries.* NORDUnet A/S. ISBN 87-990712-0-7.

Lindley, J. (1978). Report on the North West Network Service. *Regional Network Activities,* session in *Networkshop 2,* 69–84, Liverpool (April).

Linington, P.F. (1983). Fundamentals of the Layer Service Definitions and Protocol Specifications. *Proc. IEEE,* **71**, No.12, 1334.

Linington, P.F., and Spratt, E.B. (1991). *The Objectives of and Services Offered by the UK Academic Networking Programme.* CB(91) REV4.1 Annex 4, 29/30 January 1991.

Linington, P.F., Slater, J.B., and Spratt, E.B. (1992). *The Criteria and Methodology to be used in the Selection of Sites for Connection to Wide Area Networks.* ACN(92)8.

Lukasik, S.J. (2010). Why the ARPANET was built. *IEEE Annals of the History of Computing* (to appear). Accessed at http://www.cistp.gatech.edu/publications/files/ARPANETv8.pdf March 2010.

McConachie, M.A. (1978). MIDNET — Midlands Universities. *Regional Network Activities,* session in *Networkshop 2,* 57–8, Liverpool (April).

Metcalfe, R.M., and Boggs, D.R. (1976). Ethernet: Distributed Packet Switching for Computer Networks. *Communications of the ACM,* **19**, No.7, 395–403.

Naughton, J. (2000). *A Brief History of the Future: the origins of the Internet.* Phoenix. ISBN 0-75381-093-X (2005).

Needham, R.M., and Herbert, A.J. (1982). *The Cambridge Distributed Computing System.* Addison-Wesley. ISBN 0-201-14092-6.

Network Unit (1979). *Final Report.* March.

Oppen, D.C., and Dalal, Y.K. (1983). The Clearinghouse: a decentralized agent for locating named objects in a distributed environment. *ACM Transactions on Information Systems,* **1**, No.3, 230–53 (July).

Orwell, G. (1949). *1984.* (George Orwell was the pseudonym of Eric Arthur Blair.)

Partridge, C. (2008). The Technical Development of Internet Email. *IEEE Annals of the History of Computing,* **30,** No.2, 3–29 (April-June).

Perlman, R. (1985). An Algorithm for Distributed Computation of a Spanning Tree in an Extended LAN. *ACM SIGCOMM Computer Communication Review,* **15,** No.4, 44-53.

Perkins, C., and Cooper, C. (1999). *MECCANO Experience with the TEN-155 Managed Bandwidth Service Alpha Test:Report to QUANTUM Project.* ERCIM TEN-155 Alpha Test Report.

Perry, D. (2009). Green Data Centres. *JANET News,* No.9, 7 (September). (Accessed at http://www.ja.net/documents/publications/news/news-9.pdf January 2010.)

Peterson, L.L., and Davie, B.S. (2003). *Computer Networks: A Systems Approach* (3rd ed.). Morgan Kaufmann. ISBN 1-55860-833-8.

Phillips, I., Parish, D., Sandford, M., Bashir, O., and Pagonis, A. (2006). Architecture for the Management and Presentation of Communication Network Performance Data. *IEEE Transactions on Instrumentation and Measurement,* **55,** No.3, pp.931–8 (June). (Accessed via https://dspace.lboro.ac.uk/dspace-jspui/handle/2134/2578 and http://hdl.handle.net/2134/2578 June 2009.)

Powell, C.J. ('Kit') (1980). Evolution of networks using standard protocols. *Computer Communications,* **3,** No.3, 117–22 (June).

Rice, J.D. (1978). The North-West Universities Network. *Interlinking of Computer Networks,* Proceedings of the NATO Advanced Study Institute, Bonas, France, 365–83. D. Reidel (1979). ISBN 90-277-0979-3.

Rivest, R.L., Shamir, A., and Adleman, A. (1978). A Method for Obtaining Digital Signatures and Public-Key Cryptosystems. *Communcations of the ACM,* **21,** No.2, pp.120–26. (Accessed at http://theory.lcs.mit.edu/~rivest/rsapaper.pdf October 2009.)

Roberts, L.G. (1978). *The Evolution of Packet Switching.* IEEE invited paper, November. (Accessed at http://www.packet.cc/files/ev-packet-sw.html May 2008.)

Roberts, L.G. (1999). *Internet Chronology.* (Accessed at http://www.ziplink.net/~lroberts/InternetChronology.html May 2008.)

Rosenbrock, H.H. (1975). *Engineering Computing Requirements: Technical Group Report.* Report to Engineering Board of SRC. (Accessed at http://www.chilton-computing.org.uk/acd/icf/eb_report/overview.htm August 2008; see also http://www.chilton-computing.org.uk/acd/sus/perq_history/part_1/c2.htm August 2008)

Rosner, R.A. (1978). Networking among U.K. Universities. *Interlinking of Computer Networks,* Proceedings of the NATO Advanced Study Institute, Bonas, France, 353–63. D. Reidel (1979). ISBN 90-277-0979-3.

Rosner, R.A. (1996). *In the beginning: an affectionate memoir.* UCISA.

Salford CNTR (2008). Details of the Claude Chappe mechanical semaphore or optical telegraph network in France may be found at the Computer Networking and Telecommunications Research website at the University of Salford: http://www.cntr.salford.ac.uk/comms/telegraphy.php and http://www.cntr.salford.ac.uk/comms/ebirth.php, both accessed September 2008.

Shoch, J.F. (1978). Inter-Network Naming, Addressing, and Routing. *Proceedings of Compcon78,* 72–9.

SRI (1982). *Internet Protocol Transition Workbook.* March. Published by the Network Information Center, SRI International, Menlo Park, CA 94025, USA.

Stephens, C. (1997). HUMBUL Updated: Gateway to Humanities Resources on the Web. *Computers & Texts*, **15**, 23 (August). (Accessed at http://users.ox.ac.uk/~ctitext2/publish/comtxt/ct15/stephens.html February 2010.)

Swift, J. (1726). *Gulliver's Travels*.

Temple, S. (1984). *The Design of a Ring Communication Network*. PhD Thesis; Technical Report No 52, University of Cambridge Computer Laboratory.

Thomas, J. (1978). SWUCN. *Regional Network Activities*, session in *Networkshop 2*, 59–67, Liverpool (April).

UKERNA (2001). *Report of Quality of Service Think Tank*. (Accessed at http://webarchive.ja.net/development/qos/qos-tt-report.pdf August 2009.)

UKERNA (2004). *SuperJANET5:An Architecture for Diversity*. (Accessed at http://www.webarchive.ja.net/sj5/requirementsanalysis/an-architecture-for-diversity.pdf May 2010.)

Universities Funding Council (1991a). *The Structure of UKERNA*.

Universities Funding Council (1991b). *Opportunities and Challenges for the Networking Programme in the 1990s*.

Want, R., Hopper, A., Falcão, V., and Gibbons, J. (1992). The active badge location system. *ACM Transactions on Information Systems*, **10**, No.1 (January).

Weinstein, C.J., and Forgie, J.W. (1983). Experience with Speech Communication in Packet Networks. *IEEE J. Selected Areas in Communications*, **SAC-1**, No.6, 963–80 (December).

Wells, M. (1973). *Report of the Network Working Party*. Computer Board paper, widely circulated for comment. (Sometimes referred to as 'Wells 1'.)

Wells, M. (1975). *Network Working Party Report*. Computer Board paper, widely circulated for comment. (Sometimes referred to as 'Wells 2'.)

Wells, M. (1984). The JANET project. *University Computing*, **6**, 56–62.

Wells, M. (1986). A progress report on JANET. *University Computing*, **8**, 146–53.

Wells, M. (1988). Access to JANET. *University Computing*, **10**, 149–53.

Wilkes, M.V. (1975). Communication using a Digital Ring. *Proceedings of the PACNET Conference, Sendai, Japan*.

Wilkes, M.V., and Needham, R.M. (1979). *The Cambridge CAP Computer and its Operating System*. North Holland. ISBN 0-444-00357-6.

Zhang, L., Deering, S., Estrin, D., Shenker, S., and Zappala, D. (1993). RSVP: A New Resource ReSerVation Protocol. *IEEE Network*, **7**, No.5, 8–18 (September).

Chronology

Year	Development	Organisation & funding	International	Background
1965				Donald Davies coins the term 'packet switching'.
1966		Computer Board created to fund central computing for UK universities.		
1967	Need for SWUCN identified.		ARPANET construction begins.	
1968				Multimedia communications demonstration by Doug Engelbart at Fall Joint Computer Conference, San Francisco.
1969	Design and implementation of SWUCN begins. METRONET begins operation.		ARPANET first four nodes (IMPs) installed.	National computing services begin at MRCC and ULCC, based on CDC6600 systems.
1970	RCOnet proposed.			
1971	RCOnet development begins.		Agreement reached to mount EIN. Aloha network begins operation at the University of Hawaii.	British Post Office announces intention to mount EPSS and invites participation. ARPANET TIP introduced.
1972	DTI/CCTA/ICL GANNET project begins.	North-West Network consortium formed.	ARPANET expands to 30 nodes during the year.	At Xerox PARC, Alan Kay articulates concept of 'Dynabook', the precursor of the laptop (from ideas dating back to 1968).

CHRONOLOGY | 381

Year	Development	Organisation & funding	International	Background
1973	RCOnet begins operation. CYGNET operational.	Computer Board working party visits US ARPANET. First Wells Report.	UCL ARPANET node (TIP) installed, providing 9.6 Kbps UK-US link; TIP linked to RHEL IBM 360/195, making it a host on ARPANET.	Initial discussion of proposal to split ARPANET NCP into TCP and IP. British Steel begins consideration of an enterprise network. Initial development of Ethernet at Xerox PARC.
1974	SWUCN begins operation. Development of SRCnet begins. Collaborative development of EPSS begins.			TCP/IP proposal published by Kahn & Cerf. Intel 8080 and Motorola 6800 appear.
1975		Second Wells Report.	EIN begins experimental operation.	CCITT begins discussions which will lead to X.25. Cambridge ring design published. MOS Technology 6502 appears.
1976	North-West Network collaboration with ICL begins, based on GANNET.	The Network Unit of the Computer Board and Research Councils established; Mervyn Williams appointed Head.		SRC ICF created. NCCN convened. Development of Ethernet reported by Metcalf and Boggs. Zilog Z80 appears. Diffie-Hellman shared secret key algorithm published. DES encryption standard published by NBS.

Year	Development	Organisation & funding	International	Background
1977	First issue of Network News published. First Networkshop held in September at Glasgow University.		First implementations of TCP/IP demonstrated by Stanford University, BBN and UCL.	Prototype Cambridge ring begins operation. SRC DCS programme begins. EPSS operation begins.
1978	North-West Network service begins.	DCPU established.	UK-US link via UCL moves to use of SATNET at 20Kbps. IPSS service begins. Euronet begins operation.	NCCN reports. British Steel network operational. Provisional specification of X.25, together with initial versions of X.3, X.28 and X.29, published by CCITT. Initial specification of TCP and IP published. Motorola 6809 appears. RSA public key encryption algorithm published.
1979	X.25-based MidNET complete.	JNT established; Roland Rosner appointed Head. Support for UK Coloured Books made mandatory for computer systems funded by the Computer Board.		Usenet news begins in US. Motorola 68000 appears.
1980	Initial specification of the JNT PAD published.			PSS trial service opens. BT created from the telecommunications part of the Post Office. DIX Ethernet version 1.0 published.

CHRONOLOGY 383

Year	Development	Organisation & funding	International	Background
1981	Proposal mooted that SERCnet should form wide-area backbone of academic network.	NMC convened.	UCL UK-US link begins using 48Kbps IPSS link.	SRC becomes SERC.
	Construction of METRONET II begins, based on X.25.			PSS full service begins; EPSS closed.
	First discussion of the need for name registration in UK academic networking.			BT partially privatised but government retains majority interest and thus control.
				Apollo token ring launched.
				CSNET funded by NSF.
				BITNET begins.
1982	Demonstration of pre-production JNT PAD by Camtec at Nottingham Networkshop.	Proposal to base wide-area component of academic network on SERCnet agreed by Computer Board and SERC.	ARPANET changes to IP.	Mercury Communications (a subsidiary of Cable & Wireless) granted a telecommunications licence: duopoly created.
	NRS design study begins at Salford University.	JANET network operations structure defined: Network Executive, assisted by NCC and NOCs.		DIX Ethernet version 2.0 published.
				Initial version of IEEE 802.3 10Mbps CSMA/CD published.

Year	Development	Organisation & funding	International	Background
1983	'JANET' name comes into use.	Barrie Charles appointed Acting Head of JNT until Peter Linington appointed Head of JNT and NE.	UCL establishes '.UK' as third top-level Internet domain.	ANSI publishes initial version of FDDI MAC standard.
	NRS implementation begins.		EARN begins operation in Europe.	Paul Mockapetris' implementation of DNS appears.
	University of Kent begins providing access to Usenet news.	David Hartley, as Chair of NMC, informs BT of Computer Board and Research Councils' intent to construct a private academic network.	ARPANET and MILNET split.	
		Mike Wells appointed Director of Networking.		
		First two staff members appointed to NE, Ian Smith and Keith Mainwaring.		
		NMC reconstituted as NAC.		
1984	JANET officially begins operation on 1 April.		Joint funding of UK-US link begins: DTI/SERC (Alvey)/Computer Board in UK; NASA, DARPA and NSF in US.	Privatisation of BT completed (Telecommunications Act), remaining government holding sold, and monopoly formally ended.
	RCOnet begins migration to X.25.			
	NRS trial begins.			Discussion and planning of NSFnet begin.
	Upgrade of JANET trunk network from 9.6 Kbps to 48 Kbps begins.			CCITT publishes H.120 video compression standard.
	Camtec CR-X.25 gateway and CR PAD become available.			*Data Protection Act 1984* (privacy protection).

CHRONOLOGY 385

Year	Development	Organisation & funding	International	Background
1985	Full connectivity achieved for all JANET sites, i.e., universities and Research Councils.		EARN begins operation in UK.	IBM token ring becomes available.
	NRS prototype service begins.			IEEE 802.5 4Mbps token ring standard published.
				Bridge spanning tree algorithm (developed at DEC) published by Radia Perlman.
				SONET/SDH standards begin to emerge.
				Interception of Communications Act 1985 (regulation of law enforcement).
1986	NRS full service begins at Salford University.		NSFnet backbone begins operation.	QPSX proposed as IEEE 802.6 MAN standard.
	JANET backbone upgraded to 256Kbps.		RARE established.	Work on JPEG compression standard begins.
1987	UKnet formed and contracted to provide Usenet news feed to JANET.	Bob Cooper appointed Head of JNT.		ANSI publishes initial FDDI MAC standard.
		JNT, UCL, ULCC & USC ISI propose JANET–NSFnet link.		
1988	FDDI Advisory Group convened.	UFC succeeds UGC.		ISO MPEG compression work begins.
		PCFC created.		Morris worm infects Internet.
		NSF/CB/SERC agree to fund new US-UK link (using TAT8).		CERT/CC set up at CMU.

Year	Development	Organisation & funding	International	Background
1989	Mailbase (University of Newcastle) becomes available on trial.	Funding for JANET Mk II and the High Performance LAN Initiative approved.	64Kbps link between JANET (ULCC) and NSFnet (Princeton) comes into operation.	Basis for Web proposed by Tim Berners-Lee at CERN.
		Approval given to begin planning SuperJANET.	First meeting of RIPE.	Bellcore publishes definition of SMDS.
				BITNET merges with CSNET.
1990	FDDI pilot begins.	NASG established to advise Computer Board and Research Councils on future management structure for JANET.	UK-US link capacity upgraded to 256Kbps.	UK 'duopoly' review.
	DoDAG convened.		IXI pilot begins.	IEEE 802.6 MAN standard published (technology now referred to as DQDB).
	SuperJANET project team formed.	KPMG assesses options for legal organisation of The JNT Association.		CCITT H.261 video compression standard published.
				Twisted-pair Ethernet standardised.
				Computer Misuse Act 1990 (unauthorised access).

CHRONOLOGY 387

Year	Development	Organisation & funding	International	Background
1991	Shoestring project demonstrates feasibility of IP-over-X.25.	Computer Board approves IP-over-X.25 pilot.	UK-US link capacity upgraded to 512Kbps.	ISC succeeds Computer Board.
	JIPS begins.	Additional £9M for DES approved in PES for SuperJANET.	IXI becomes EMPB, operating at 2Mbps.	UK 'duopoly' ended.
	JANET DNS mounted by NOSC at ULCC.	UFC approves £20M for SuperJANET project.	CSNET ceases operation.	CERN Web server appears.
	FDDI deployment begins.	PCFC funds the connection of a number of polytechnics and colleges to JANET.		
	Mailbase begins full service.	Willie Black appointed Head of JNT.		
		KPMG recommends form of organisation to be adopted for The JNT Association.		
		University of Kent recommends basis for an SLA for The JNT Association.		
		NASG recommends to CB that The JNT Association be set up on the basis proposed by KPMG and the University of Kent.		
1992	MBone mounted over JIPS.	ISC approves the setting up of The JNT Association.	EBONE begins operation.	Polytechnics become universities.
	MBone video experiments begin over JANET; audio-video-casts of IETF meetings available over JANET.	KPMG prepares business case for The JNT Association for the Treasury.		HEFCs (HEFCE, HEFCW and SHEFC) created to succeed UFC and PCFC.
		SuperJANET RFP issued, followed by ITT.		Initial JPEG compression standard published.
		Decision to award 4-year (1993–97) SuperJANET contract to BT approved by UFC.		

Year	Development	Organisation & funding	International	Background
1993	IP traffic over JANET begins to exceed X.25 traffic.	SuperJANET contract for £18M signed by UFC and BT.	UK-US link operating at 1.5Mbps, then 2Mbps.	ISC becomes JISC, responsible to the HEFCs and DELNI.
	Native IP links introduced in JANET.	HEFCs come into operation and ratify BT SuperJANET contract to 1997.		DANTE launched.
	'Little-endian' (Internet) naming accepted.	The JNT Association is incorporated by the HEFCs and OST (for the RCs) as a not-for-profit company limited by guarantee.		Switched Ethernet developed.
	SuperJANET pilot IP, ATM and video networks established.			IntServ and RSVP concepts published.
	First demonstrations of SuperJANET take place.	NAC reconstituted as JISC ACN.		
	SuperJANET IP pilot network incorporated into JIPS.	TAU set up at the University of Kent.		
	30-site pilot IP-over-SMDS network begins operation.	JISC confirms IP as the preferred protocol for JANET; ISO CONS no longer strategic.		
1994	SMDS coverage increased to 50 sites and incorporated into JIPS.	The JNT Association begins trading as UKERNA with responsibility for JANET from 1 April and David Hartley as CEO.	EARN merges with RARE to form TERENA.	Athens identity management system developed.
	SuperJANET ATM network expanded to 14 sites.	JISC approves SuperJANET II funds to expand SMDS coverage.		BBSRC, PPARC and EPSRC created.
	SuperJANET wins BCS Award; award ceremony carried on ATM-based video.	Programme to establish MANs/RNs begins.		JPEG becomes an ISO standard.
	JANET CERT created.	ACN reviews JANET connection policy in context of NHS.		
	Web-based Mailbase interface introduced.			
	TACS begins operation.			
	LINX created.			

Year	Development	Organisation & funding	International	Background
1995	Last issue (44) of *Network News* published.	Market testing of UKERNA begins.	TEN-34 proposed to EU under Framework IV Programme.	CCLRC created, with responsibility for DL and RAL.
	Pilot SJVCN begins.	Revision of JANET AUP; creation of separate Connection Policy.	UK-US link capacity now 4Mbps.	Altavista Web search engine made available by Digital.
	Pilot ATM-over-SDH network established.		NSFnet upgraded to 155Mbps (vBNS).	100 Mbps Ethernet standardised.
	Mixed IP data and CBR video established over ATM network.		Access to Pan-European ATM Pilot network provided.	Preliminary specification of IPv6 circulated.
1996	First issue of *UKERNA News* published.	Market testing of UKERNA completed.	JAMES pilot network begins.	UK IWF formed.
	SMVCN video codecs procured (M-JPEG over ATM); pilot begins.	BT SuperJANET contract extended to 1998.	UK-US link capacity now 20Mbps.	ITU-T VCEG publish H.263 and H.323.
	Videoconferencing strategy developed.		6bone launched.	IPv6 support introduced in Linux.
	Athens identity management service introduced.			*Defamation Act 1996* (content liability).
	FEnet96 project begins, funded by FEFCW to connect FE colleges in Wales (WelshNet).			

Year	Development	Organisation & funding	International	Background
1997	JANET X.25 service ceases; Coloured Books support ceases; NRS closed.	JOD created as part of NOSC.	JANET joins 6bone.	RSVP defined.
	SMVCN full service begins.	Geoff McMullen appointed CEO of UKERNA.	TEN-34 begins service, operated by JANET NOSC.	
	Videoconferencing strategy approved by JCN; JVCSS started.	3-year SuperJANET III contract (1998–2001) awarded to Cable & Wireless.	QUANTUM proposed to EU to develop and upgrade TEN-34 (TEN-155).	
	SJVCN replaced by JVCS using ISDN; JVCSS becomes part of JVCS managed by University of Edinburgh.	HEFCE announces initiative to connect 72 selected English FE colleges to JANET.	UK-US link capacity now 45Mbps.	
	Survey of videoconferencing equipment conducted by University of Newcastle and UCL.	SFEFC and DELNI request UKERNA to establish programmes to connect FE colleges in Scotland and Northern Ireland.		
1998	SuperJANET III begins service.	HEFCW commissions UKERNA to report on videoconferencing needs of HE and FE in Wales.	JAMES pilot ends.	DiffServ defined.
	Consideration of JANET use in the context of schools begins.	Institutional charging introduced for transatlantic use.	TEN-155 begins service.	1 Gbps Ethernet standardised.
	VCAS begins (provided by University of Newcastle and UCL).		UK-US link capacity now 90Mbps.	RFC 2460 ratified by IETF as IPv6 standard.
			NSFnet/vBNS becomes Internet2, upgraded to 622Mbps and operated by UCAID.	Google Web search engine service launched.
				Data Protection Act 1998 (privacy protection).

Year	Development	Organisation & funding	International	Background
1999	ATM MBS service launched.	English LEAs begin forming schools RBCs.	UK-US link capacity now 310Mbps.	1Gbps Ethernet products launched.
	SMDS phased out.			
	JANET X.500 directory service closed.			
	Planning for SuperJANET4 begins.			
	JANET allocated a block of IPv6 addresses.			
	Formation of wireless advisory group mooted.			
2000	Cisco GSR 12016 routers chosen for SuperJANET4.	MCI WorldCom chosen to supply SuperJANET4 transmission infrastructure on a 5-year contract (2000–05).	Bermuda2 (2000–02) UK-Internet2 IPv6 trial project begins.	Wide-area 1Gbps Ethernet products launched.
	Mailbase replaced by JISCMAIL (based on LISTSERV at RAL).		JANET peers with Abilene at 34 Mbps.	ITU-T publish revised H.323.
	VCAS becomes VTAS, with expanded remit.	WVN project obtains matching EU funds.	GN1 planning for GÉANT begins.	IETF RFC 3261 SIP accepted for use in cellphone service.
	Programme to develop IP-based videoconferencing announced, based on H.323.	Follett review of JISC.		*Regulation of Investigatory Powers Act 2000* (regulation of unlawful interception of a network).
	QoS Think Tank convened.			

CHRONOLOGY | 391

Year	Development	Organisation & funding	International	Background
2001	SuperJANET4 begins service at 2.5Gbps.	Robin Arak appointed CEO of UKERNA.	GÉANT begins service at 10Gbps.	5-year UK e-Science Programme begins.
	QoS Think Tank reports.	NHS-HE Forum created.	UK-US link capacity now 930Mbps.	International Review of UK Research in Computer Science published.
	WVN begins IP-based service.		Abilene peering upgraded to 155Mbps.	
	1st SuperJANET5 strategy workshop held in Warwick.		Optical network service development programmes established in Canada (CANARIE), The Netherlands (NetherLight) and USA (StarLight, STARTAP).	Anti-Terrorism, Crime and Security Act 2001 (retention of records).
	C2k deployment to primary schools begins in Northern Ireland.			
2002	SuperJANET4 backbone upgraded to 10Gbps.	Governance of UKERNA amended to restructure and broaden membership of The JNT Association.	6NET (2002-05) EU IPv6 trial project begins.	Teleglobe and KPNQwest go into administration; WorldCom's financial viability is in doubt.
	Initial IP-based JVCS begins, covering whole of UK.	Case for UKLight approved by JCSR.	UK-US link upgraded to 5Gbps.	Initial version of 10 Gbps Ethernet standard published.
	JANET X.400 support ceases.		Abilene upgraded from 2.5Gbps to 10Gbps.	
	LLNW begins operation.			Electronic Commerce (EC Directive) Regulations 2002 (content liability).
	Spark project begins in Scotland.			
	2nd SuperJANET5 strategy workshop held in Glasgow.			
	Wireless Advisory Group convened.			
	QoS trial launched.			
	Model architecture for JANET-NHS interconnection proposed by Andrew Cormack.			

Year	Development	Organisation & funding	International	Background
2003	JVCS IP-based service launched.	Funding announced by Department of Education and Skills for JANET to provide RBC interconnect, to be completed by 2005.		First version of Shibboleth released by Internet2.
	Schools start to use videoconferencing.			*Communication Act 2003* (stealing communications service).
	JANET RBC interconnect begins service.	UKLight funding approved by HEFCE.		*Sexual Offences Act 2003* (investigation of indecent or obscene material).
	SSDN begins operation.			
	C2k post-primary deployment begins in Northern Ireland.			
	SuperJANET5 requirements formulated.			
	JANET IPv6 dual-support deployed.			

Year	Development	Organisation & funding	International	Background
2004	MAN/RN programme completed with establishment of NIRAN.	Bob Day appointed Acting CEO of UKERNA.	GN2 planning of GÉANT2 begins.	BT selected to provide N3 (next generation NHS network).
	UKLight begins operation to StarLight and NetherLight.	SuperJANET4 contract with WorldCom extended for a year (to end 2006).		
	JVCS booking service upgraded to cope with growth.	SuperJANET5 transmission platform procurement begins.		
	WVN extended to an initial group of Welsh-medium schools.	RFI for SuperJANET5 routers issued, followed by trials of products from Chiaro, Cisco and Juniper.		
	One-year schools videoconferencing pilot begins in England.	JISC funds 3-year SDSS project at University of Edinburgh.		
	AGSC set up at Manchester University.	Malcolm Teague appointed NHS-HE Co-ordinator.		
	Schools content distribution begins (e.g., Pathé News archive).			
	SuperJANET5 strategic requirements and architecture defined.			
	JANET Web Filtering Service launched.			
	JANET Roaming architecture proposed.			
	Pilot JVCS conferencing on demand begins.			

Year	Development	Organisation & funding	International	Background
2005	JVCS introduces conferencing on demand service (IVC).	Tim Marshall appointed CEO of UKERNA.	GÉANT2 service begins.	AHRC succeeds AHRB.
	JVCS automated quality assessment introduced.	Contract for SuperJANET5 transmission platform using DWDM placed with Verizon Business.		SFEFC and SHEFC merge to form SFC.
	UK-wide one-year schools videoconferencing pilot undertaken; LGfL mounts videoconference with 120 participating venues.			
	JVCS extended nation-wide to whole education sector.			
	Access to UKLight extended geographically.			
	One-year JANET roaming trial launched.			

Year	Development	Organisation & funding	International	Background
2006	Welsh RNs begin using LLNW infrastructure.	Continued funding for schools interconnect announced; schools sector becomes JANET funding partner.	6bone ends.	*Police and Justice Act 2006* (amended to regulate unauthorised interference).
	UK Access Management Federation launched, based on SDSS.			*Terrorism Act 2006* (content liability).
	JRS launched.	Planning for Glow begins in Scotland.		
	Pilot JANET VRVS and Access Grid gateway service begins.	Funding for UKLight dark-fibre segment (Aurora) approved by HEFCE.		
	Trial provision of selected Freeview TV channels over JANET launched.	NTL (Virgin Media) selected to supply Aurora infrastructure.		
	Preliminary specification of N3-JANET gateway published.	NOSC, now called the NOC, transfers from Univerity of London to merge with UKERNA.		
	Pilot trial undertaken of dedicated wavelength, 10Gbps ethernet circuit between RAL and CERN via TVN, SuperJANET5, GÉANT2 and SWITCH.	Contract to supply Juniper T640 routers awarded to Lucent.		

Year	Development	Organisation & funding	International	Background
2007	SuperJANET5 begins service.	JANET NOC moves from ULCC in Guilford Street to new premises in Grays Inn Road.		CCLRC and PPARC merge to form STFC.
	JANET achieves e-Government Shared Services Award 2007.			EVO begins to replace VRVS.
	Last issue (39) of *UKERNA News* published; first issue of *JANET News* published.	UKERNA moves to new building, Lumen House.		
	Verizon conducts trials of additional WDM channels operating at 40Gbps; 40Gbps link begins service in JANET core.	The JNT Association adopts JANET(UK) as its new trading name.		
	JANET Lightpath service established.	Welsh Assembly Government announces funding for PSBA Network.		
	JVCS Booking Service starts to use UK Access Management Federation.	IPv6 incorporated into JANET backbone SLA.		
	JANET txt service launched.			
	JANET QoS trial ends; QoS support not introduced.			
	JANET Training introduces use of virtual machine systems.			
	Aurora begins operation.			
	JANET-N3 gateway begins trial operation.			

Year	Development	Organisation & funding	International	Background
2008	First phase of Welsh PSBA Network comes into operation at 2.5Gbps, provided by LLNW.	Funds obtained to extend Aurora. IPv6 incorporated into JANET SLA.	GÉANT funded for a further 4 years; GN3 begins.	Athens identity management service ceases (in UK education and research sector).
	Glow deployment begins.			*Criminal Justice and Immigration Act 2008* (investigation of indecent or obscene material).
	Core of JANET backbone upgraded to 40Gbps.			
	Support for JANET txt incorporated into Moodle VLE.			
	HIPNet use of Aurora begins.			
	Library services, VLE and JVCS sessions successfully trialled over JANET-N3 gateway.			
2009	Use of 100Gbps successfully tested in JANET core transmission platform.	JANET celebrates 25 years of service. Announcement of planned closure of JANET Usenet news service in July 2010.	3rd generation of GÉANT begins service.	*Data Retention (EC Directive) Regulations 2009* (retention of records by ISPs – voice and email).
	Optical TDM demonstrated by an international consortium using Aurora.			

Index

6bone 295, 300
6NET 295

A

AARNet 279
Aberdeen Metropolitan Area Network. *See* AbMAN
Abilene 205, 227, 229, 230
AbMAN (Aberdeen Metropolitan Area Network) 214, 226, 307
Access Grid 114, 249, 250, 252, 281
Access Grid Support Centre. *See* AGSC
ACN (Advisory Committee on Networking) 166, 282
Addressing 44
Advisory Committee on Networking. *See* ACN
AFRC (Agricultural and Food Research Council) 161
Agricultural and Food Research Council. *See* AFRC
AGSC (Access Grid Support Centre) 250
AHRC (Arts and Humanities Research Council) 161
AltaVista 119, 173
Alty, J.L. 305
Arak, Robin 308
ARPANET i, 1, 17, 18, 40, 42, 43, 45, 48, 54, 57, 107, 111, 158, 288, 290
  adoption of IP 55
  adoption of TCP/IP 55, 90, 107, 108, 109
  becomes Internet 107
  connection to SRCnet 51
  creation of 6
  email protocol 53
  evolution of internetworking concept 106
  experiments in packetized audio 140
  hosts on 257
  link to UCL 19, 60, 123, 124
  mail addressing 54
  mail headers 53
  MILNET split 107, 108, 154
  naming scheme 55, 115
  real significance of 19
  should JANET have copied? 292
  success gains international attention 13
  success of packet-switching 18
  transatlantic connections 124
  UK delegation to visit 13
  use of email 46
  uses little-endian convention 54
Arts and Humanities Research Council. *See* AHRC
Asynchronous Transfer Mode. *See* ATM
ATAG (ATM Technical Advisory Group) 180

Athens 83, 269
  funded by JISC 269
Atlas 5, 8, 9, 58
Atlas Centre 30, 58, 253
Atlas Laboratory 5, 9, 28, 29, 253
ATM (Asynchronous Transfer Mode) 140, 146, 180, 182, 183, 186–191, 193, 195, 200–203, 205, 206, 210, 212, 221, 222, 225, 227, 229, 233, 244–246, 252, 288, 292, 296, 297, 300
  video over ATM 148
ATM Technical Advisory Group. *See* ATAG

B

Babbage, Charles 3
Baran, Paul 6
Barber, Derek 9, 37, 123
Barry, Peter 61
Bartlett, Keith 37, 41
BBSRC (Biotechnology and Biological Sciences Research Council) 161
Becta (British Education Communications Technology Agency) 218, 270, 277
Bennett, Chris 53
BERKOM project 171
Bermuda 2 295
Berners-Lee, Tim 110, 136
BGP (Border Gateway Protocol) 186
Biotechnology and Biological Sciences Research Council. *See* BBSRC
BITNET i, 109, 127, 129, 130, 287
Black, G 305
Black, Willie 181, 308
Border Gateway Protocol. *See* BGP
Bray MP, Rt Hon Jeremy 174
British Education Communications Technology Agency. *See* Becta
British Standards Institution. *See* BSI
British Telecom. *See* BT
British Universities Network Interface Equipment. *See* BUNNIE
broadband ISDN 99, 137, 146, 149, 150, 171, 173, 180, 200, 202, 288, 289
  ATM standardised for 150
  circuit- or packet-oriented 137
  signalling 149
Bryant, Paul 29
BSI (British Standards Institution) 39
BT (British Telecom) ii, 36, 43, 44, 55, 57, 66, 68, 70, 71, 72, 80, 87, 92, 100, 119, 126, 127, 138, 172, 175, 181–184, 186, 187, 189, 190, 193, 197, 200, 201, 212, 221, 259, 260, 262, 293, 327
  awarded contract for N3 283
  awarded contract for SuperJANET 182
  charging 44
  offers SMDS 151
  ultimatum received from NMC 72
BUNNIE (British Universities Network Interface Equipment) 15, 18
Burren, John 29, 175, 183

C

C&NLMAN (Cumbria and North Lancashire Metropolitan Area Network) 213, 214, 307
Cable & Wireless 221, 230
  awarded contract for SuperJANET III 201
  manages interior core of SuperJANET III 201
Camtec 61, 66, 90
CANARIE 230
CCITT (Comité Consultatif International de Télégraphique et Téléphonique) 37, 39, 40, 42, 47, 55, 78, 91, 100, 116, 126, 144, 149, 150, 180, 288, 292, 321
CCLRC (Council for the Central Laboratory of the Research Councils) 161
CCTA (Central Computing and Telecommunications Agency) 55
CDC 5, 10
Central Computing and Telecommunications Agency. *See* CCTA
Central Laboratory of the Research Councils. *See* CLRC
Cerf, Vint 107
CERT (Computer Emergency Response Team) 168, 252, 268
CETL4HealthNE 284
Charles, Barrie 58, 71, 81, 308
Charles, Janet 32, 58
Cheney, Chris 65, 180, 183
CineGrid 282
Clarke, Charles 217, 248
Clarke, Peter 230
Classroom 2000 Project (C2k) 218, 323
Clearinghouse 115
CLEO (Cumbria and Lancashire Education Online) 213, 323
CLRC (Central Laboratory of the Research Councils) 161
ClydeNET (Clyde Regional Network) 214, 307
Clyde Regional Network. *See* ClydeNET
Clyne, Les 58, 180, 183
Cohen, Danny 107
Coloured Books 25, 26, 41, 62, 95, 109, 113, 116, 131, 285, 288, 292
  Blue Book 50, 87
  Fawn Book 49
  Green Book 48
  Grey Book 53, 63, 113, 117, 118, 119
  no longer supported 119
  Orange Book 88
  Peach Book 92
  Pink Book 92
  Red Book 52, 208
  White Book 113
  Yellow Book 48
Comité Consultatif International de Télégraphique et Téléphonique. *See* CCITT
Computer Board of the Universities and Research Councils 4, 5, 8, 11, 13, 15, 16, 19, 22, 23–27, 29–34, 57–59, 66, 68–72, 74, 76, 81, 83–86, 91, 93, 99, 110, 111, 125, 126, 129, 159, 160, 163–166, 172, 174, 262, 263, 269, 287–290, 293, 302, 305
  endorses replacement of JNT 162
  expenditure 68, 69
  funding by 9, 306
  succeeded by ISC 160
Computer Emergency Response Team. *See* CERT
Computer Security Incident Response Team. *See* CSIRT
Constable, Geoff 284
Cooperation for OSI Networking in Europe. *See* COSINE
Cooper, Bob 58, 81, 164, 172, 174, 175, 181, 308, 309
Cormack, Andrew 269, 283
COSINE (Cooperation for OSI Networking in Europe) 128, 199
COST Project 9
Council for the Central Laboratory of the Research Councils. *See* CCLRC
Craigie, Jim 58
CSIRT (Computer Security Incident Response Team) 252, 268
CSNET 108, 109, 111, 129
  connected to ARPANET using TCP/IP 108
Cumbria and Lancashire Education Online. *See* CLEO
Cumbria and North Lancashire Metropolitan Area Network. *See* C&NLMAN
CYGNET 26, 306

D

Dallas, Ian 88
Daniels, Trevor 29
DANTE (Delivery of Advanced Network Technology to Europe) 123, 129, 130, 199, 205, 243, 268, 295
Daresbury Laboratory 5, 9, 19, 22, 28–30, 58, 71, 72, 82, 93, 161, 190, 231
  regionalisation 9
DARPA (Defense Advanced Research Projects Agency) 108
Data Communications Protocol Unit. *See* DCPU
Davies, Donald i, 6, 7, 9, 11, 123
Day, Bob 111, 183, 308
DCPU (Data Communications Protocol Unit) 37, 40, 50, 52, 60, 72
  File Transfer Protocol Implementers Group 50
Deering, Steve 110
Defense Advanced Research Projects Agency. *See* DARPA
DELNI (Department for Employment and Learning, Northern Ireland) 161, 166, 213
Demon 126
Dense WDM 225, 234, 241, 243
Department for Employment and Learning, Northern Ireland. *See* DELNI
Department of Defense Advisory Group. *See* DoDAG
DIAL (Dual Interconnect Across London) 198
DiffServ 148, 155, 227, 229, 290, 295, 296, 297
Director of Networking 71
Distributed Queue Dual Bus. *See* DQDB
DNS (Domain Name Service) 55, 111, 115, 116, 118, 296
DoDAG (Department of Defense Advisory Group) 111, 113
Domain Name Service. *See* DNS
Donald Davies 6
DPS/6 39

DPS/20 39
DQDB (Distributed Queue Dual Bus) 100, 150, 194
Dual Interconnect Across London. *See* DIAL
Dyer, John 183
Dynabook 136

E

E2BN (East of England Broadband Network) 323
EARN (European Academic Research Network) i, 109, 127, 129–131, 287
EaStMAN (Edinbrugh and Stirling Metropolitan Area Network) 214
EaStMAN (Edinburgh and Stirling Metropolitan Area Network) 226, 307
East Midlands Broadband Consortium. *See* EMBC
East Midlands Metropolitan Area Network. *See* EMMAN
EastNet (East of England Network) 213, 214, 223, 231, 307
East of England Broadband Network. *See* E2BN
East of England Network. *See* EastNet
EBONE 128, 132
ECMA (European Computer Manufacturers Association) 91
Economic and Social Research Council. *See* ESRC
Edinburgh and Stirling Metropolitan Area Network. *See* EaStMAN
Edinburgh Regional Computer Centre. *See* ERCC
Edinburgh University Computing Services. *See* EUCS
EIN (European Informatics Network) 9, 13, 17, 123, 126
email 45
   big-endian vs little-endian 54, 103, 117, 118, 292
EMBC (East Midlands Broadband Consortium) 323
EMMAN (East Midlands Metropolitan Area Network) 307
EMPB (European Multi-Protocol Backbone) 128, 129
Enabling Virtual Organizations. *See* EVO
Engelbart, Doug 7
Engineering and Physical Sciences Research Council. *See* EPSRC
EPSRC (Engineering and Physical Sciences Research Council) 161, 231, 279, 280
EPSS (Experimental Packet Switched Service) 9, 13, 17, 18, 22, 26, 28, 31, 37, 40, 43, 44, 47, 60, 84, 262
   protocols 40
   terminal access protocol 47
   User Forum 37, 40, 43
   User Forum study groups 40
ERCC (Edinburgh Regional Computer Centre) 5, 71, 72
ERCIM (European Research Consortium for Informatics and Mathematics) 205
ERSC (Economic and Social Research Council) 161
e-Science 229, 230, 234, 244, 249, 250, 282
Ethernet 35, 86, 88–92, 97, 102, 103, 109, 114, 130, 143, 151, 152, 172, 182, 183, 190, 206, 207, 242, 269
   1Gbps 151

   10Gbps 152
   carrier Ethernet 300
   Ethernet over SDH 298
   Ethernet over WDM 298
   gigabit Ethernet 210, 221, 289
   removal of distance limit 298
   shared Ethernet 104, 151
   switched Ethernet 104, 141, 151, 252, 299
   wide-area Ethernet 299
EUCS (Edinburgh University Computing Services) 111, 117
EU Framework IV 199
EU Framework V 203, 205
EU Framework 6 243
EU Framework 7 243
EUnet 130
EuroCERT 268
Euronet 126
EuropaNet 129
European Academic Research Network. *See* EARN
European Computer Manufacturers Association. *See* ECMA
European Informatics Network. *See* EIN
European Multi-Protocol Backbone. *See* EMPB
European Research Consortium for Informatics and Mathematics. *See* ERCIM
EVO (Enabling Virtual Organizations) 281
Experimental Packet Switched Service. *See* EPSS

F

FaTMAN (Fife and Tayside Metropolitan Area Network) 214, 226, 307
fat pipe
   upgraded to 1.5Mbps T1 link 126
FDDI (Fiber Distributed Data Interface) 98, 102, 151, 162, 172, 177, 180, 181, 183, 184, 190, 206, 212, 252, 306
   achievements of programme 104
   Advisory Group 99, 101, 104
   assessment of bridge and router support 103
FDDI-II 99
Federal Information Processing Standards. *See* FIPS
Fiber Distributed Data Interface. *See* FDDI
Fife and Tayside Metropolitan Area Network. *See* FaTMAN
FIPS (Federal Information Processing Standards) 123
Flowers, Brian 4
Flowers Report 23, 289
Follett, Brian 215
Follett Report 216
funding
   as part of the national research and education budget 4
   by Computer Board 5, 34, 71, 126, 159
   by JISC
      terms of 166
   by JNT 125
   by MoD 125
   by NERC 159
   by NSF 126
   by SERC 126, 159

by SRC 5
central funding model 33
creation of new councils 161
for JANET to provide RBC interconnect 217
for science 3
for US link 126
free at point of use principle 262
how much for SuperJANET 174
of JANET connections 260
polytechnics 58, 93
top-sliced 4, 33

G

Gandalf 65
GANNET 26, 27, 52, 60, 306
GBnet 82
GÉANT 227, 229, 243, 250, 261
Generalised MPLS. *See* GMPLS
Gigabit Testbed Initiative 171, 208
GlobalSign 270
Glow 218, 323, 324
GMPLS (Generalised MPLS) 298
Gold Network Service 119
Google 173
Gopher 113
GOSIP (Government Open Systems Interconnection Profile) 55
Government Open Systems Interconnection Profile. *See* GOSIP
Grapevine 115
Grays Inn Road 232
Grid computing 208, 249, 269
Guy, Mike 65

H

H.261 187, 245, 246
H.263 146, 149, 227
H.323 149, 248, 280, 281
Hartley, David 69, 70, 72, 162, 164, 181, 199, 308
  becomes UKERNA's first Chief Executive 165
HEAnet 227, 261
Heard, Ken 58
HEFCE (Higher Education Funding Council for England) 161, 166, 231, 279
  programme to fund and rationalise connections for English FE colleges 201
HEFCW (Higher Education Funding Council for Wales) 161, 164, 166, 247
Heterogeneous IP Networks. *See* HIPNet
Higher Education Funding Council for England. *See* HEFCE
Higher Education Funding Council for Wales. *See* HEFCW
High Performance LAN Initiative 96, 110, 135, 172, 178, 211, 222
  approved 94
  funding approved 99
HIPNet (Heterogeneous IP Networks) 280
Hopper, Andrew 69
Horne, Dr Nigel W. 174

Howat, George 117
Humanities Bulletin Board. *See* HUMBUL
HUMBUL (Humanities Bulletin Board) 83
Husband, Tom 308
Hutton, James 180

I

IBM 5
ICF (Interactive Computing Facility) 30, 31, 51, 64, 69, 251
ICMP (Internet Control and Monitoring Protocol) 295
IETF (Internet Engineering Task Force) 40, 91, 109, 111, 114, 117, 124, 133, 148, 150, 180, 186, 289, 295
  110
IGMP (Internet Group Management Protocol) 295
Information Systems Committee. *See* ISC
Information Technology Advisory Board. *See* ITAB
Information Technology Standards Unit. *See* ITSU
Interactive Computing Facility. *See* ICF
Interactive Terminal Protocol. *See* ITP
International Organisation for Standardisation. *See* ISO
International Packet Switched Service. *See* IPSS
International X.25 Infrastructure. *See* IXI
Internet 123, 129
  architecture able to encompass LANs 292
  ARPANET becomes 154
  birth 108
  emergence of architecture 106
  integrated services architecture 141
  mail gateway from JANET 124
  origins in government-sponsored research 294
  rapid expansion in early 1990s 154
  security not addressed in original architecture 257
  UK access for education and research 124
  Web aids growth 136
Internet2 123, 205, 230, 261, 270, 294
Internet Activities Board 109
Internet Architecture Board 109
Internet Control and Monitoring Protocol. *See* ICMP
Internet Engineering Task Force. *See* IETF
Internet Group Management Protocol. *See* IGMP
Internet Protocol. *See* IP (Internet Protocol)
Internet Research Task Force. *See* IRTF
Internet Watch Foundation. *See* IWF
IntServ 147, 148, 227, 295, 296
Inuk 281
IP (Internet Protocol) 107
  confirmed as preferred JANET protocol 116
  IP Technical Advisory Group (IPTAG) 113
  principal network protocol for JANET 184
IPSS (International Packet Switched Service) 85, 123, 126
  made irrelevant by Internet 262
IPTV 173, 281
IRTF (Internet Research Task Force) 109
ISC (Information Systems Committee) 84, 112, 116, 129, 161, 164, 181, 183

replaces Computer Board 160
ISO (International Organisation for Standardisation) 39, 41, 43, 50, 52, 55, 60, 78, 91, 92, 99, 112, 113, 116–118, 120, 128, 129, 133, 145, 289, 292
ITAB (Information Technology Advisory Board) 174
ITP (Interactive Terminal Protocol) 47
ITSU (Information Technology Standards Unit) 41
IWF (Internet Watch Foundation) 274
IXI (International X.25 Infrastructure) 114, 128, 129, 199

J

Jacobson, Van 110, 113
JAMES (Joint ATM Experiment on European Services) 200
JANET 81, 83, 119, 126, 159, 306, 307, 308
 adoption of IP 77, 109, 116, 134
 and the Millennium Bug 205
 approval to phase out use of SMDS 210
 assembling the network 72
 assigning addresses for hosts 80
 awards
  British Computer Society Award 198, 244
  e-Government Shared Services Award 303
 backbone 70, 75, 93, 94, 111, 114, 133, 155, 193, 196, 198, 200–202, 211, 222, 224, 225, 226, 228, 229–231, 234, 235, 237–239, 242, 243, 252, 284, 296, 323
 campus access 89
 changing technological focus 142
 community 158, 270, 325
 connection of polytechnics 93
 connection of schools sector 216
 connection to IXI 128
 connection to US 229
  capacity to US doubled 199
  mail gateway to early Internet 124
  NSFnet connection 126
 costs and charging 292
 early services 80
 evolution of three-tier AS hierarchical structure 155
 extension of wide area paradigm to local context 85
 information dissemination 83
 integration of Ethernet LAN access 91
 interconnection with NHS network 255
 interconnect with local authorities 218
 international connectivity 123
 largest SMDS IP network 186
 legality of JANET 74
 Location Awareness trials 301
 naming of 75
 Network Control Centre (NCC) 71
 Network Operations Centres 71
 newsletter 81
 object of envy 105
 operations and support development 251, 252
 Optical Development Programme 298
 pilot VRVS service 281
 policy 167
  Acceptable Use Policy (AUP) 82, 119, 167, 168, 213, 260, 261, 263–265
  Connection Policy 167, 168, 186, 260, 261, 263
  Connection Policy in respect of NHS organisations 282
  Connection Policy review & development 261
  network standards 59
  Security Policy 264, 265
 pressure to support Internet protocols 102
 problems of providing peering capacity and peering resilience 198
 protocols 37, 40, 41
 quarter century 290
 requirement of monitoring 236
 service level agreement 166–168, 218, 261, 263, 268, 296
 to provide interconnect amongst RBCs 248
 trademark of higher education funding councils 164
 types of connection 261
  Interconnect 261
  Primary connection 261
  Proxy connection 261
  Sponsored connection 261
 VPNs 297
 X.25 37
 X.25 service closes 85
JANET Aurora 255, 279, 302
JANET IP Service. *See* JIPS
JANET Liaison Desk 251
JANET Lightpath 242, 251, 279, 297, 298, 300
JANET Mk II 94, 110, 114, 126, 135, 172, 177, 178, 181, 187
 funding obtained 94
 upgrades tail off in anticipation of SuperJANET 95
JANET News 82
JANET NHS Gateway 282
JANET-NSFnet gateway 125
JANET Operations Desk. *See* JOD
JANET Roaming 279
JANET Service Desk 252
JANET Training 300
JANET txt 277, 278
JANET(UK) 119, 253, 303
 funding member of IWF 274
 Heads of 308
 new trading name of JNT Association 82
JANET Videoconferencing Booking Service 271
JANET Videoconferencing Service. *See* JVCS
JANET Videoconferencing Switching Service. *See* JVCSS
JANET Web Filtering Service 274, 275
JCSR (JISC Committee for Support of Research) 210, 231, 279
JIF (Joint Infrastructure Fund) 231
JIPS (JANET IP Service) 102, 111–114, 116, 117, 136, 165, 184, 186, 188, 232, 251
JISC Committee for Support of Research. *See* JCSR
JISC (Joint Information Systems Committee) 84, 110, 116, 126, 164, 166, 167, 184, 194, 200, 211, 213, 234, 270, 279, 282, 283
 funding of regional networks 213

successor to ISC 161
JISCmail 84
JNT 22, 34, 35, 40, 52, 53, 57, 58, 251, 309
    achieves full complement of staff 58
    addressing authority 64
    budget 58
    consequences of establishing JIPS 116
    creation of 1, 34
    Director of Networking 71, 72
    discusses need for registration 81
    discussions with Post Office 68
    established 57, 59
    Heads of 308
    initiates discussion of regulation and security 265
    Joint Head of the JNT and Network Executive 72
    naming authority 64
    operational inconveniences 159
    original team 58
    policy 58
    remit 57
    replacement 162
    responsibility for protocols for academia 60
    responsible for assigning addresses within the network 80
    sets in motion PAD facilities 65
    sponsors development of Z80 PAD 65
    standards-based network components as policy 58
    superseded by UKERNA 81
JNT Association, The 253. *See also* JANET(UK), UKERNA
    begins operation 165
    Chairs of the Board 308
    creation approved 164
    Heads of 308
    incorporated 164
    naming of 164
    restructuring of membership 216
    staff transfer from SERC 165
    UKERNA adopted as trading name 164
JNT PAD 61, 62, 64–66, 85
Job Transfer and Manipulation Protocol. *See* JTMP
JOD (JANET Operations Desk) 252
Joint ATM Experiment on European Services. *See* JAMES
Joint Information Systems Committee. *See* JISC
Joint Infrastructure Fund. *See* JIF
Joint Network Team. *See* JNT
Joint Photographic Experts Group. *See* JPEG
Joint Working Party on Networks 68, 71, 165
    report 68
JPEG (Joint Photographic Experts Group) 145
JTMP (Job Transfer and Manipulation Protocol) 52, 59, 60
JVCS (JANET Videoconferencing Service) 187, 247, 248, 280, 284, 302
JVCSS (JANET Videoconferencing Switching Service) 246

K

Kahn, Bob 107

Kay, Alan 136
Kennisnet 218
Kentish Metropolitan Area Network. *See* KentMAN
Kent MAN (Kentish Metropolitan Area Network) 307
Kingston Communications 36, 327
Kingston-upon-Hull 36, 258
Kirstein, Peter 8, 17, 23, 40, 107, 110, 114, 120, 123, 124, 287
Kleinrock, Leonard 6
KPMG International 162, 163, 164, 174
KPNQwest 229

L

Label Distribution Protocol. *See* LDP
LAN. *See also* Ethernet
    Apollo token ring 86
    applied to campus access 89
    cabling 89
    Cambridge Backbone Ring 88
    Cambridge Ring 87–92, 100
    Orwell Ring 87
    Pierce loop 86
    Söderblom 86
Larmouth, John 52, 80
Laws, John 40
layering
    in Internet protocols 40
    in OSI 40
    link layer 42
    network layer 41
    song 310
    X.25 layers 42
LDP (Label Distribution Protocol) 206
Learning and Teaching Scotland. *See* LTS
Learning Network South East. *See* LeNSE
legislation
    Anti-Terrorism, Crime and Security Act 2001 328
    Communications Act 2003 325
    Computer Misuse Act 1990 325
    Criminal Justice and Immigration Act 2008 327
    Data Protection Act 1984 327
    Data Protection Act 1998 327
    Data Retention (EC Directive) Regulations 2009 328
    Defamation Act 1996 326
    Electronic Commerce (EC Directive) Regulations 2002 326
    European Convention on Human Rights 327
    Human Rights Act 1998 327
    Interception of Communications Act 1985 327
    Police and Criminal Evidence Act 1984 327
    Police and Justice Act 2006 325
    Protection of Children Act 1978 327
    Regulation of Investigatory Powers Act 2000 325, 327
    Sexual Offences Act 2003 327
    Telecommunication Bill 1980 68, 70
    Telecommunications Act 1981 72
    Telecommunications Act 1984 73

Index   405

Terrorism Act 2006 326
LeNSE (Learning Network South East) 155, 214, 307
Level 3 Communications 230
LGfL (London Grid for Learning) 323
Licklider, J.C.R. 6
Lifelong Learning Network in Wales. *See* LLNW
Linington, Peter 37, 72, 81, 162, 251, 308, 309
LINX (London Internet Exchange) 126, 198
LLNW (Lifelong Learning Network in Wales) 216, 217, 323
LMN (London Metropolitan Network) 214, 307
Location Independent Networking 279
London Grid for Learning. *See* LGfL
London Internet Exchange. *See* LINX
London Metropolitan Network. *See* LMN
LTS (Learning and Teaching Scotland) 218
Lucent
  chosen as SuperJANET5 backbone router supplier 240
Lumen House 253

M

Mailbase 84
Mainwaring, Keith 71
Manning, Geoff 70, 71, 181
MANs (Metropolitan Area Networks) 212
Marshall, Tim 308
MBone 114, 246, 300
  experimental MBone mounted over JIPS 114
  videoconferencing tools 250
McClure, Roger 308
McCrindle, Fred 216
McMullen, Geoff 308
MECCANO (Multimedia Education and Conferencing Collaboration over ATM Networks and Others) 205
Mercury 73, 94, 138, 172, 175, 327
METRONET 25, 306
Metropolitan Area Networks. *See* MANs
Michie, Donald 3
Microsoft Network 274
MidMAN (Midlands MAN) 214, 307
MidNET 27, 306
Millennium Bug, The 204
MILNET 107, 108, 154
Ministry of Defence. *See* MoD
Mockapetris, Paul 115
MoD (Ministry of Defence) 17, 123, 124, 161, 287
Monopoly, Post Office 28, 31, 33, 36, 39, 40, 42, 57–59, 68, 70, 72, 73, 138, 258, 259
Moore's Law 299
Morris, Chris 31
Morris, Ernest 165, 308
Morris worm 266, 268, 272
Motion Picture Experts Group. *See* MPEG
MPEG (Motion Picture Experts Group) 146
MPLS (Multiprotocol Label Switching) 155, 206, 229, 271, 290, 297, 298, 300
multicast 110, 113, 114, 290
Multimedia Education and Conferencing Collaboration over ATM Networks and Others. *See* MECCANO

Multiprotocol Label Switching. *See* MPLS

N

NAC (Network Advisory Committee) 71, 129, 160, 166
Name Registration Scheme. *See* NRS
NASG (Networking Association Steering Group) 162–165
National Bureau of Standards. *See* NBS
National Committee on Computer Networks. *See* NCCN
National Grid for Learning in Wales. *See* NGfL Cymru
National Information on Software and Services. *See* NISS
National Institute of Standards and Technology. *See* NIST
national network
  adoption of protocols prioritised 18
  assembling the network 72
  construction recommended 15
  involvement of Computer Board and Research Councils 22
  operational requirement drafted 70
  providing flexibility to accommodate evolution 32
  provisioning strategy recommended 15
  recommendation to examine feasibility of 13
National Physical Laboratory. *See* NPL
National Science Foundation. *See* NSF
Natural Environment Research Council. *See* NERC
NBS (National Bureau of Standards) 123
NCCN (National Committee on Computer Networks) 57
NCP 42
Nelson, Dorothy 261
NERC (Natural Environment Research Council) 9, 31, 68–72, 93, 159
Netcomm 2000 94
NetherLight 231, 242
Net North West. *See* NNW
Network Advisory Committee. *See* NAC
Networked Information Services Project. *See* NISP
Network Executive 69, 71, 83, 129, 157, 158, 211, 251, 309
  Joint Head of the JNT and Network Executive 72
  responsibilities 69
Network Independent File Transfer Protocol. *See* NIFTP
Networking Association
  agreed successor to JNT 162
  naming of 164
  suggested organisational forms 162
Networking Association Steering Group. *See* NASG
Network Management Committee. *See* NMC
Network News 81
  first issue 34
Network News Transfer Protocol. *See* NNTP
Network Operations and Service Centre. *See* NOSC
Networkshop 34, 61, 62, 65, 66, 83, 252

Durham, 1980 65
first meeting 61
importance of 61
list of venues 1977-2010 319
Nottingham, 1982 62, 66
Reading, 1988 83, 99
Network Unit 22, 58–60, 64, 68, 71, 75, 81, 105
   accommodation 32
   Final Report 34
   Heads of 308
   set up 31
   staff 31
   terms of reference 32
Network Working Party 15, 19
NGfL Cymru 217, 323
NG (Northern Grid) 323
NHS
   N3 282, 283
   NHS-HE Co-ordinator 283
   NHS-HE Forum 283
   NHSnet 282
NIFTP (Network Independent File Transfer Protocol) 50–53, 59, 60, 87, 118
NIRAN (Northern Ireland Regional Area Network) 213, 214, 219, 291, 307, 323
NISP (Networked Information Services Project) 83
NISS (National Information on Software and Services) 83, 269
NIST (National Institute of Standards and Technology) 123
NMC (Network Management Committee) 69–71, 165, 199, 251
   ultimatum to BT 72
NNTP (Network News Transfer Protocol) 82
   82
NNW (Net North West) 214, 231, 307
NORDUnet 128
NorMAN (North East Metropolitan Area Network) 214, 307
North East Metropolitan Area Network. *See* NorMAN
Northern Grid. *See* NG
Northern Ireland Regional Area Network. *See* NIRAN
North Wales MAN. *See* NWMAN
North West Learning Grid. *See* NWLG
NOSC (Network Operations and Service Centre) 112, 128, 165, 184, 199, 201, 202, 213, 232, 251
   merges with UKERNA 252
NPL (National Physical Laboratory) i, 6, 7, 9, 11, 37, 60, 64, 108, 123, 124, 126
NRS (Name Registration Scheme) 55, 65, 80, 81, 115, 116, 118
NSF (National Science Foundation) i, 108, 119, 123, 126, 171, 205, 287, 288
NSFnet 108, 112, 119, 125, 126, 288, 298
NUNET 47, 306
NWLG (North West Learning Grid) 323
NWMAN (North Wales MAN) 217, 307

O

Office of Science and Technology 164, 231

O'Reilly, John 7
OSI Model 39, 40, 128, 129, 131, 133
   compliance an EU procurement requirement 55

P

Packet Assembler Disassembler. *See* PAD
Packet Switched Service. *See* PSS
packet-switching 3, 39, 69, 72, 106, 110, 136, 293
   forerunners 6
   growing international interest 37
   standardisation 39
   successful in experimental network 18
   term coined 7
   used in ARPANET 7
PAD (Packet Assembler Disassembler) 48, 49, 62, 64–66, 76, 84, 85, 89, 90, 108, 118
   JNT PAD 61, 62, 64, 66, 85
   JNT specification 65
PageOne 277
Particle Physics and Astronomy Research Council. *See* PPARC
PBB-TE 299
PCFC (Polytechnics and Colleges Funding Council) 58, 93, 110, 114, 160, 161, 163, 261, 262
PDH (Plesiochronous Digital Hierarchy) 142, 171, 179, 182–184, 186, 187, 189, 190, 192, 193, 195, 200, 298, 321, 322
Peatfield, Tony 70, 71
Peters, Geoff 308
Pipex 126
Planetlab 301
Plesiochronous Digital Hierarchy. *See* PDH
polytechnics
   become universities 114
   connections upgraded to 64kbps 114
   connection to JANET 93
Polytechnics and Colleges Funding Council. *See* PCFC
Postel, Jon 107, 115, 120
Post Office 5, 9, 11, 13, 17, 22, 24–26, 28, 31–33, 36, 37, 40, 42, 57, 60, 61, 64, 66, 68, 124, 126, 173
PPARC (Particle Physics and Astronomy Research Council) 161, 231
protocols. *See also* Coloured Books
   adoption made a priority 18
   Coloured Books 41, 312
   concept 36
   convergence during 1970s and 1980s 28
   email 53, 118
      addressing 54
      format 53
   file transfer 49
   IP 107
   JANET protocols 41
   layering 39
      in OSI 40
   ST-II 147
   terminal access 47
   triple X set 47
PSBA Network 167, 217, 323
PSINet 82

PSS (Packet Switched Service) 11, 27, 28, 33, 36, 37, 40, 42–45, 48, 50, 60, 62, 64–66, 68, 70–72, 80, 84, 85, 93, 119, 123, 126, 127, 266, 293
 academic gateways implemented 84
 based on X.25 64
 begins 84
 billing 262
 opens for service 66
 paying for 84
 unacceptable for national network 68
 use dwindles as JANET comes into operation 85
 use of triple X 48
 User Forum 37
  High Level Protocol Group 50
 uses X.25 42

Q

QoS (Quality of Service) 141, 146, 148, 227, 244, 289, 290, 295–297
 Think Tank 227, 248, 296
Quality Network Technology for User-oriented Multimedia. *See* QUANTUM
Quality of Service. *See* QoS
QUANTUM (Quality Network Technology for User-oriented Multimedia) 205

R

RAL (Rutherford Appleton Laboratory) 58, 69, 71, 72, 82–84, 93, 111, 129, 131, 157, 159, 161, 175, 181, 184, 188, 189, 200, 205, 209, 213, 231, 242, 251
 gateway between JANET and EARN 127
RARE (Réseaux Associés pour la Recherche Européenne) 127, 129, 132
RCOnet 23, 25, 26, 47, 60, 61, 306
Read, Malcolm 164
Real Time Protocol. *See* RTP
regionalisation 4, 290
 8
regional networks
 Acceptable Use Policy 213
 beginnings of 23
 funding of 213
 nature of 213
 re-emergence during 1990s 211
 traffic routing 213
Remote Job Entry. *See* RJE (Remote Job Entry)
Research Councils, The 261
Réseaux Associés pour la Recherche Européenne. *See* RARE
Réseaux IP Européens. *See* RIPE
Resource Reservation Signalling Protocol. *See* RSVP
RFC 801 107
RFC 1613 117
RFC 3261 150
RHEL (Rutherford High Energy Laboratory) 5, 8–10, 28–30, 106, 124, 233
RIPE (Réseaux IP Européens) 128

RJE (Remote Job Entry) 5, 10, 18, 23–26, 28, 32, 36, 40, 45, 46, 51, 52, 60, 85, 124, 127, 176, 285
Roberts, Larry 7, 293
Rogerson, Duncan 238
Rosenbrock, H.H.
 Rosenbrock Report 29
Rosner, Roland 31, 58, 70–72, 75, 81, 282, 308
 34
Royal Signals and Radar Establishment. *See* RSRE
RSRE (Royal Signals and Radar Establishment) 40, 60, 161
RSVP (Resource Reservation Signalling Protocol) 147, 148, 206
RTP (Real Time Protocol) 148, 150
Rutherford Appleton Laboratory. *See* RAL
Rutherford High Energy Laboratory. *See* RHEL

S

SAML (Security Assertion Markup Language) 270
Sankar, James 279
Scicon Consultancy International 69
Science and Engineering Research Council. *See* SERC
Science and Technology Facilities Council. *See* STFC
Science Research Council. *See* SRC
Science Research Investment Fund. *See* SRIF
Scottish Funding Council. *See* SFC
Scottish Further Education Funding Council. *See* SFEFC
Scottish Higher Education Funding Council. *See* SHEFC
Scottish MANS Videoconferencing Network. *See* SMVCN
Scottish Schools Digital Network. *See* SSDN
SDH (Synchronous Digital Hierarchy) 142–144, 151, 173, 178, 180–183, 189, 190, 192, 193, 199, 200, 221, 223, 224, 234, 243, 288, 298, 299, 321, 322
SDSS (Shibboleth Development and Support Services) 270
Security Assertion Markup Language. *See* SAML
SEGfL (South East Grid for Learning) 323
SERCnet 66, 70, 71, 72, 74, 79, 81, 84, 85, 135, 158, 251, 260, 285, 286. *See also* SRCnet
 suggestion that it be handed over to JNT 70
SERC (Science and Engineering Research Council) 66, 69–71, 86, 88, 93, 125, 126, 129, 159, 161, 162, 164, 165, 179, 181, 263, 287
Session Initiation Protocol. *See* SIP
SFC (Scottish Funding Council) 162
SFEFC (Scottish Further Education Funding Council) 162
Shaw, Sandy 117
SHEFC (Scottish Higher Education Funding Council) 161, 162, 164, 166
Shibboleth 270
Shibboleth Development and Support Services. *See* SDSS
Shoch, John 115
Shoestring 111, 116, 128, 232, 291
Simple Mail Transfer Protocol. *See* SMTP
Simple Screen Management Protocol. *See* SSMP

SIP (Session Initiation Protocol) 150, 281
SMDS (Switched Multi-megabit Data Service) 101,
  150, 151, 182–186, 193, 196, 200, 201, 221, 286
  approval to phase out use within JANET 210
  based on compatibility with DQDB 101
  European SMDS Interest Group (ESIG) 186
Smith, Ian 71, 251
SMTP (Simple Mail Transfer Protocol)
  53, 113, 118
SMVCN (Scottish MANS Videoconferencing
  Network) 246, 248
South East Grid for Learning. *See* SEGfL
South Wales MAN. *See* SWMAN
South West England Regional Network.
  See SWERN
South West Grid for Learning. *See* SWGfL
South West Universities Computer Network.
  See SWUCN
Space Physics Analysis Network. *See* SPAN
spam 265
SPAN (Space Physics Analysis Network) 131, 287
Spark 218
Spratt, Brian 162, 305
Sprint 230
SRCnet 45–47, 50, 51, 60, 62–64, 66, 69, 80, 84, 233,
  287
  alternative link layer 43
  email 40
  naming conventions 45
  renamed SERCnet 66
SRC (Science Research Council) 5, 8, 9, 15, 19, 22,
  23, 28–33, 50, 51, 58, 66, 76, 87, 259, 287, 289
  Astronomy, Space and Radio Board 31
  Distributed Computing System research
    programme 87
  Engineering Board 29, 30
  Facilities Committee for Computing (FCC) 66,
    69, 71, 165
  regionalisation 8
  renamed SERC 66
SRIF (Science Research Investment Fund) 231
SSDN (Scottish Schools Digital Network) 218, 324
SSMP (Simple Screen Management Protocol)
  49
StarLight 230, 231
Starlink 31, 130
STFC (Science and Technology Facilities Council) 161
SuperJANET ii, 95, 96, 104, 110, 112, 114, 117, 118,
  130, 135, 138, 143, 151, 159, 162, 172, 205, 208,
  225, 227, 232, 286, 288, 290, 291, 293
  achieved original strategic goals 252
  awarded British Computer Society Award 198
  birth of concept 171
  case made 93
  choice of initial sites 183
  concept defined 133
  concept given form 180
  contract awarded to BT 182
  dependency on single supplier 230
  early application demonstrations 187
  emergence of backbone 193
  exemplar applications 175
  fibre infrastructure 177
  first public event 187
  formal Invitation to Tender 181
  funding approved 178
  how much would it cost 174
  international access 198
  IP service development 184
  network topology 181
  realisation of vision 133
  trademark of higher education funding councils
    164
SuperJANET II 188, 193, 225
SuperJANET III 146, 200, 202, 212, 221, 222,
  225–227, 229, 247, 248, 289, 296
  architecture 201
  contract awarded to Cable & Wireless 201
  decision to use ATM 202
  international access 205
  Managed Bandwidth Service (MBS) 203, 205
  technology developments 206
SuperJANET4 133, 155, 169, 212, 213, 217, 222,
  225–227, 229, 233, 239, 242, 248, 250, 282, 287,
  289, 296, 297
  an IP network 222
  backbone upgraded to 10Gbps 225
  contract extended 232
  gigabit routers 221
  responsibility back in the hands of the
    community 226
  test network 229
  topology 224
  undisturbed by failure of WorldCom 230
  WorldCom chosen as supplier 224
SuperJANET5 133, 213, 231, 251, 289
  achievement of original goals of SuperJANET
    232
  developing strategic requirements 234
  development of transmission core platform 243
  eliminate single points of failure 235
  first pilot service 242
  flexible transmission platform 236
  Lucent chosen as backbone router supplier 240
  operational requirement and Invitation to
    Tender 240
  planning and architecture development 233
  requirements summary 234
  transmission infrastructure 243
  Verizon chosen as infrastructure supplier 240
SURFnet 230, 250
SWERN (South West England Regional Network)
  214, 307
SWGfL (South West Grid for Learning) 323
SWITCH 242
Switched Multi-megabit Data Service. *See* SMDS
SWMAN (South Wales MAN) 217, 307
SWUCN (South West Universities Computer
  Network) 24, 25, 47, 52, 60, 306
Synchronous Digital Hierarchy. *See* SDH

T

TACS (Terminal Access Conversion Service) 118
TAT-8 124, 287
TAU (Technical Advisory Unit) 166
Taylor, B.R. 305
TCP/IP 90, 91, 92, 95, 107, 109, 110, 113, 116–118,
  124, 131, 136, 252, 287–289, 292
  spread of support 108

INDEX 409

TDM (Time Division Multiplexing) 142, 143, 146, 179, 207, 234, 235, 280, 288, 293
Teague, Malcolm 283
Technical Advisory Unit. *See* TAU
technology drivers 139
   controlling and scheduling packet traffic 140
   deployment of optical fibre 139
   digitisation and encoding of audio 140
   switched Ethernet 141
Telecity 198, 241, 242, 243, 284
Teleglobe 229
Telehouse 126, 198, 206, 241, 242, 243
TELNET 118, 119, 124, 125
   118
TEN-34 126, 130, 132, 199, 205, 287
TEN-155 203, 205
TERENA (Trans-European Research and Education Networking Association) 129, 130, 205, 230, 268
   Mobility Task Force 279
terminal access 46
   protocols 47
   teletype terminals 46
   visual terminals 47
Terminal Access Conversion Service. *See* TACS
Thames Valley Network. *See* TVN
Time Division Multiplexing. *See* TDM
Trans-European Research and Education Networking Association. *See* TERENA
Transport Service 43
   Yellow Book 43
Transport Service Implementers Group 81
Turing, Alan 3
TVN (Thames Valley Network) 213, 214, 242, 307

U

UBE (unsolicited bulk email) 265
UCAID (University Corporation for Advanced Internet Development) 205
UCAS (Universities and Colleges Admissions Service) 118
UCCS (University of Cambridge Computing Service) 65, 69
UCL (University College London) 8, 10, 17, 19, 46, 51, 53, 54, 60, 72, 87, 106, 107, 110, 111, 114, 116, 124, 145, 175, 184, 187–190, 198, 205, 230, 245, 246, 279, 282, 287, 295
UDP (User Datagram Protocol) 148
UFC (Universities Funding Council) 93, 99, 160–165, 174, 178, 181, 182, 189, 190
   ceases to exist 189
UGC (University Grants Committee) 4, 160, 164
UHIMI (University of the Highlands and Islands Millennium Institute) 214, 307
UHI (University of the Highlands and Islands) 226
UK Access Management Federation for Education and Research 270, 276, 279
UKERNA News 81
UKERNA (United Kingdom Education and Research Network Association) 81, 82, 181, 201
   adopted as trading name of JNT Association 164

   agrees with UCAID to promote collaboration between Europe and the US 205
   conducts procurement for Scottish MANS Videoconferencing Network 246
   contract with regional networks 213
   defines formal OR for SuperJANET5 238
   discussions with BT 126
   establishes trust framework with GlobalSign 270
   explores provision of TV-style content over JANET 281
   first appearance of term 164
   governance and operation 166
   Heads of 308
   joins 6bone 295
   managed router service 202
   manages HEFCE programme of funding and rationalising connections for the English FE colleges 201
   market testing 168
   merges with NOSC 252
   moves out of Atlas Centre 253
   name credited to David Hartley and Bob Cooper 164
   not-for-profit organisation 166
   objectives 165
   publishes SuperJANET5 Invitation to Tender 240
   responsible for managing SuperJANET4 IP routers 222
   steps to guard against supplier failure 230
   subcontracted to administer LLNW procurement 217
   trademark of higher education funding councils 164
   undertakes study of combined needs of FE and HE sectors in Wales 247
UKLight 230, 231, 242, 250, 255, 277, 279, 297, 302
UKnet 82, 126
ULCC (University of London Computer Centre) 5, 10, 22–25, 65, 70, 71, 72, 85, 111, 112, 114, 116, 125, 126, 128, 165, 184, 198, 199, 201, 232, 251
UMRCC (University of Manchester Regional Computer Centre) 5, 22, 23, 26, 27, 72, 251
United Kingdom Education and Research Network Association. *See* UKERNA
Universities and Colleges Admissions Service. *See* UCAS
Universities Funding Council. *See* UFC
University College London. *See* UCL
University Corporation for Advanced Internet Development. *See* UCAID
University Grants Committee. *See* UGC
University of Cambridge Computing Service. *See* UCCS
University of London Computer Centre. *See* ULCC
University of Manchester Regional Computer Centre. *See* UMRCC
University of the Highlands and Islands. *See* UHI
University of the Highlands and Islands Millennium Institute. *See* UHIMI
unsolicited bulk email. *See* UBE
Usenet 54, 82, 130, 275
   EUnet 82
   Usenet News 82, 273
      closure of JANET service 82

User Datagram Protocol. *See* UDP
User Support Workshop 252

V

VCAS (Videoconferencing Advisory Service) 247
VCEG (Video Coding Experts Group) 146
Verdon, F.P. 305
Verizon
  chosen as SuperJANET5 infrastructure supplier 240
Video Coding Experts Group. *See* VCEG
videoconferencing 113, 114, 137, 139, 144, 146, 150, 173, 183, 187, 217, 218, 246, 296
  development of on JANET 244
  extension to NHS sites 252
  IP-based 227
  largest schools event to date 249
  pilot service uses ATM 190
  provision for colleges 201
  real-time 228
  SuperJANET VideoConferencing Network (SJVCN) 244, 246, 247
  Vic 145
  video-based services 280
Videoconferencing Advisory Service. *See* VCAS
Video Technology Advisory Service. *See* VTAS
Virtual Rooms Videoconferencing System. *See* VRVS
von Neumann, John 3
VRVS (Virtual Rooms Videoconferencing System) 250, 281
VTAS (Video Technology Advisory Service) 247

W

Walkinshaw, W. 305
Wave Division Multiplexing. *See* WDM
WDM (Wave Division Multiplexing) 143, 173, 179, 207, 225, 230, 234, 236, 241, 243, 284, 288, 289, 293, 298
Wellcome Trust 231
Wells, Mike 8, 13, 14, 18, 26, 54, 71, 72, 251, 302, 305, 309
Wells Reports 13, 19, 22, 31, 32, 34
  first 8, 14, 19, 290, 291
  second 19, 22, 31
WelshNet 246, 247
  prelude to WVN 245
Welsh Video Network. *See* WVN
West Midlands Regional Broadband Consortium. *See* WMnet
Weston, Sue 99, 102
Wilbur, Steve 175, 183
Wilkes, Maurice 3
Williams, Mervyn 31, 58, 308
Williams, P.E. 305
Wireless Advisory Group 278
WMnet (West Midlands Regional Broadband Consortium) 323
Wood, Shirley 81, 251
WorldCom 224, 225, 230
  chosen as SuperJANET4 supplier 224
  revelations of financial status 230
World Wide Web 110, 113, 116, 134, 136, 218, 227, 249, 258, 266, 277, 288, 292
  continued exponential growth of Internet 136
WVN (Welsh Video Network) 216, 245, 247, 248, 252, 284

X

X.25 25–27, 37, 42–44, 47, 48, 55, 59, 60–64, 66, 72, 85, 89, 90, 92, 105, 108, 111, 112, 114, 116–118, 120, 126, 128, 129, 132, 149, 152, 179, 180, 182, 184, 186–188, 206, 251, 271, 273, 285, 292, 300, 306
  basis of JANET for first decade 37
  closure of service 85, 119
  IP over X.25 95
  operating over SMDS 188
  over Ethernet 92
  protocol of choice for MidNET 27
  to be used in academic network 26
  TS29 48
  X.25 switch specification drawn up 64
X.29 117–119
X.400 113, 116, 118–120, 292
X.500 55, 120
Xerox PARC 49, 55, 86, 108, 115, 136, 208
XOT (X.25 over TCP) 117, 188, 271
X-Windows 49, 110

Y

YBTS (Yellow Book Transport Service) 43, 44, 48, 50, 59, 60, 66, 85, 88, 91, 266
  addressing 44
Yellow Book Transport Service. *See* YBTS
YHGfL (Yorkshire & Humberside Grid for Learning) 323
YHMAN (Yorkshire and Humberside Metropolitan Area Network) 214, 231, 307
Yorkshire and Humberside Metropolitan Area Network. *See* YHMAN
Yorkshire & Humberside Grid for Learning. *See* YHGfL

Z

Zacharov, B. 305

Biographical Note

Chris Cooper holds a degree in mathematics and theoretical physics from the University of Cambridge and a PhD in theoretical particle physics from the University of London. From 1974 until 2001 he worked at the Rutherford Appleton Laboratory (RAL) before gaining a chair at Oxford Brookes University and becoming network strategist to JANET. For 25 years from 1980 his research interests were in multiservice networks. He has implemented, taught, researched and consulted internationally on networking.

Chris has worked on projects in the UK, Europe and the USA, and has consulted in the UK and USA. He was one of the earliest users of the UCL ARPANET link in 1973 and contributed to various aspects of the development of SRCnet and JANET, from participation in the definition and implementation of the Coloured Books in the early 1980s, to participation in the SuperJANET programme from 1990 to 2007, specifically the original SuperJANET, SuperJANET4 and SuperJANET5. From 1981 to 1983 he participated in Project Universe, a £5m early multiservice and distributed systems network project using linked satellite and Cambridge Ring networks – a collaboration consisting of BT Research Laboratories, Cambridge University, GEC Marconi Research Centre, Logica Ltd, Loughborough University, RAL and UCL. Subsequently he worked as a consultant on the Unidata programme, based at UCAR in Boulder, Colorado, where he participated in establishing the nationwide communication infrastructure for atmospheric science, based on two-way Ku-band satellite communication, one-way broadcast satellite distribution of meteorological data and the early Internet at the time of the creation of NSFnet. He subsequently helped to establish the MSc programme at Oxford Brookes University in High Speed Networking and Distributed Systems in 1992, on which he taught until 2008. He has participated in EPSRC and EU RACE and ACTS projects, latterly as co-investigator on the UK EPSRC-funded project Open Overlays, in which Oxford Brookes University and Lancaster University developed dynamically reconfigurable middleware in support of a service-oriented distributed systems architecture applicable to Grid computing and Web services.

Chris has served on EPSRC, NERC and JISC committees for research support and funding in the UK, including the e-Science programme. He has reviewed for all of these, the US NSF and the Royal Dutch Academy of Science, and for the IEEE Journal on Selected Areas in Communications, Electronics Letters, Computer Communications and IEE Communications.